D1163903

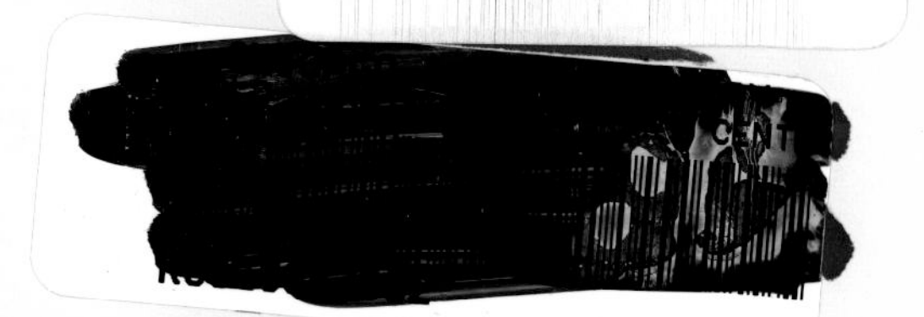

# LUNAR MINERALOGY

# LUNAR MINERALOGY

JUDITH W. FRONDEL
Honorary Research Associate
Department of Geological Sciences
Harvard University
Cambridge, Massachussetts

A WILEY-INTERSCIENCE PUBLICATION

JOHN WILEY & SONS, New York · London · Sydney · Toronto

***Library of Congress Cataloging in Publication Data:***

Frondel, Judith Weiss, 1912-
Lunar mineralogy.

"A Wiley-Interscience publication."
Includes bibliographical references and index.
1. Lunar mineralogy. I. Title.

QB592.F76 549.9′99′1 75-9786
ISBN 0-471-28289-8

Printed in the United States of America

10 9 8 7 6 5 4 3 2 1

TO CLIFF | WHOSE ENTHUSIASM FOR THE APOLLO PROGRAM WAS RESPONSIBLE FOR THIS VOLUME

TO CLIFF

WHOSE ENTHUSIASM
FOR THE
APOLLO PROGRAM
WAS RESPONSIBLE
FOR THIS VOLUME

# PREFACE

On July 21, 1969, Apollo 11 astronauts Neil Armstrong and Edwin Aldrin stepped onto the Moon. Within hours, in a record-breaking long distance telephone call, they were greeted by the President of the United States of America. "This is the greatest day since creation," he declared. In contrast to this enthusiasm, immediately following the opening of the first lunar sample boxes, a member of the Lunar Sample Preliminary Examination Team received a letter from a private citizen with a bitter complaint that many billions of dollars had been spent to obtain a few pounds of rocks "just like those" in any terrestrial backyard.

Despite many similarities, however, the lunar rocks are not exactly like terrestrial ones. Many lunar research scientists have noted that the rocks' individual components—the lunar minerals—are trying to tell us something with their form, chemistry, and evidence of a unique environment. With the Apollo program completed, no new lunar material will be added in the foreseeable future to the specimens in storage at the Johnson Space Center in Houston and among the samples distributed to the investigative teams throughout the world. It follows, then, that this precious hoard has been, is being, and will continue to be studied most intensively.

The greater part of the data on the lunar material already has been obtained and published in various journals and special publications. This volume has a double purpose: to present a reasonably compact record of the data that have appeared up to and including the Fifth Lunar Science Conference held at Houston in March 1974, and to serve as an aid to future mineralogical investigations.

The work is an outgrowth of two editions of a *Glossary of Lunar Minerals*, which was supported (NASA contract NAS-9-1800) and distributed by NASA, in xeroxed copy, to the Principal Investigators of the Apollo program. The book has been organized generally according to Dana's *System of*

*Mineralogy*. It contains a summary description, with extensive reference to the literature, of all the lunar minerals that have been found in the Apollo 11 through 17 samples, as well as the Luna 16 and 20 missions from the USSR.

I wish to thank my husband, Clifford Frondel, for his reading of the manuscript. Any improvement in the work, as it progressed, was due to his criticism. Any failures in the work are solely my responsibility. Mrs. Kathy Dwyer's cheerful deciphering of my handwriting and her expert typing are much appreciated. I am greatly indebted to NASA of Johnson Space Center, and particularly to Dr. Robert Laughan and Dr. Uel Clanton, for the generous gift of many NASA photographs. To the NASA effort that made the Apollo program possible, to the astronauts who brought back the lunar material, and more specifically, to those many kind people who offered their data, illustrations, comments, and encouragement, I am deeply grateful.

JUDITH W. FRONDEL

*Cambridge, Massachusetts*
*February 1975*

# CONTENTS

# LUNAR MINERALOGY

# INTRODUCTION

On July 25, 1969, the opening of the first rock box returned from the Apollo 11 mission was televised nationally (Figure I–1). The whole world (the news media, especially) expected immediate answers to the questions of the Moon's origin. As the box came into view on the television screen, the dusty lunar samples were seen to resemble a batch of burned baked potatoes (Figure I–2). At what was to be the moment of revelation, the television program was cut off abruptly. Perhaps a hasty curtain was being pulled down on a disappointing sight.

The disappointment, however, was short-lived. The next day, as the first carefully cleaned rock in its vacuum chamber was moved up to the camera ports, some questions were quickly answered. The rock was crystalline. It was unmistakably a medium-grained basalt, indicating that it had solidified from a magma (Figure I–3). Therefore, regardless of mode of origin, some of the Moon, at one time, had been hot. Despite some differences in amounts, the major mineral constituents of the rock were similar to the major minerals in terrestrial basalts. Thus it could be concluded further that the material of the Moon and the material of Earth are related.

More than 5 years have passed, and almost 850 lbs. of lunar samples have been returned to Earth. More than 800 scientists, representing approximately 180 investigative teams, have worked intensively to categorize only 10% of the returned material. Several hundred kilograms of lunar material, consisting of perhaps 20,000 individually identified samples, still are available for study. A large portion of the collected samples will be kept on reserve in the Johnson Space Center in Houston. Investigations will continue,[1,2] although dimimished monetary support will reduce the number of investigators and slow the pace of research.

## Geochemical and Geological History of the Lunar Rocks

Many results have been obtained already, and many problems resolved; nevertheless the debate over lunar origin continues. Of the three basic ideas

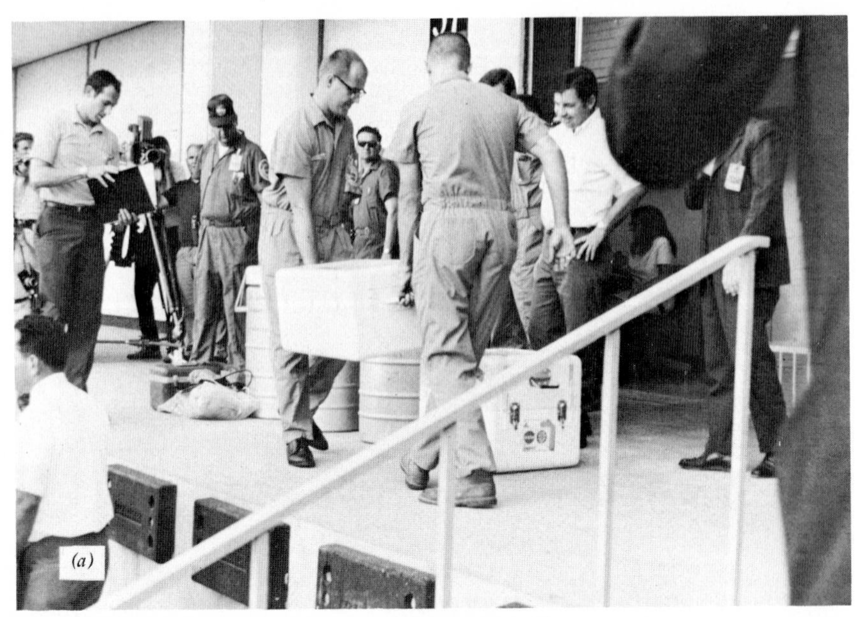

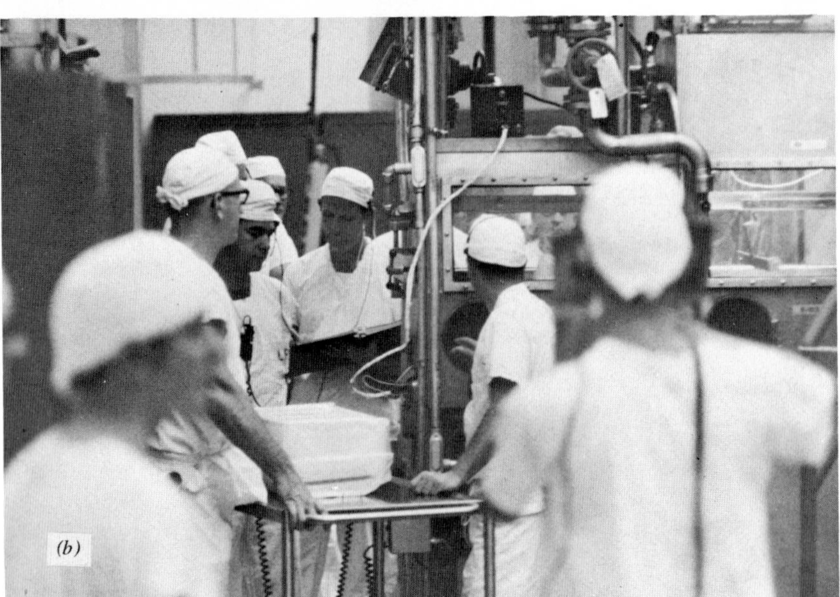

Fig. I–1 (*a*) Delivery of the first Moon rocks to the Lunar Receiving Laboratory in Houston, on July 25, 1969. (*b*) Preparing to open the first lunar rock box, July 26, 1969. (*c*) Monitoring the opening of the rock box. In-house television at the Lunar Receiving Laboratory. Photographs courtesy of C. Frondel, Harvard University, Cambridge, Mass.

Fig. I–1 (*continued*).

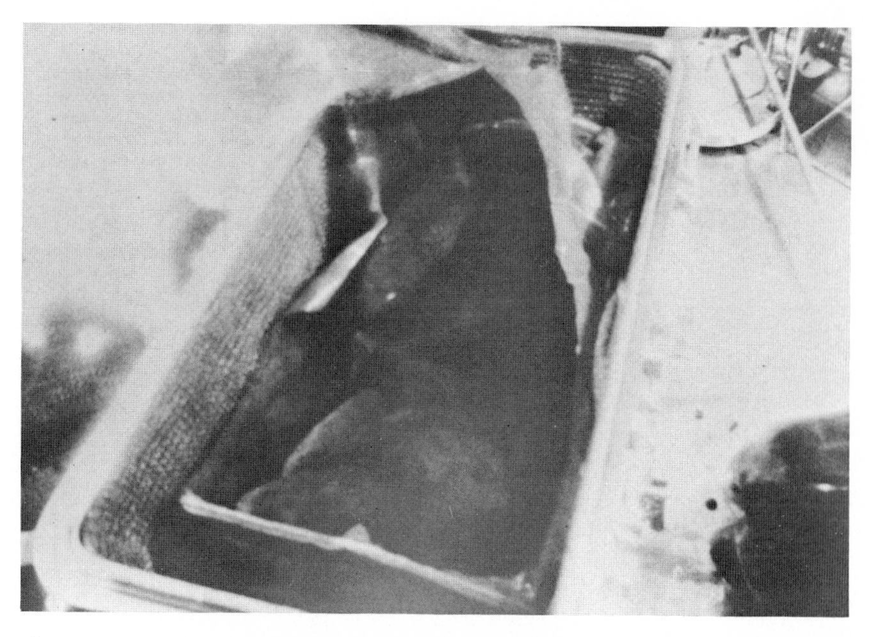

Fig. I–2 The first view of the lunar rocks. NASA photograph S-69-45002.

Fig. I–3 Medium-grained basalt 10003, the first cleaned lunar rock to be observed. Photograph courtesy of C. Frondel, Harvard University, Cambridge, Mass.

on the Moon's beginnings—accretion, capture, or fission—the third has been largely discredited. From present geochemical as well as geophysical data, it appears that the Moon formed under conditions different from those of the formation of Earth and the meteorites. The Moon (quite possibly the whole Moon and not just the outer shell) is depleted in all elements more volatile than iron (e.g., sodium, potassium, rubidium, cesium, zinc) and is enriched in refractory elements. If one model of lunar formation is considered (i.e., condensation during cooling in a disc-shaped nebula), the earliest condensates were compounds of calcium, aluminum, titanium, and refractory trace elements such as zirconium, tungsten, vanadium, thorium, barium, strontium, and rare earth elements (REE). High initial temperature of condensation, radioactivity 16 times that of chondritic meteorites, and rapid accretion led to early and extensive differentiation. Both the fractional crystallization of the whole Moon and later partial melting of the approximately 250 km of the outer shell occurred early in lunar history. Present geophysical data are consistent with a lunar interior that is currently hot and may be virtually barren of iron.[3]

The accretion of a Moon from refractory material already depleted in volatile elements is assumed to have been homogeneous. This assumption is based on the good correlation between ratios of volatile to nonvolatile elements in both highland and mare rocks. Some additional unique geochemical features of the Moon are high abundances of Mg and Cr in aluminous crustal rocks, high Cr/Ni ratios, and Eu anomalies. Two models for the deep lunar interior (below 1000 km) are offered. The core is formed by an immiscible Fe-FeS liquid that sank, removing most siderophile and chalcophile elements; or, if melting did not extend below 1000 km, the center is primitive unfractionated material, now partially molten because of trapped K, U, and Th.[4]

The europium deficiency is due in part to fractionation of calcic plagioclase.[5] The Eu depletion of Luna 20 samples has been found to be less than that for Apollo 12 mare basalts, and this has been explained by the higher plagioclase contents of the highlands material in Luna 20 samples. The negative Eu anomaly shown by these samples nevertheless argues for the Moon as a whole being depleted in Eu. Additional mechanisms have been suggested to bring about this depletion (e.g., partial melting, fractional crystallization, or vaporization of Eu[6]). At low oxygen fugacity and low confining pressure, europium oxide is extremely volatile relative to other rare earth elements.[5] From the low $f_{O_2}$ values obtained experimentally for the bulk lunar rocks, mineral separates, and even for phenocrysts, it appears that lunar magmas were in reduced states throughout their cooling history.[7]

Another major difference between the Moon and Earth is the almost total lack of lunar atmosphere. The atmosphere of the Moon today is so tenuous that it can be regarded as a collisionless gas.[8] Mechanisms for producing the present rarefied atmosphere are degassing of the Moon, meteoritic vaporization, neutralization of solar wind ions, and possibly spallation products of cosmic ray bombardment. Native He, $^{40}Ar$, and Ne have been identified. Helium and neon must come mainly from the solar wind, while $^{40}Ar$ results from the decay of $^{40}K$ within the Moon.[9] Rare molecules of water in the lunar atmosphere have been reported. Although their origin has been suspect, it has been suggested that trace amounts of water vapor could be produced under lunar conditions from the $H_2$ of the solar wind and the oxygen in the lunar soil.[10]

The Earth is active and alive. Earthquakes and volcanoes are dramatic evidence that processes responsible for the terrestrial igneous rocks are still at work. Wind and water, with help from gravity and temperature changes, break up, transport, and lay down material to make the sediments. Tectonic processes, involving high temperatures and pressures, transform the igneous and sedimentary rocks into metamorphic rocks.

Compared with Earth, the Moon is a quiet place. No winds or waters move about on its surface. The force of gravity, albeit one-sixth that of Earth, causes some movement of material. Large and small avalanches have left their traces and indicate slope failures unassociated with impact or volcanic events.[11] Tracks caused by rolling and bouncing of boulders are common, and some soil slumping on lunar slopes could be triggered by thermal stress.[12] Apollo seismic experiments record many small moonquakes (at approximately 800 km below the lunar surface) that appear to have a tidal association.[2] Other small seismic events, termed thermal moonquakes, may be related to diurnal temperature changes that induce fracturing or movement of rock along zones of weakness.[12] The impact of an occasional meteorite (with a mass range of 100 g to 1000 kg) also has been recorded.[13] Cosmic rays hit the Moon. The solar wind bathes it in a stream of ions and permanently implants some of the ions in the lunar soil.[8]

The Moon's sculptured landscape and cratered surface, as well as the lunar rocks, are records of an earlier active time. Layering on the Moon has been produced in ways very different from the terrestrial processes of sedimentation. The stratigraphy, observed in core samples, indicates that layering in the lunar regolith has been formed by complicated processes of soil excavation, redistribution, and mixing, all due largely to impact.[14] Large-scale layering, such as that seen on the walls of Hadley Rille, the Apollo 15 site, has been produced by successive lava flows. It has been suggested, however, that although the lineation observed on some of the lunar slopes may represent true geologic structure, some of the photographically recorded lineaments may be optical illusions (i.e., enhancement by oblique incident sunlight on fortuitous alignments of small topographic irregularities).[15]

Some investigators state that the Moon has had no volcanic activity. Others claim that the lunar history is similar to Earth's continuous record of volcanism and chemical differentiation, with lunar volcanism active in the recent past. There is, nevertheless, almost unanimous agreement that the dark mare regions on the Moon are underlain by extensive lava flows.[2] Basalts flooding the Imbrium Basin are similar in extent to the Columbia River basalts in the State of Washington.[16] Filling of the lunar maria involved volcanism and tectonism along pre-mare–basement fractures. These processes formed such features as mare ridges, fault systems, sinuous rilles that may be collapsed lava tubes, and volcanic flows.[17] It is reported that some individual fissures have been identified (photographically) as single eruptive sources of lava.[16] Although the large lunar craters have been formed by impact, small craters containing domical structures are probably of volcanic origin.[18] The various surface features may have undergone a more or less continuous process of formation and modification during evolution of the lunar maria.[17] Suggested dates for three major episodes of eruption of the basalts are: Phase I, 3.0 ± 0.4 billion years ago, Phase II, 2.7 ± 0.3 billion years ago, and Phase III, 2.5 ± 0.3 billion years ago.[16] Other investigators claim that the filling of the mare basins occurred largely between 3.2 to 3.8 billion years ago.[2]

Despite strong circumstantial evidence for rocks with ages of 4.5 to 4.6 billion years, only a few Apollo samples have crystallization dates lying between 4.0 and 4.6 billion years ago. It is believed that the formation of almost all the cratering on the Moon had occurred by 4 billion years ago.[2] The period of high meteorite influx produced high-grade metamorphic recrystallization of the rocks over wide areas of the Moon. The temperatures achieved by the impacts were variable, but some were high enough to cause varying degrees of partial melting of precursor rocks such as the highland breccias.[19] Total and partial melting, a relatively insignificant process on Earth, is important in the petrogenesis of the lunar highlands, where there is strong evidence of mixing and melting.[21] Metamorphism by impact is recorded in the rocks by mechanical twinning in minerals (e.g., in ilmenite, pyroxene, and plagioclase), fracturing of mineral grains (Figure I–4), partial or complete vitrification of mineral grains (e.g., production of thetomorphic plagioclase glass, and melting of the entire rock to an inhomogeneous melt).[21]

On Earth, field geologists try to find outcrops and sample bedrock wherever and whenever it is possible to do so. Mapping of and collecting from float are done only as a last resort. It has been stated that nowhere on the Moon have the crystalline rocks been found as part of a large *in situ* formation;[22] nevertheless, what may represent bedrock has been sampled (e.g., mare basalt from Hadley Rille). Photographically recorded, the exposures are described as a 5-meter thick regolith at the top of the rille, below which bedrock extends to about 55 meters; talus covers the remaining approximately 300 meters to the bottom. Blocks in the talus are up to 30 meters in diameter. Samples of mare basalt were collected from an exposure at the lip of the rille as well as from the regolith and the rims of two small craters nearby. Although correlation of the sampling locality with rock units observed across the rille is uncertain, it seems likely that most of the collected samples of mare basalt originated in one or more of the thick massive units.[23]

According to one theory, the compositions of the mare basalts were determined principally by partial melting processes in their source regions, followed by fractional crystallization near the surface. The generation of mare basalts is believed to have been caused by heating from U and Th in the interior (i.e., below 400 km) that was unaffected by early lunar differentiation. Later differentiation extracted the U, Th, and K by means of the generated basaltic magmas. These then were intruded upward into the

Fig. I–4 Fractured plagioclase grains in cataclastic anorthosite 60215,13. NASA photograph S-72-43966, courtesy of R. Laughan, Johnson Space Center, Houston.

outer 400 km and only a small proportion of the magmas reached the surface.[24]

The lunar mare-basalt samples have been grouped in various investigations by grain size (e.g., fine-, medium-, and coarse-grained basalts), by texture (ophitic, porphyritic, etc.), and by some dominant mineral components (e.g., olivine, ilmenite, and cristobalite basalts). These groups are not mutually exclusive, and different investigators have used differing names for the same rock.* There are also nonmare basalts (found in terra regions) that differ both in mineralogy and, possibly, origin.[25] The basalts have been classified also by their chemistry. Analyses reveal that some basalts are enriched in potassium, rare-earth elements, and phosphorus. These rocks have been designated by the acronym KREEP basalts, and some have been called, also, "gray mottled" basalts.[26] Basalts with an $Al_2O_3$ content of 20 to 24% have been termed high-aluminum basalts. (Some of these are the nonmare basalts, known also as highland basalts.)

* In Appendix II all rock specimens referred to in this book are listed by specimen number and the various descriptive names.

The lunar anorthositic rocks, crystalline rocks containing more that 24% $Al_2O_3$, have a mineral composition dominated by highly calcic plagioclase.[27] These rocks are primarily from the lunar highlands. The anorthositic rocks vary somewhat in their mineralogy, and attempts have been made to classify them on the basis of analogous terrestrial rocks. Hence names such as troctolite, norite, and troctolitic anorthosite have been used and grouped together under the heading of ANT (anorthositic noritic troctolite) rocks.

Some of all the lunar rock types may show the effects of shock metamorphism or impact, but the breccias predominate among the lunar metamorphic rocks. All degrees of brecciation, recrystallization, and even complete melting have been observed. From a study of the metamorphism of Apollo 14 breccias, Warner[28] has classified breccias largely on the basis of texture of the matrix, the amount of glass in the matrix, and the amount of glass clasts in the rocks. Other classifications of breccias have been made, and all attempt to show the range of metamorphism. The products of intense thermal metamorphism have been termed "melt rocks." These could have been produced either by partial melting of a single rock type or by melting with or without differentiation of heterogeneous rocks or breccias.[29]

Little of the collected material represents rocks from the lunar depths. Hence a dunite clast from breccia 72417 has been studied with great interest. The coarse grain size, simplectic intergrowths, and composition (~ 95% olivine) of the clast suggest that it did not form as a near-surface cumulate. It is believed the dunite was excavated from depth by a cratering event and subjected to shock that caused deformation and recrystallization of the olivine. Later the dunite was incorporated into the blue-gray matrix of the breccia. It has been concluded, tentatively, that the dunite clast (possible age ~ 4.6 billion years) is part of a very early differentiate, derived from the upper lunar mantle and so depleted in trace elements (including U and Th) that it can not represent the sources for the younger basaltic magmas. It is suggested that the magma producing the dunite was formed during early lunar differentiation and was associated with gravitational settling.[30]

The Apollo 15 green glass (Figure I–5) and the Apollo 17 orange glass (and soil) are of special interest because of their homogeneous, ultramafic compositions. Many explanations for the origin of the glasses have been offered. Some suggestions are that the glasses were formed by lava fountaining,[31] that the glasses are remains of a projectile that formed a primary crater,[32] or that the glasses represent a deep pyroxenite layer that has undergone partial melting because of a large impact.[33]

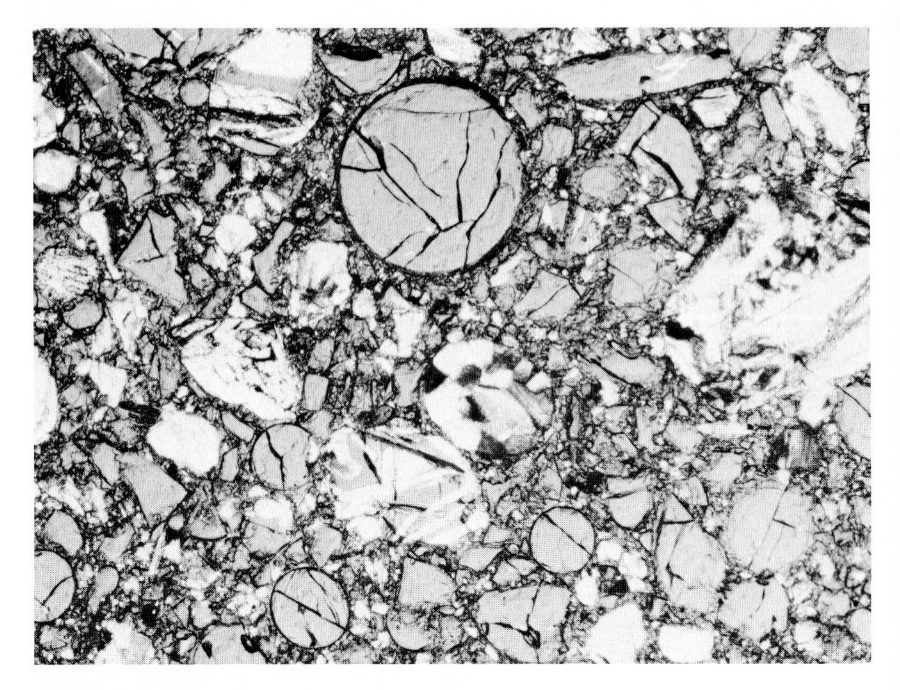

Fig. I–5 Green glass (spherical globules) in thin section of pyroclastic rock 15427. NASA photograph S-71-53148, courtesy of R. Laughan, Johnson Space Center, Houston.

No large granite masses, so common on Earth, have been found on the Moon. The presence of a lunar "granite" has been claimed on the basis of mare-basalt mesostasis areas rich in the KREEP component (i.e., mesostasis containing barian feldspar, phosphates, tranquillityite, and zirconolite). In Apollo 14 breccias some fragments have been termed rhyolite and tranquillityite-bearing granophyre (Figure I–6).[34] These areas of mesostasis can hardly be termed granite and at best can be called a "granitic component" only on the basis of their chemistry. In breccia 12013 the granitic component is essentially bimineralic with dominant potash feldspar plus silica.[35]

A satisfactory classification of the lunar rocks must be left to the petrologists who are interested even more in the mode of origin of the rocks than in their names. *Lunar Science*, by S. R. Taylor (to be published by Pergamon Press), will contain such a classification.

## *References*

1. Lunar Science Institute, *Post Apollo Lun. Sci.*, 36, 62.
2. Calio, *Apollo 17 Prelim. Sci. Rep.*, xv–xvii.

3. Anderson, *Lun. Sci. IV, Abstr.*, 40–42.
4. Taylor and Jakeš, *Lun. Sci. V, Abstr.*, 786.
5. Biggar et al, *Suppl.* **2,** 636.
6. Nguyen, L. D., et al, *Lun. Sci. IV, Abstr.*, 561.
7. Sato and Hickling, *Lun. Sci. IV, Abstr.*, 650–651.
8. Hodges et al, *Lun. Sci. V, Abstr.*,343–344.
9. ———, *Lun. Sci. IV, Abstr.*, 374.
10. Cadenhead et al, *Lun. Sci. IV, Abstr.*, 110.
11. Howard, *Lun. Sci. IV, Abstr.*, 386.
12. Duennebier and Hutton, *J. Geophys. Res.* **79,** 4351.
13. Latham et al, *Lun. Sci. IV, Abstr.*, 458.
14. Fleischer and Hart, *Apollo 15 Lunar Samples*, 371.
15. Wolfe and Bailey, *Suppl.* **3,** 24.
16. Schaber, *Suppl.* **4,** 73, 76, 91.
17. Young et al, *Apollo 17 Prelim. Sci. Rep.*, 31–10, 31–11.
18. El-Baz, *Apollo 17 Prelim. Sci. Rep.*, 32–12.
19. Bence et al, *Suppl.* **4,** 597.
20. Warner et al, *Lun. Sci. V, Abstr.*, 823.
21. Chao et al, *Suppl.* **1,** 295–296.
22. Gold, *Lun. Sci. IV, Abstr.*, 296.
23. Howard et al, *Suppl.* **3,** 7, 8, 12.
24. Ringwood et al, *Lun. Sci. V, Abstr.*, 636–638.
25. Brown et al, *Lun. Sci. V, Abstr.*, 89.
26. Anderson et al, *Suppl.* **2,** 431.
27. Bansal et al, *Lun. Sci. IV, Abstr.*, 48.
28. Warner, *Suppl.* **3,** 624–630.
29. Dowty et al, *Lun. Sci. V, Abstr.*, 174–175.
30. Albee et al, *Lun. Sci. V, Abstr.*, 5.
31. McKay, *EOS* **54,** 600.
32. Carr and Meyer, *Apollo 15 Lunar Samples*, 49.
33. Prinz et al, *EOS* **54,** 605.
34. Brown et al, *Suppl.* **3,** 154–156.
35. Drake et al, *EPSL* **9,** 121.

## Comparison of Lunar and Terrestrial Mineralogy and Relation of Lunar Mineralogy to Lunar Environment

Of the approximately 100 described lunar phases, about 60 are valid minerals, 14 are tentatively identified, and the remaining phases are either unidentified or doubtful. There are more than 2200 known terrestrial minerals. Mineralogically speaking, the Moon is a poor place.

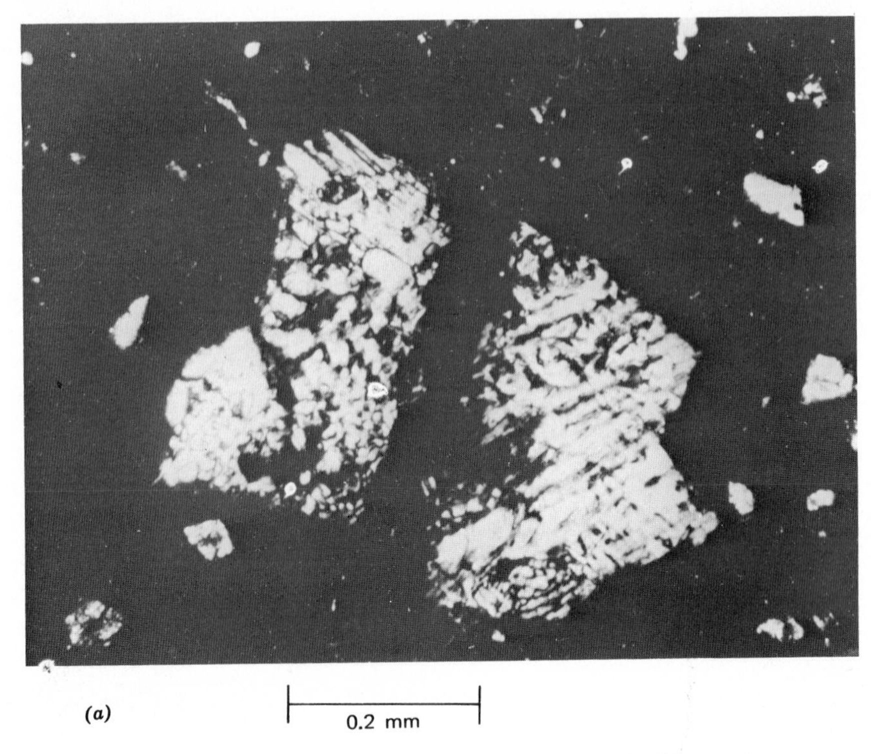


Fig. I–6 (*a*) Irregular rhyolite fragment in 14306,55. Note Na-rich margin, pyroxene needles, and black cubes of Fe metal. From *Suppl.* **3,** 822. (*b*) Granophyric aggregate of quartz and potassium feldspar from polymict breccia 14306,53. From *Suppl.* **3,** 822. Photograph courtesy of J. V. Smith, Department of Geophysical Sciences, University of Chicago.

Since there is no water on the Moon, it can have no minerals formed from alteration of material by weathering processes or hydration. Water is one of the extremely important ingredients in terrestrial regional metamorphism. Water aids in the recrystallization of one phase to enother. The Moon lacks, for example, the amphiboles, the micas, the serpentine minerals, and the clay minerals so plentiful on Earth. And without an atmosphere, no products of oxidation can form on the lunar surface.

Compared with Earth, the Moon has relatively low pressures and temperatures. The contorted and folded rocks of Earth are evidence of the high pressures and temperatures attained during periods of orogeny. Phase changes, recrystallization of original minerals, or disappearance of some phases were brought about by tectonic processes. In early lunar history, tectonic processes were at work also, but with a Moon depleted in elements more volatile than iron, the variety of minerals composing the rocks was limited.

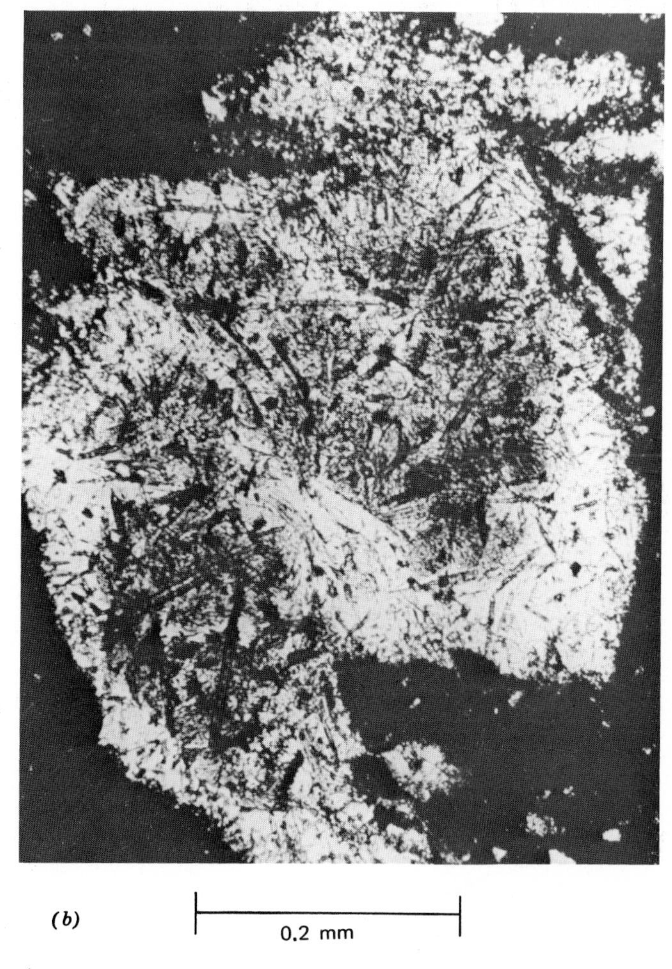


Fig. I–6 (*continued*).

The Moon has no granites nor pegmatites, those terrestrial treasure houses for both miner and mineralogist. From the pegmatites come many of Earth's spectacular specimens of amazing variety, form, beauty, and, often, size.

The Moon has a limited number of "spectacular" specimens. Worthy of mention are coarse-grained basalt 12052 and porphyritic basalt 15495. In 12052, honey-brown pyroxene crystals 3 mm long and 1 mm wide form clusters of blades, radiating from common centers (Figure I–7). A few of the pyroxene crystals are as much as 8 mm long. Interstitial feldspar crystals occur between the pyroxene blades.[1] Acicular ilmenite crystals 5

Fig. I–7 Porphyritic basalt 12052,10 with vugs containing pyroxene crystals. Scale is in centimeters. NASA photograph S-70-44633, courtesy of R. Laughan, Johnson Space Center, Houston.

mm long and 0.5 mm wide were observed lining vugs in the same rock.[2] In 15495, euhedral pyroxene prisms (15 × 2 mm), are exposed in vugs (Figure I–8). The crystals have dark brownish-green exteriors and green cores; they are well formed, with sharp, clean faces.[3]

Some of the Moon's best crystals, minerals that most probably have crystallized from a vapor phase, must be viewed with the aid of scanning electron microscopy (SEM). In vuggy breccias, euhedral crystals of iron growing on plagioclase or pyroxene crystals have been observed. The iron crystals are rare, but where they do occur they are relatively abundant.[4]

These minute crystals, together with similarly small but well-formed apatite and whitlockite crystals, must be considered to be the Moon's gems.

Fig. I–8 Porphyritic basalt 15495 with euhedral pyroxene prisms exposed in vugs. NASA photograph S-71-44206 courtesy of R. Laughan, Johnson Space Center, Houston.

There are no lunar diamonds. Carbon and very high temperatures and pressures are needed to form diamonds. There are no rubies or sapphires on the Moon. The lunar corundum that occurs is of insignificant amount and suspicious origin. Only one garnet crystal has been found in all the examined lunar samples. Zircons are not uncommon in the mesostasis of some lunar rocks, but the zircon crystals are too small and too few to have other than petrogenic importance.

Metallic iron is widespread on the Moon, but in quantity the iron is only a minor accessory mineral, and much of it is a meteoritic addition to the lunar material. There are no gold, no silver, or other precious metals, except traces in parts per million (ppm) or parts per billion (ppb) in some rocks. A trace or two of native copper occurs together with some minor sulfides. One phosphide, schreibersite, is another minor accessory mineral. It is found together with the carbide cohenite, and both are probably meteoritic contributions to the Moon. There are no halides or (authentic) carbonates. There is only a limited phosphate mineralogy that occurs in the mesostasis of rocks with a KREEP component.

Except for ubiquitous and plentiful ilmenite, simple oxides are represented on the Moon mainly by the accessory minerals rutile and baddeleyite. A number of $SiO_2$ polymorphs, cristobalite, tridymite, and quartz, occur in very small amounts. Zirconolite and armalcolite, although minor in amount, are interesting multiple oxide phases. Detailed investigation of the former has revealed its structural formula (a problem of terrestrial mineralogical origin) and, perhaps, resolved the controversy over its name (*zirkelite* has priority over *zirconolite*—see Chapter 5). Armalcolite is a new mineral, although it is a lunar analog of a terrestrial phase. Spinels are important lunar multiple oxide phases. They are abundant accessories in the lunar rocks and have interesting chemical compositions (differing somewhat from terrestrial spinels) that are of considerable petrogenic significance.

Silicates form the bulk of the lunar rocks, but again, the variety is restricted. The feldspars and pyroxenes are the chief rock components. The feldspars are primarily plagioclase and mainly highly calcic plagioclase of very narrow compositional range. Potash feldspars occur in small amounts in the mesostasis of some of the rocks, especially those with the so-called granitic component. The pyroxenes are limited to the Mg, Fe, and Ca varieties. One unusual iron-rich pyroxenoid, not found in Earth, occurs as a late-forming phase. It is the new mineral, pyroxferroite. The olivines compose the third large rock-forming group of lunar minerals. Their compositions are commonly magnesium-rich, with iron-rich fayalite occuring only as a late accessory mineral. One complicated silicate of iron, titanium, and zirconium is the new mineral tranquillityite, a minor accessory having no terrestrial analog.

The lunar minerals have no great variety and, obviously, no potential for economic exploitation. Their importance lies in the information they yield concerning the processes of lunar rock formation, and hence, the origin and history of the Moon.

## *References*

1. Heiken and Anderson, *Lunar Sample Information Catalog, Apollo 12*, 150.
2. Lindsay, *Lunar Sample Information Catalog, Apollo 12*, 150.
3. Reid and Bass, *Lunar Sample Information Catalog, Apollo 15*, 225.
4. Clanton et al, *Lun. Sci. IV, Abstr.*, 143.

# CHAPTER ONE | NATIVE ELEMENTS

## Metallic Iron: Kamacite and Taenite, Doubtful Cr-Fe Metal, Unidentified Phase

Before the first lunar landing it was expected that large quantities of meteoritic metal would be found on the Moon's surface. The Apollo 11 samples, however, contained only a minor amount of metallic iron, and much of it appeared to be of lunar rather than extralunar origin. Subsequent missions have yielded much more meteoritic iron, especially from samples of highland rocks, but the individual particles are small (generally $\sim$ 125 to 250 $\mu$). No large masses of metal, similar to many meteorite specimens found on Earth, have been discovered on the Moon.

In soil samples from the highland sites of Apollo 14, 16, and 17, and Luna 20, meteoritic metal is more than 75% of the total metal fragment occurrence. Much of this metal has lost its meteoritic microstructure (e.g., Widmanstätten structure) by shock processes from cratering events and meteorite impacts that served to liberate the metal from the rock. Some metallic particles, with both meteoritic structure and composition show no signs of having been reprocessed. The metal appears to have been liberated from its host rock without suffering undue shock effects, probably because the more brittle silicates tend to crack instead of the more ductile metal.[1,2] In some rocks (e.g., KREEP basalt 14310 and anorthositic gabbro 68415) there may be a mixture of meteoritic and lunar metal, supporting a possible remelt hypothesis for the origin of these rocks.[3] Even though about 80% of the metallic particles in the Apollo 12 soil probably were derived from the igneous rocks, the iron particles larger than 125 $\mu$ are of meteoritic origin. These meteoritic inclusions may have been added by low-velocity impacts in the lunar soil during major meteoritic bombardment on the Moon.[4]

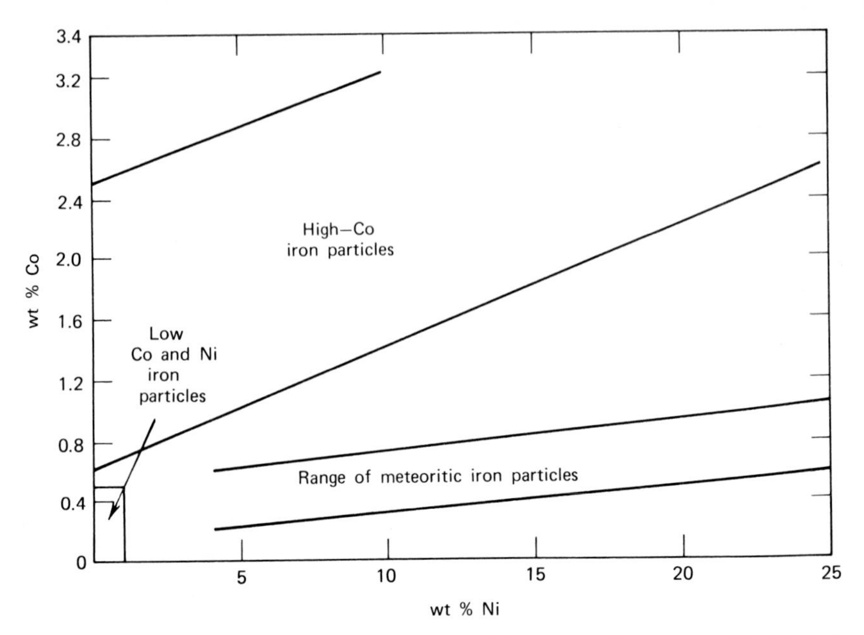


Fig. 1–1 Diagram of Co/Ni in lunar iron particles summarizes suggested criteria for determining meteoritic origin of the particles. Adapted from Goldstein and Axon, *Apollo 15 Lunar Samples*, 81.

Criteria for determining the meteoritic origin of iron particles have been based on their Co/Ni ratios (Ni content $> 4\%$; Co content $< 1\%$: Figure 1–1).[4] It has been advised, however, that these criteria be used as the source of strong suggestions rather than definitive indications.[3]

Much of the metallic iron from the mare basalts of Apollo 11, 12, 15, and 17 apparently originated in the igneous rocks. Some of these iron particles may be the result of reduction of iron-bearing silicates or glass during shock and/or thermal metamorphism.[5] Some kamacite may have come from a breakdown of fayalite-rich olivine (e.g., in basalt 14053, pure iron is associated with glass mainly of an MgO and $SiO_2$ composition).[6] Some euhedral iron crystals found in vugs of Apollo 14, 15, and 16 breccias appear to have grown in a vapor phase close to equilibrium conditions.[7]

Kamacite ($\alpha$-iron) is stable at low temperatures and has low Ni content. Taenite ($\gamma$-nickel-iron), which has a high Ni content, is the high-temperature form.[1] If the amount of Ni, however, is greater than 27%, the taenite is stable at low temperatures.[8] At intermediate temperatures both the $\alpha$ and $\gamma$ phases are stable and may coexist. If kamacite is reheated to this region of intermediate temperatures, small laths of nickel-rich $\gamma$ iron may exsolve from the $\alpha$-iron parent. Conventionally, this $\gamma$-iron phase is termed "isothermal taenite."[1] Both kamacite and taenite have been found among the metallic iron particles of meteoritic origin and of those indigenous to the Moon. In addition, many two-phase particles of $\alpha + \gamma$ iron have been observed. Some have structures that could be termed "isothermal taenite," but many others show chemically zoned taenite such as could have formed with the growth of kamacite as the parent taenite slowly cooled through the region of intermediate temperatures. Detailed studies of the microstructure of these lunar iron particles permit estimation of their equilibrium temperatures.[1]

*References*

1. Goldstein and Axon, *Suppl.* **4,** 751, 756–757, 769–770.
2. Goldstein et al, *Suppl.* **3,** 1037, 1062.
3. Taylor et al, *Suppl.* **4,** 826–827.
4. Goldstein and Yakowitz, *Suppl.* **2,** 179, 189–190.
5. Goldstein et al, Preprint, *Metal Silicate Relationships in Apollo 17 Soils.*
6. El Goresy et al, *EPSL* **13,** 121, 127.
7. Clanton et al, *Suppl.* **4,** 925.
8. Goldstein and Ogilvie, *Trans. AIME* **233,** 2083.

## KAMACITE $\alpha$-Fe or $\alpha$-(Ni,Fe)

*Synonymy*

Cobaltian metallic nickel-iron; Keil et al, *Suppl.* **2,** 332.
Cobaltian Ni-Fe; Prinz et al, *Meteorit.* **6,** 301.
Cobalt nickel-iron; Simpson and Bowie, *Suppl.* **2,** 207.
Fe-metal; El Goresy et al, *Suppl.* **3,** 334.
Iron; Brown et al, *Suppl.* **3,** 153.
Iron-nickel; Frondel et al, *Suppl.* **1,** 467.
Metallic Fe; Roedder and Weiblen, *Suppl.* **3,** 259.
Metallic FeNi (in part); El Goresy et al, *Suppl.* **2,** 220.
Metallic iron; Haggerty, *Suppl.* **3,** 307.
Native iron; Bailey et al, *Suppl.* **1,** 179.
Nickeliferrous Fe; Cameron, *Suppl.* **1,** 239.
Pure Fe; El Goresy et al, *EPSL* **13,** 121.
Ti- and Cr-bearing metallic iron; Melson et al, *Lun. Sci. III, Rev. Abstr.,* 536.

*Occurrence and Form*

In the crystalline rocks kamacite is found as rounded blebs with occasionally poorly developed polygonal outlines.[1] Although kamacite grains have

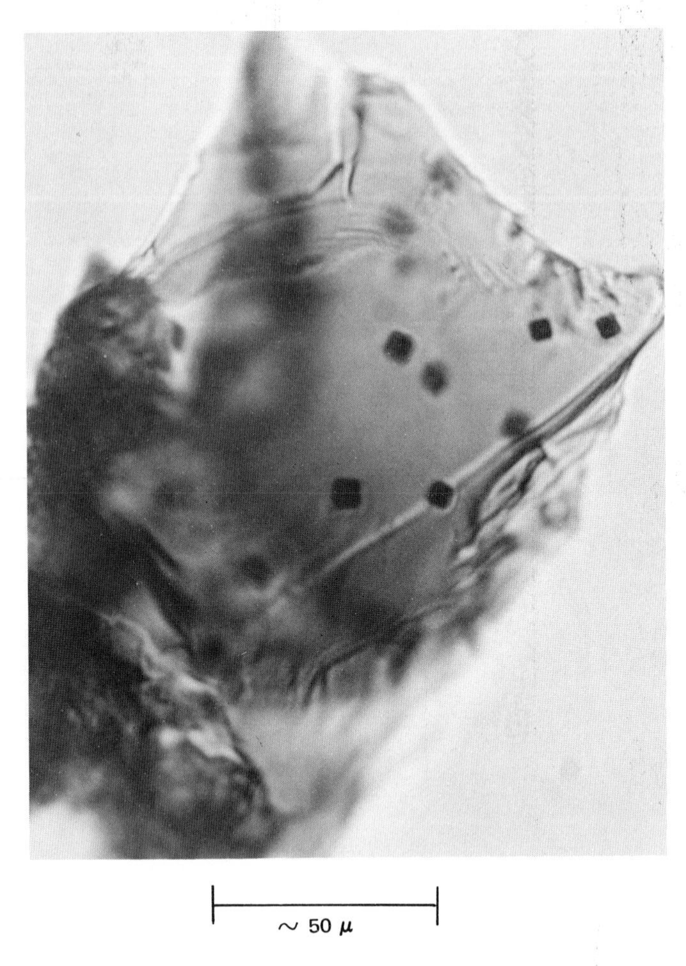


Fig. 1–2 Cubes of iron (or Ni-Fe) in a glass fragment from a fines sample. From *Suppl.* **1,** 464. Photograph courtesy of C. Frondel, Harvard University, Cambridge, Mass.

been observed isolated from troilite,[2] in general these two phases are closely associated or intergrown, suggesting formation from an iron sulfide melt that separated immiscibly from the silicate magma.[3] Like troilite, kamacite is a late-stage crystallization product commonly associated with other residual products in the mesostasis areas outlined by feldspar–pyroxene frameworks.[4]

In the breccias and soil, the iron occurs as dendrites, droplets, and irregular fragments, as well as spheroidal-discoidal or spindle-shaped bodies. Some irregular grains are intergrown with plagioclase and pyroxene. Thin

Fig. 1–3 Neumann bands in meteoritic kamacite. From *Suppl.* **2,** 183. Photograph courtesy of J. I. Goldstein, Department of Metallurgy and Materials Science, Lehigh University, Bethlehem, Pa.

films or spheroidal particles have been observed in crustiform glass, and in one glass fragment there are minute cubes of kamacite (Figure 1–2).[5] From fines sample 10084, the kamacite inclusions are less than 30 $\mu$ in size, often rounded, and swathed in troilite; the microstructure indicates shock hardening.[6] From fines sample 10085-4 there are also a few angular or wirelike fragments of a meteoritic kamacite with well-developed Neumann bands.[7] From 10085-17 M, among several fragments of meteorites, a single crystal of kamacite was found containing a complex array of Neumann and other deformation bands. The flowed Neumann bands are evidence of postshock plastic deformation (Figure 1–3).[8] Also from 10085 there is a nickel-iron pellet weighing 88 mg and measuring $3.5 \times 3.2 \times 2.3$ mm (Figure 1–4*a*). It is one of the largest meteorite inclusions in the lunar soil. The pellet consists of approximately 85% kamacite and 15% troilite, estimated from the pellet's density of 7.36. The internal structure of the pellet (Figure 1–4*b*) is a dendritic intergrowth of iron and troilite.[9]

Fig. 1–4 (*a*) Nickel-iron pellet from fines sample 10085. From *Suppl.* **1**, 658. (*b*) Polished section of nickel-iron pellet showing dendritic agregate of taenite (white) and troilite (gray). From *Suppl.* **1**, 659. Photographs courtesy of B. Mason, Smithsonian Institution, Washington, D.C.

The association of metallic iron with troilite, which is pronounced in Apollo 11 rocks, is less common in Apollo 12 and 14 samples.[10] The metal phase in Apollo 12 samples has a complicated paragenesis; for example, in porphyritic basalt 12063,9, the iron occurs with the oxides, with pyroxene and plagioclase, with troilite and ilmenite in the fayalite–cristobalite–glass assemblage, and in the end stage as a eutectic intergrowth with troilite. In porphyritic basalt 12004,11, kamacite (and taenite also) occurs as inclusions in early formed olivine.[11] In KREEP basalt 14310, the iron is

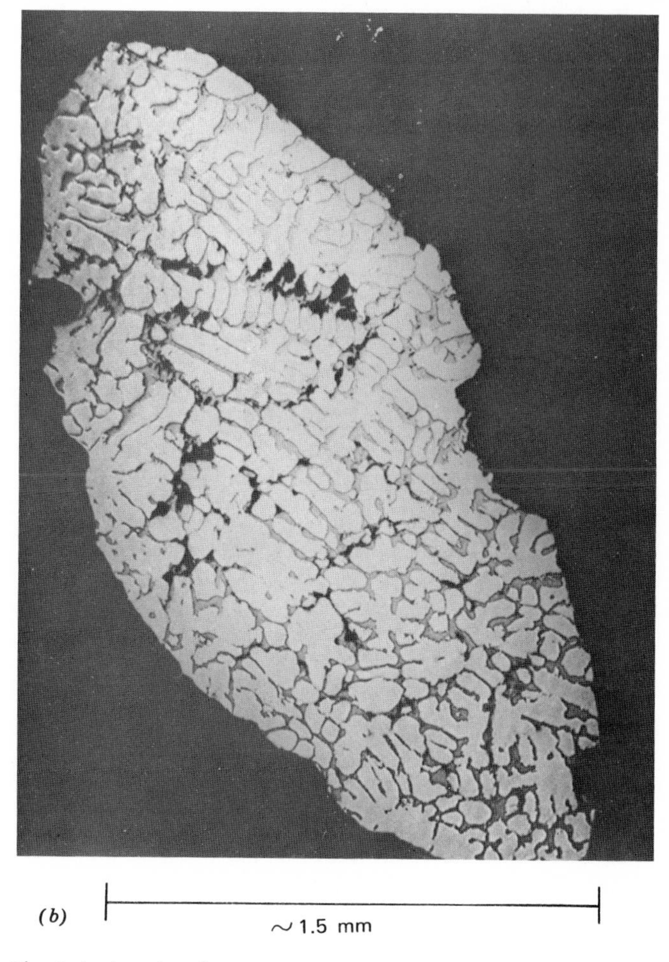


Fig. 1–4 (*continued*).

found both as discrete, irregular grains and as inclusions in troilite. The metal–troilite aggregates range widely in their Fe/FeS ratios. Metal precipitation began at the onset of pyroxene crystallization and continued throughout the crystallization sequence, with the last metal being formed by subsolidus reduction of ulvöspinel. Metal is abundant in the mesostasis.[12]

In clasts of breccia 14068, next to plagioclase the most important mineral inclusion is kamacite. It occurs as subrounded blebs with splattery outlines and stringers of metal and smaller satellite blebs surrounding them

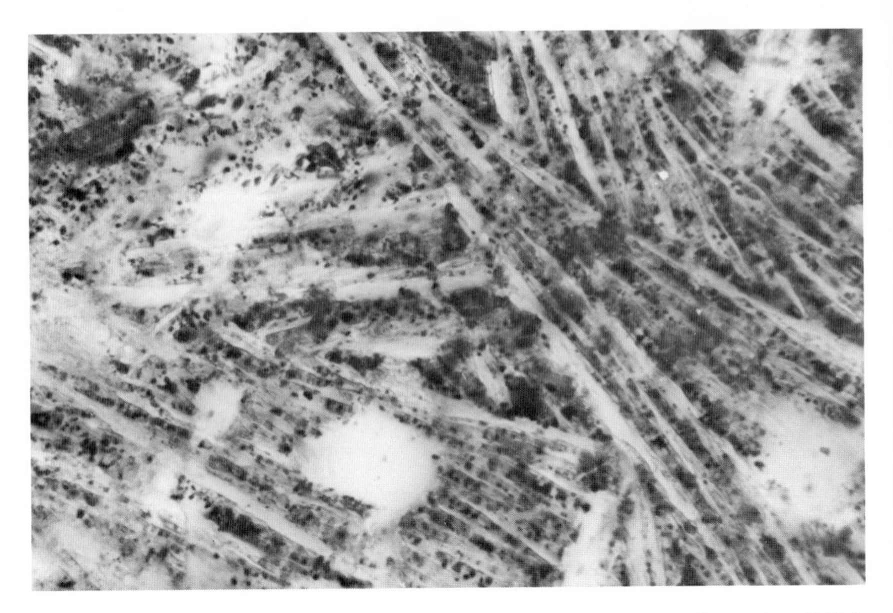

Fig. 1–5 Abundant subrounded blebs of kamacite in thin section of breccia 14068,7. NASA photograph S-71-40812, courtesy of R. Laughan, Johnson Space Center, Houston.

(Figure 1–5).[13] Kamacite is present in soil sample 15602 as discrete rectangular grains up to 400 $\mu$, as interlacing stringers along silicate grain boundaries, and as spherules ($< 1$–$20\ \mu$) in eutectic intergrowth with troilite.[14] In some of the Apollo 16 cataclastic and highly metamorphosed breccias, the kamacite is found as large rounded blebs up to 1 mm in size in the silicate matrix, or as local concentrations of grains filling the interstices between the silicates. Many of the metal blebs are rimmed by small troilite grains.[15] In gabbroic anorthosite 68415, kamacite occurs typically as blebs in the residual glass. Less commonly the iron is included within bordering plagioclase and pyroxene. The composition of the iron is consistent with that of meteoritic origin, and its spherical or globular shape suggests separation due to liquid immiscibility.[16]

In some Luna 20 samples needle-shaped[17] or rodlike iron crystals (0.5 $\mu$ wide and nearly 100 $\mu$ long) have been observed as inclusions with preferred orientations in plagioclase crystals (Figure 1–6).[18] The preferred orientations suggest that the iron crystals were exsolved and aligned by crystallographic control of the host plagioclase.[17]

Kamacite crystals of remarkable perfection have been found in vugs of breccias from Apollo 14, 15, and 16 samples. Only a small proportion of the breccias contain the iron crystals, but they are abundant where they do

Fig. 1–6 Oriented crystals of metallic iron in Luna 20 plagioclase. Reflected light; width of inclusions = 0.14 $\mu$. From *Geochim. Cosmochim. Acta* **37**, 758. Photograph courtesy of P. Bell, Geophysical Laboratory, Washington, D. C.

occur, and they appear to have grown in a vapor phase on a substrate of plagioclase and pyroxene crystals (Figure 1–7). The iron crystals are from 1 to 17 $\mu$, with dimensions along their crystallographic axis ranging from less than 1 to 10 $\mu$. Observation of these tiny crystals, by means of stereophotographs taken with a scanning electron microcsope has permitted the identification of the following habits (Figures 1–8 – 1–10):

1. In one group (from Apollo 14, 15, and 16 breccias) the trapezohedron $\{hll\}$ predominates and is modified with smaller cube faces $\{100\}$ (Figure 1–8).
2. Another group (found in Apollo 15 breccias) has the cube $\{100\}$ as the dominant form with the trapezohedron $\{hll\}$ and tetrahedron $\{hk0\}$ faces smaller and about equally developed.
3. Still another group (from Apollo 14 and 15 breccias) has the octahedron $\{111\}$ dominant with smaller but equally well-developed cube $\{100\}$ and dodecahedron $\{110\}$ faces (Figure 1–9).[19]

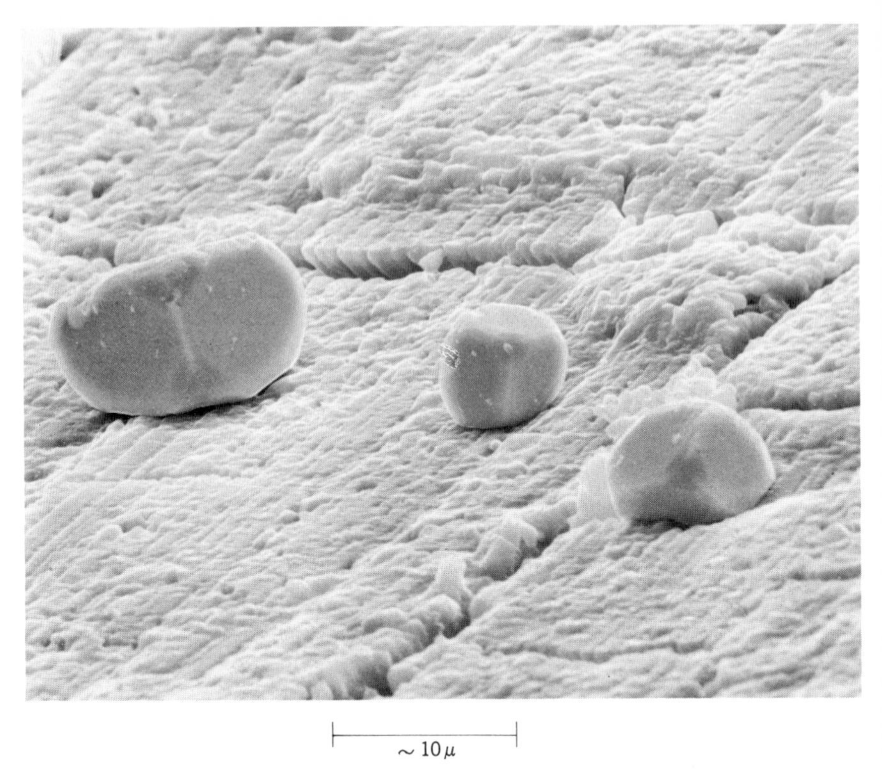


Fig. 1–7 SEM photograph of iron crystals in a grain of pyroxene from breccia 15402. The iron crystallized after the stepped pyroxene overgrowth was formed. From *Suppl.* **4,** 927. NASA photograph S-72-53357, courtesy of U. Clanton, Johnson Space Center, Houston.

4. From breccia 67482 one iron crystal has the trapezohedron {*hll*} dominant, with modifications by the cube {100} and trisoctahedron {*hkl*} faces (Figure 1–10).[20]

Although much kamacite occurs as single crystals, it is not uncommonly found as part of two-phase or multiphase assemblages. In Apollo 14 and to a lesser extent Apollo 16 samples, kamacite with troilite (Figure 1–11) and kamacite with cohenite have been observed. From all Apollo sites, and the Luna 20 sites as well, iron-troilite-schreibersite spherules have been found. The internal metallographic structure of these spherules is consistent with the rapid resolidification of molten droplets of iron-sulphur-phosphorus liquids.[21] In a number of Apollo 16 soil samples, single crystals of kamacite are the predominant iron phase. Some of the crystals contain

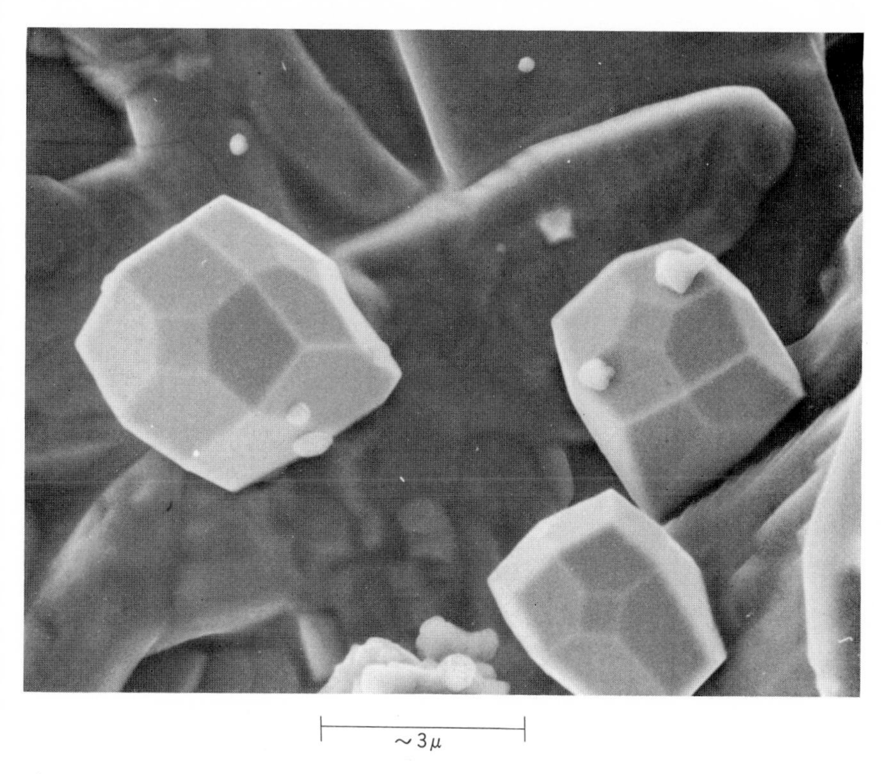


Fig. 1–8 SEM photograph of iron crystals showing dominant trapezohedral faces and smaller cube faces. From *Lun. Sci. IV. Abstr.*, 144. NASA photograph S-72-55208, courtesy of U. Clanton, Johnson Space Center, Houston.

an occasional troilite grain or incomplete troilite rim. In about 30% of the kamacite particles, however, there are some small rounded or lamellar grains of schreibersite.[22] Two-phase particles of kamacite and taenite have been noted in the soils of all the lunar missions and have been studied in great detail by Goldstein and Yakowitz, Goldstein et al, Goldstein and Axon, Goldstein and Blau, and Axon and Goldstein.[23–26]

## *References*

1. Bailey et al, *Suppl.* **1,** 179.
2. Brown et al, *Suppl.* **1,** 206.
3. Skinner, *Suppl.* **1,** 894.
4. Haggerty et al, *Suppl.* **1,** 528.
5. Frondel et al, *Suppl.* **1,** 467.
6. Adler et al, *Suppl.* **1,** 92.

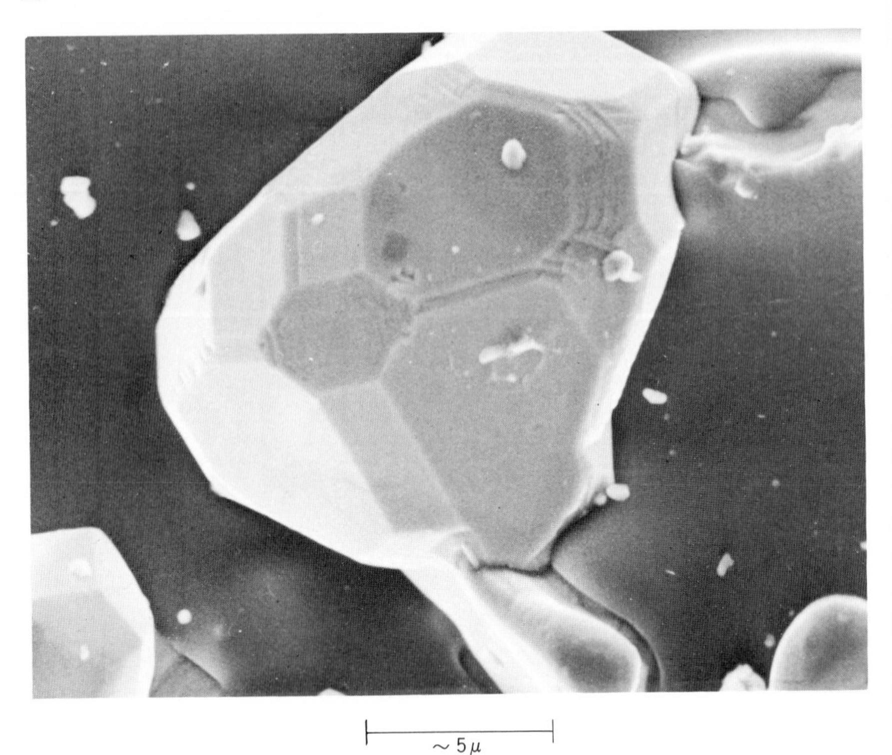


Fig. 1–9 SEM photograph of an iron (?) crystal coated with iron sulfide, presumably troilite. The dominant form is the octahedron. The cube and dodecahedron faces are smaller and less well developed. From *Lun. Sci. IV*, *Abstr.*, 145. NASA photograph S-72-53687, courtesy of U. Clanton, Johnson Space Center, Houston.

7. Agrell et al, *Sci.* **167,** 586.
8. Goldstein et al, *Suppl.* **1,** 509–511.
9. Mason et al, *Suppl.* **1,** 658–660.
10. El Goresy et al, *Suppl.* **3,** 334.
11. Taylor et al, *Suppl.* **2,** 863.
12. Ridley et al, *Suppl.* **3,** 163–165.
13. Helz, *Suppl.* **3,** 871–872.
14. Haggerty, *Apollo 15 Lunar Samples*, 85.
15. El Goresy et al, *Suppl.* **4,** 737–738.
16. Gancarz et al, *EPSL* **16,** 316.
17. Bell and Mao, *Geochim. Cosmochim. Acta* **37,** 757.
18. Brett et al, *Geochim. Cosmochim. Acta* **37,** 770.
19. Clanton et al, *Suppl.* **4,** 925–926, 931.

Fig. 1–10 SEM photograph of an rion crystal showing combination of trapezohedron {*hll*} (*n*), cube {100} (*a*), and trisoctahedron {*hhl*} (*p*) faces. Note the two sets of trapezohedron faces. NASA photograph S-73-35143, courtesy of U. Clanton, Johnson Space Center, Houston.

20. ———, *Lun. Sci. V*, *Abstr.*, 126.
21. Goldstein and Axon, *Suppl.* **4,** 760, 768–769.
22. Reed and Taylor, *Meteorit.* **9,** 24–25.
23. Goldstein and Yakowitz, *Suppl.* **2,** 186–187.
24. Goldstein et al, *Suppl.* **3,** 1041–1042.
25. Goldstein and Blau, *Geochim. Cosmochim. Acta* **37,** 848–849.
26. Axon and Goldstein, *EPSL* **18,** 173–180.

## *Optics*

The kamacite (as observed in samples exposed only to dry nitrogen at the Lunar Receiving Laboratory) is bright silver white with a brilliant metallic luster. Parts of material exposed to the terrestrial atmosphere for some

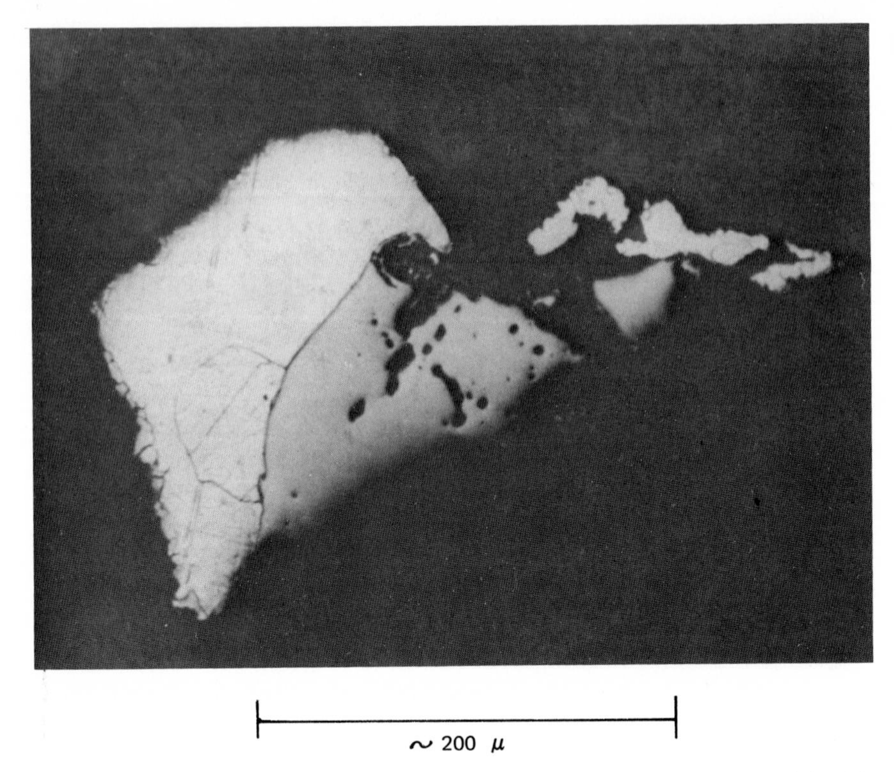


Fig. 1–11 Low-nickel $\alpha$ phase (kamacite) associated with massive sulfide in particle 22.9. Etched 1% Nital. From *Suppl.* **3,** 1051. Photograph courtesy of J. I. Goldstein, Department of Metallurgy and Materials Science, Lehigh University, Bethlehem, Pa.

months acquired a faint brown to bronzy tarnish or became rusty. Aggregates of irregular grains have a rough surface and an iron-black color.[1]

*Reference*

1. Frondel et al, *Suppl.* **1,** 467.

*Chemical composition*

Although kamacite (i.e., $\alpha$-[Fe,Ni], with space group $F_{m3m}$) may contain as much as 25% Ni,[1] the kamacite particles on the Moon generally have a nickel content below 10%. Lunar kamacite can be divided into various groups based on the amounts of Co and Ni (Figure 1-1).

Some kamacite is virtually pure iron or extremely low in both Ni and Co. Either the vapor-phase deposited kamacite crystals (seen in breccia vugs) have no Ni and Co, or these elements occur in amounts less than about 0.5%.[2] No Ni has been detected in the kamacite globules included in troilite from basalts 10044-50 and 10058-22 (Table 1-1, [1]).[3] In basalt 14053, the iron phase (interpreted as a breakdown of fayalite-rich olivine) contains less than 0.02% Ni.[4] In the same rock, kamacite associated with reduction of late-stage Fe-Ti-Cr oxide grains, contains 1.4% Ti and 0.5% Cr. It is possible, however, that these contents are due to smearing of adjacent grains, or secondary X-ray fluorescence may be involved.[5] In fine-grained plagioclase aggregates of anorthosite 60015, there are 15-$\mu$ grains of kamacite in which the Co content is only 0.20 to 0.25% and the Ni content is 0.10 to 0.15%. This metal may have originated by reduction of Fe, Co, and Ni ions in the shock-produced liquid of plagioclase composition; apparently it crystallized from the plagioclase melt as a relatively late interstitial phase.[6]

Kamacite, with low to moderate Ni content and Co generally about 1 to 3%, could be termed cobaltian kamacite. This type has been found in porphyritic basalt 12021 and ophitic basalt 12051,59 (Table 1-1; [3], [4]). Another type of cobaltian kamacite has a high Co content and a medium Ni content. Such kamacite has been found in $\alpha + \gamma$ iron particles from Apollo 15 soil samples (Table 1-1; [6], [7]).[7] Kamacite with 5.9% Co and 8% Ni occurs in mare basalt 15076,12, but this rock also contains kamacite with only 0.8% Co and 0.2% Ni.[8]

Kamacite of meteoritic origin, according to the criteria of Goldstein and Yakowitz,[9] has Co $< 1\%$ and Ni $> 4\%$. Kamacite with such a composition is abundant in Apollo 14 and 16 samples (Table 1-1; [8]–[11]). The phosphorus content of kamacite–schreibersite particles in microbreccia 66055 is appreciably higher than that associated with meteoritic metal (which ordinarily has about 1% P). Many of the composite metal blebs have a P content exceeding 2%, and in one particle the bulk P content is 7%. This argues against meteoritic origin for the particles, although the Ni-Co concentrations in the kamacite (Table 1-1 [5]) fall within the limits for meteoritic metal. Hence it remains to be seen whether such Ni-Co values can be produced also under lunar conditions.[10]

In an enstatite chondrite from soil sample 15602,29, kamacite grains (400 $\mu$), with an Ni content of 6.4 to 6.9%, have a relatively constant Si content of 2.9 to 3.2%. This amount is characteristic of highly reduced enstatite chondrites to which the Fe-Ni-Si solid solution is unique.[11]

## *References*

1. Strunz, *Tabellen*, 96.
2. Clanton et al, *Lun. Sci. IV*, *Abstr.*, 143.
3. Cameron, *Suppl.* **1**, 237–239.

**Table 1-1 Analyses**

| | [1] 10058 | [2] 10084 | [3] 12021 | [4] 12051,59 | [5] 66055 | [6] 15261 | [7] 15071 |
|---|---|---|---|---|---|---|---|
| Si | 0.12 | — | 0.16 | — | — | — | — |
| Ti | — | — | 0.03 | — | — | — | — |
| Cr | 0.09 | — | 0.05 | — | — | — | 0.46 |
| Fe | 98.95† | 99.2 | 97.54 | 95.7 | 94.1 | 88.4† | 82.74† |
| Ni | — | 0.1 | 0.61 | 1.72 | 4.4 | 4.7 | 5.0 |
| Co | 0.80 | 0.7 | 1.24 | 2.59 | 0.56 | 6.9 | 11.8 |
| Total | 100.0* | 100.0 | 99.70* | 100.01 | 99.25* | 100.0* | 100.0* |

*References*

[1] Brown et al, *Suppl.* **1,** 206. From a medium-grained basalt. *Analysis includes Ca = 0.04%. †Fe by difference. Kamacite.

[2] Agrell et al, *Suppl.* **1,** 112. From a fines sample. Kamacite.

[3] Weill et al, *Suppl.* **2,** 417. From a porphyritic basalt. *Analysis includes Mn = 0.07%. Cobaltian kamacite.

[4] Keil et al, *Suppl.* **2,** 331. From an ophitic basalt. Cobaltian kamacite.

[5] McKay et al, *Suppl.* **4,** 814. From a microbreccia. *Analysis includes P = 0.19%. Kamacite.

[6] Axon and Goldstein, *EPSL* **18,** 175. From a soil sample. *Analysis includes P = < 0.02%. †Fe by difference. From a composite $\alpha + \gamma$ iron particle. Cobaltian kamacite.

[7] ———, *ibid*, 176. From a soil sample. *Analysis includes P = < 0.02%. †Fe by difference. From a composite $\alpha + \gamma$ iron particle. Cobaltian kamacite.

4. El Goresy, *EPSL* **13,** 121, 127.
5. Melson et al, *Lun. Sci. III, Rev. Abstr.*, 536.
6. Sclar and Bauer, *Lun. Sci. V*, Abstr., 687.
7. Axon and Goldstein, *EPSL* **18,** 175–176.
8. Brown et al, *Suppl.* **3,** 153.
9. Goldstein and Yakowitz, *Suppl.* **2,** 186–187.
10. McKay et al, *Suppl.* **4,** 814–815, 817–818.
11. Haggerty, *Apollo 15 Lunar Samples*, 85.

## TAENITE γ-(Fe,Ni)

*Synonymy*

FeNi; Taylor et al, *Suppl.* **2,** 863.
Iron-nickel; Battey et al, *Lun. Sci. III, Rev. Abstr.*, 46.

**of Lunar Iron**

| | [8] 66095,89 | [9] 60601 | [10] 14276 | [11] 68415 | [12] 10085 | [13] 12004,11 | [14] 15261 |
|---|---|---|---|---|---|---|---|
| Si | — | — | — | — | 0.14 | — | — |
| Ti | — | — | — | — | 0.05 | — | — |
| Cr | — | — | — | — | 0.04 | <0.03 | — |
| Fe | 94.3 | 93.2 | 91.71 | 89.60 | 85.82† | 67.4 | 37.8† |
| Ni | 5.51 | 6.0 | 7.45 | 9.18 | 13.4 | 26.7 | 60.0 |
| Co | 0.37 | 0.4 | — | — | 0.48 | 2.37 | 1.3 |
| Total | 100.32* | 99.6* | 99.16 | 98.78 | 100.00* | 96.47 | 100.0* |

[8] El Goresy et al, *EPSL* **18,** 412. From a metamorphosed breccia. *Analysis includes P = 0.14%. Kamacite.
[9] Wlotzka et al, *Lun. Sci. IV, Abstr.*, 789. From a fines sample. *Analysis includes: Cu = 340 ppm, Ir = 1.45 ppm, W = 80 ppm, Au = 1.20 ppm, As = 22 ppm. Kamacite.
[10] Gancarz et al, *EPSL* **16,** 308–309. From a KREEP basalt. Kamacite.
[11] ———, *ibid*, 308. From a gabbroic anorthosite. Kamacite.
[12] Brown et al, *Suppl.* **1,** 206. From a soil sample. *Analysis includes Mn = 0.07%. †Fe by difference. Taenite.
[13] Taylor et al, *Suppl.* **2,** 864. From a porphyritic basalt. Inclusion in olivine. Taenite.
[14] Axon and Goldstein, *EPSL* **18,** 185. From a soil sample. *Analysis includes P = < 0.02%. †Fe by difference. From a composite $\alpha + \gamma$ iron particle. Taenite.

Metallic FeNi (in part); El Goresy et al, *Suppl.* **2,** 220.
Metallic Fe-Ni-Co alloy; Dowty et al, *Suppl.* **4,** 437.
Metallic NiFe; Ridley et al, *Meteorit.* **6,** 304.
NiFe; Chao et al; *Suppl.* **3,** 656.
NiFe-metal; Taylor et al; *Suppl.* **3,** 1002.
Nickel-iron; Ridley et al, *Suppl.* **3,** 163.

*Occurrence and Form*

In fines sample 10085-17M, taenite occurs as rounded particles about 5 $\mu$ in diameter,[1] or as minute spheres (1 $\mu$ or less) concentrated at the periphery of 1-mm glassy spherules.[2] A nickel-iron pellet from 10085 contains dendritic intergrowths of troilite and taenite.[3] Taenite associated with troilite has been found also in KREEP basalt 14310 and anorthositic gabbro 68415.[4]

Taenite has been observed in Apollo 12 basalts either partly or wholly included in olivine that has crystallized early in the paragenetic sequence.[5] In dunite fragments 72415 to 72418, olivine grains are "dusted" with tiny ($< 1\ \mu$) inclusions of taenite and spinel.[6]

In a number of Apollo 14 and 15 fines samples, both single-phase taenite grains and composite $\alpha + \gamma$ iron particles have been found. In the two-phase assemblages there are Widmanstätten structures similar to but often on a finer scale than those found in iron meteorites. [During cooling of iron meteorites kamacite exsolves crystallographically on the (111) planes of the parent taenite, forming the Widmanstätten pattern.[7]]

*References*

1. Goldstein et al, *Suppl.* **1,** 511.
2. Brown et al, *Suppl.* **1,** 211.
3. Mason et al, *Suppl.* **1,** 659.
4. Taylor et al, *Suppl.* **4,** 826–827.
5. Brett et al, *Suppl.* **2,** 312.
6. Albee et al, *Lun. Sci. V, Abstr.*, 3.
7. Axon and Goldstein, *EPSL* **18,** 175–176.

*Chemical Composition*

Although taenite (i.e., $\gamma$-[Ni,Fe], with space group $I_{m3m}$[1]) generally contains more than 30% Ni, lunar iron with Ni as low as 13.4% has been termed taenite.[2] Taenite inclusions in fines sample 10085-17M contain about 15% Ni, which is nearly homogeneously distributed. This taenite also has a P content from 0.15 to 0.3%, and S of approximately 0.06%.[3] From 10085-4, zoned taenite (included in kamacite) has cores with 9% Ni and rims with 14% Ni.[4] In soil sample 14259,17, taenite at the edge of a kamacite fragment contains 37% Ni and 0.7% Co.[5] In particle F5-16 from soil sample 63501, the taenite (in an $\alpha + \gamma$ iron assemblage) is zoned from 15 to 25% Ni, with the Ni content at the kamacite/taenite interface indicating a final equilibrium temperature of 530 to 550°C.[6]

Taenite included in olivine has a wide range of Ni content, from about 26% (Table 1-1; [13]) up to 56%.[7] No nickel has been detected in the host olivine.[8] In crushed dunite fragments 72415,11 and 72415,12, the taenite contains 1.3 to 2.2% Co and 24.5 to 31.8% Ni. It is suggested this may indicate a primary origin of the metal and the primitive nature of the dunite.[9]

In a number of Apollo 15 soil samples single-phase taenite grains have an Ni content of 30 to 60%. In these samples, and soil sample 14003 as well, there are $\alpha + \gamma$ iron particles that lie within or near the meteoritic compositional range. Other $\alpha + \gamma$ iron particles have Ni and Co contents well outside this range. In these particles the Ni content of the taenite is high, but its Co content is always lower that than of the kamacite (Table 1-1; [6], [14]). The Co content, moreover, is highei than that of meteorites of similar structure. The source of the particles, then, may be either meteoritic or lunar; perhaps they developed during the formation of the lunar crust 1 to 20 km beneath the Moon's surface.[10]

### *References*


1. Strunz, *Tabellen*, 96.
2. Brown et al, *Suppl.* **1,** 211.
3. Goldstein et al, *Suppl.* **1,** 511.
4. Agrell et al, *Sci.* **167,** 586.
5. Battey et al, *Lun. Sci. III, Rev. Abstr.*, 46.
6. Goldstein and Axon, *Suppl.* **4,** 761.
7. Brett et al, *Suppl.* **2,** 312.
8. Boyd et al, *Lun. Sci. Conf. '71, Abstr.*, 149.
9. Albee et al, *Lun. Sci. V, Abstr.*, 3.
10. Axon and Goldstein, *EPSL* **18,** 173–176, 178–179.


## Cr-Fe Metal (doubtful)

In a special study of metallic particles in the Apollo 14 soil, one particle (#22.6) is free of Ni and Co and consists of approximately 17% Cr and 81% Fe. The particle may not belong to the lunar sample.[1]

### *Reference*


1. Goldstein et al, *Suppl.* **3,** 1056.


## Unidentified Phase

In norite 62295, within a kamacite grain, one 12-$\mu$ rounded area is composite, one half being schreibersite and the other half an unidentified anisotropic phase of weaker reflectivity and darker "lilac" hue.[1]

### *Reference*


1. Agrell et al, *Lun. Sci. IV, Abstr.*, 16.

## Copper, Brass, Tin, Tentative Nickel and Cu-Ni-Zn-Fe Metal, and Doubtful Indium and Graphite

### COPPER Cu

#### *Occurrence and Form*

In fine grained basalt 10045-35-5, rare microscopic grains of native copper occur (in association with troilite and iron) as a small segregation in troilite at a troilite-ilmenite contact (Figure 1–12). The copper appears to have crystallized as a primary phase.[1] In porphyritic basalts 12018 and 12063, the copper has been found as minute ($\sim 2\ \mu$) veins within ulvöspinel and ilmenite. Here again the copper is in direct contact with metallic FeNi and troilite.[2] A single particle of copper has been identified in ophitic basalt 12040.[3] Copper occurs also as minute angular grains in some Apollo 14 samples.[4] In basalt 14053, together with metallic Fe, troilite, an unidentified $SiO_2$ phase, and other mesostasis minerals, one small stringer was composed of metallic copper.[5]

#### *Chemical Composition*

The Ni and Zn contents of the copper from 14053 are less than 0.1%.[5] Because of the small size of the copper veins in 12018, only semiquantitive probe analysis was made. Copper is the major element, and no nickel or zinc has been detected. The phase is believed to be relatively pure copper that cannot be a contaminant, since the fresh cut surface of the specimen was ground for several microns before being polished.[2]

#### *References*

1. Simpson and Bowie, *Suppl.* **1,** 882.
2. El Goresy et al, *Suppl.* **2,** 220.
3. Walter et al, *Suppl.* **2,** 354.
4. Lunar Sample Preliminary Examination Team, *Sci.* **173,** 684.
5. Roedder and Weiblen, *Suppl.* **3,** 259.

### BRASS Cu + Zn with Minor Sn

#### *Occurrence and Form*

One brasslike grain, found in the heavy residue of coarse-grained basalt 10044, was suspected to be a contaminant.[1] Other brasslike grains, however, have been extracted from Apollo 11 fines samples and from fine-grained basalt 10017-50. In the latter sample, one sickle-shaped brass fragment had firmly adherent pieces of troilite and feldspar.[2] Similar fine-grained metallic fragments occur in breccia 12013,10.[3]

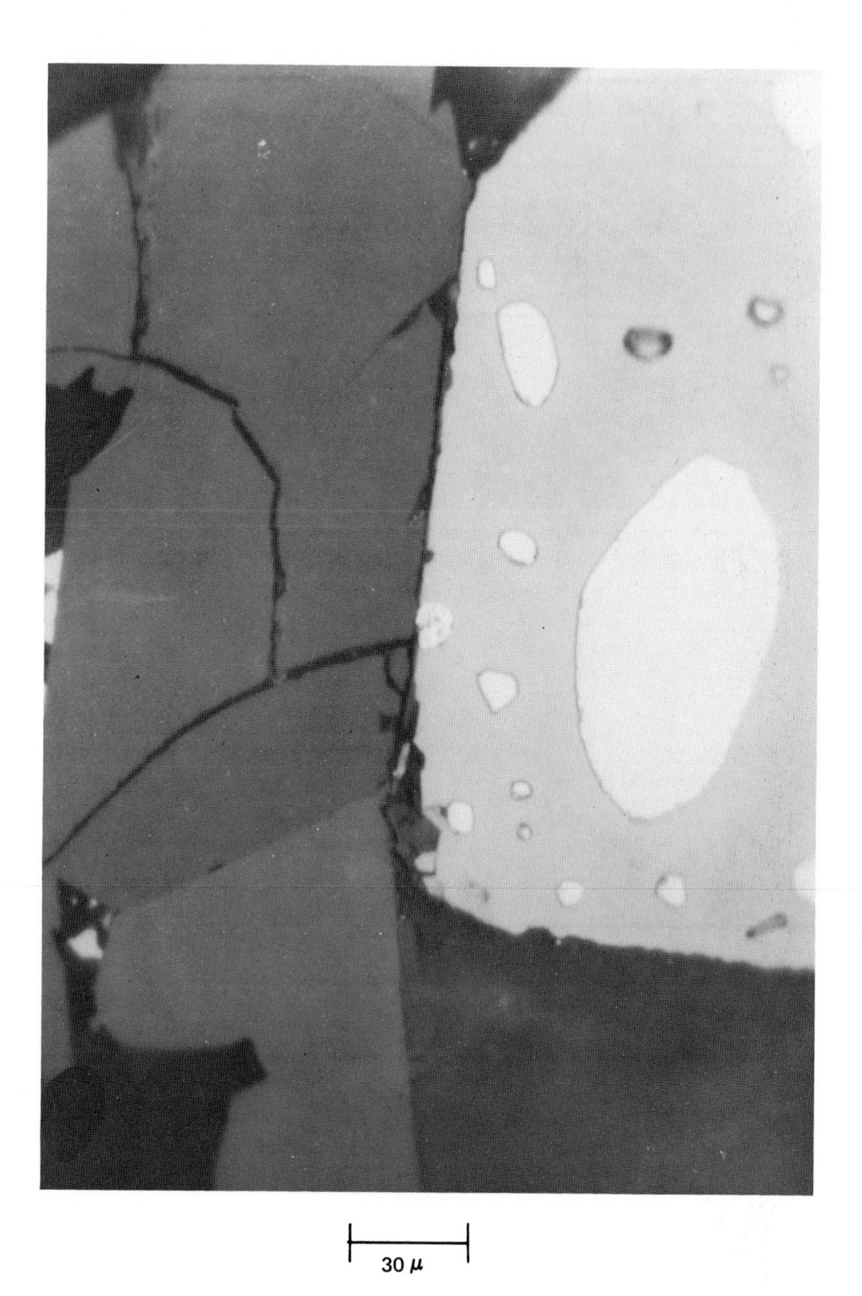


Fig. 1–12 Native copper (small white grain in center of field) and native iron (large white ellipsoid grain) together with troilite (light gray) and ilmenite (dark gray). Polished section, from fine-grained basalt 10045-35-5, viewed in plane-polarized reflected light. From *Suppl.* **1,** 881. Photograph courtesy of P. R. Simpson, Institute of Geological Sciences, London.

*Optics*

The brass fragments have a distinctive golden appearance.[2]

*Chemical Composition*

The brass grain from coarse-grained basalt 10044 has a composition of approximately $Cu_{70}Zn_{30}$, which is within the range of commercial brass, but probe analyses of the brass fragments from 10017 yielded varying proportions of Cu, Zn, and Sn:[2]

$$\begin{aligned} Cu &= 55\text{–}70\% \\ Zn &= 30\text{–}45\% \\ Sn &= 0.3\text{–}5\% \end{aligned}$$

Although terrestrial copper may contain minor Ag, As, Fe, Bi, or Sb, it never has significant amounts of Zn or Sn. The above-mentioned brass, then, appears to be truly lunar.[4]

*X-Ray Data*

Powder patterns of the various brass particles from 10017 are not identical, but they all indicate the presence of a cubic $F$ lattice with $a \sim 3.68$ Å. The powder patterns have no $\alpha$-Fe lines.[2] The powder pattern of brass particles from 12013,10 are identical to those of the other lunar brasses.[3]

*References*

1. Agrell et al, *Suppl.* **1,** 97.
2. Gay et al, *Suppl.* **1,** 483.
3. ———, *EPSL* **9,** 124.
4. *L. and T.*, 90.

## TIN Sn

The polished surface of a small iron fragment from fines sample 10084 showed areas of a more yellowish tint than the iron matrix. Microprobe analysis revealed these areas were tin with only a little iron, and the host iron contained about 1% Ni. Comparison powder photographs of $\beta$-Sn matched the extra lines in the powder pattern of the mixture. The iron fragment is believed to be of lunar origin because of its low nickel content.[1]

*Reference*

1. Gay et al, *Suppl.* **1,** 482.

## Nickel Ni (tentative)

### *Synonym*

Ni metal; Reid et al, *EPSL* **9,** 2.

### *Occurrence and Form*

It was suggested that thin veins of Ni metal, found in porphyritic basalts 12022 and 12004 associated with fractures in polished thin sections, may not be indigenous.[1] In fines sample 12003,17, however, a metallic foil a few microns thick partially surrounded an aggregate of mineral and glass fragments. It is believed that contamination is precluded because the mineral inclusions had compositions of lunar pyroxenes and feldspars.[2]

### *Optics*

The nickel foil from 12003,17 had a yellowish hue and reflectivity in oil higher than that of iron grains in the lunar soil.[2]

### *Chemical Composition*

Microprobe analysis of the nickel foil yielded a composition of almost pure Ni, with 2 to 5% Fe and 0.5% Co. Although morphologically the particle appears to have had an impact origin, the Ni content is much higher than that reported for other lunar metal particles.[2]

### *References*

1. Reid et al, *EPSL* **9**, 2.
2. Jedwab et al, *Lun. Sci. Conf.* '72, *Abstr.*, 386–387.

## Cu-Ni-Zn-Fe Metal (tentative)

Grains in porphyritic basalts 12022 and 12004, associated with fractures in polished thin sections, have a composition similar to commercial "nickel" silver":

$$\begin{aligned} Cu &= 58\% \\ Ni &= 22\% \\ Zn &= 17\% \\ Fe &= 2\% \end{aligned}$$

No source of contamination is known, but the nature of occurrence may mean that the grains are not indigenous.[1]

### *Reference*

1. Reid et al, *EPSL* **9,** 1–2.

## INDIUM In (doubtful)

Some discrete grains of indium metal were identified by X-ray diffraction from ophitic basalt 12040. Contamination from the indium vacuum seals of the lunar sample rock boxes is suspected.[1]

### *Reference*

1. Champness et al, *Suppl.* **2,** 374.

## GRAPHITE C (doubtful)

A single soft, lustrous, irregular grain, approximately 0.6 mm in diameter, was recovered from soil sample 10085. The grain gave a strong carbon signal in the microprobe. On X-ray analysis, thin flakes yielded $d_{(100)}$ = 2.12 ±0.01 Å. The $d_{(100)}$ of graphite is 2.13 Å. Quite possibly the grain is a terrestrial contaminant.[1]

### *Reference*

1. Reid et al, *Suppl.* **1,** 755.

# CHAPTER TWO | SULFIDES, PHOSPHIDE, AND CARBIDES

Numerous lunar sulfides have been identified, although their total volume in the lunar rocks and soil is very small. Troilite is the most abundant of the sulfide phases.[1] A mineral formed under strongly reducing conditions, troilite is found in meteorites and, terrestrially, in serpentinized rocks only. On the Moon it is a widespread if minor constituent.[2] Both meteoritic and lunar iron-troilite intergrowths are present in the lunar soil, and the two types of material can be differentiated by the significant Ni and P contents of the original meteoritic material.[3]

Additional lunar sulfides positively identified are bornite, cubanite, sphalerite, and niningerite. The presence of two other sulfides, chalcopyrite and mackinawite (suspect in the early days of lunar material research), probably has been substantiated. Pentlandite, talnakhite, and/or (possibly) cubanite II have been identified tentatively. It is still in doubt whether chalcocite and molybdenite are lunar materials.

The phosphide schreibersite, from lunar breccias and soils, has been used as an indicator of meteoritic contamination. However, its presence in the crystalline rocks implies that it can be also a primary lunar product.[4]

Cohenite, an iron-nickel carbide, is a very sparse accessory in the lunar rocks and soils. It is found together with Fe-metal, troilite, and schreibersite. An (unnamed) aluminum-carbide has been described,[5] and moissanite (silicon-carbide) may or may not be of lunar origin.[6]

*References*


1. Taylor and Williams, *Am. Mineral.* **58,** 952.
2. Evans, *Suppl.* **1,** 399.
3. Goldstein et al, *Suppl.* **1,** 511.
4. Grieve and Plant, *Lun. Sci. IV, Abstr.*, 318.
5. Tarasov et al, *Suppl.* **4,** 345–346.
6. Jedwab, *Lun. Sci. IV, Abstr.*, 413.

## Sulfides: Troilite, Mackinawite, Chalcopyrite, Cubanite, Bornite, Sphalerite, Niningerite, Plus (Tentative) Talnakhite, Cubanite II, Pentlandite, and (Doubtful) Chalcocite and Molybdenite

### TROILITE FeS

*Synonym*

Pyrrhotite; Skinner, *Suppl.***1**, 891.

*Occurrence and Form*

Troilite is disseminated throughout the crystalline rocks as subrounded interstitial grains, less than 50 μ and frequently less than 10 μ in size, that often contain subhedral or small rounded grains of native iron.[1] Troilite occurs also as thin stringers in ilmenite and as small blebs, up to 0.3 mm wide, that are commonly polycrystalline. Troilite has been seen on the walls of cavities forming bright metallic yellow patches, rounded as though they had solidified from a molten drop. In coarse-grained basalt 10050, troilite has appeared as single crystals implanted on pyroxene crystals lining a vug. The crystals are well developed, with bright faces that give good optical signals although the edges are somewhat rounded (Figures 2–1, 2–2). The crystals show a hexagonal habit and the forms are those that would be expected for the simple NiAs-type crystal unit cell that prevails above 320°C. Miller indices for the high-temperature unit cell are $\{10\bar{1}0\}$, $\{10\bar{1}1\}$, and $\{10\bar{1}2\}$. In coarse-grained basalt 10047, a euhedral crystal of troilite is an aggregate of subparallel crystals ranging in orientation over 5°.[2]

Rounded blebs of troilite with exsolved iron are distributed throughout the plagioclase matrix of fine-grained basalt 10069-30, whereas more irregular crystals are marginal to ilmenite or clinopyroxene grains.[3] From fines sample 10085, a nickel-iron pellet has an internal structure of a dendritic intergrowth of taenite and troilite. The troilite has inclusions of chromite and a small amount of silicate.[4] The troilite–silicate boundaries in the lunar igneous rocks are invariably irregular, and no tendency toward a spherical shape has been observed.[5]

In Apollo 11 samples troilite always is associated with kamacite, and this has been considered to be indicative of formation from an Fe-FeS liquid.[6] In Apollo 12 samples, however, the troilite more commonly is found without Fe metal and is related paragenetically to the oxides (i.e., ilmenite and the spinels).[7] The same relationship of troilite with ilmenite, spinels, and baddeleyite has been noted in Apollo 14 crystalline rocks, even though some troilite is associated with nickel-iron in a eutectic texture.[8]

Fig. 2–1 SEM photograph of euhedral troilite resting on a pyroxene substrate. From a vug in a breccia particle from fines sample 14001. NASA photograph S-74-20440, courtesy of R. Laughan, Johnson Space Center, Houston.

The troilite and iron intergrowths were apparently one of the last features to form from the lunar rocks (Figure 2–3).[9] Both phases have been noted between or attached to the various early formed Fe-Ti-Cr oxides, but more commonly they are associated with other residual products in the mesostasis. It has been suggested that the ubiquitous relationship between iron and troilite in the Apollo 11 crystalline rocks indicates that these intergrowths are a result of crystallization at a temperature above the Fe-FeS eutectic temperature of 988°C.[10] In another investigation of some Apollo 11 igneous rocks, visual examination of polished surfaces of iron-troilite intergrowths revealed that the two phases had a more or less constant ratio. This, it is advanced, precludes the possibility of their having co-crystallized from the magma, or of the iron having been formed from postmagmatic desulfurization of pyrrhotite grains. The constant-ratio intergrowth suggests the breakdown of an initial homogeneous phase

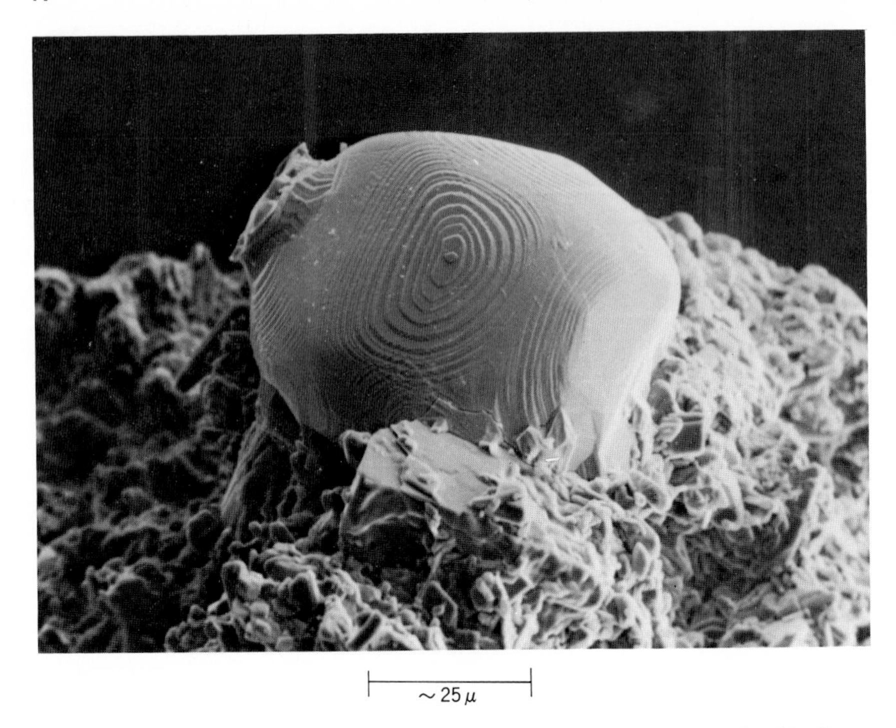


Fig. 2–2 SEM photograph of troilite crystal showing pyramidal faces outlined by linear bunching of growth lines. NASA photograph S-74-16418, courtesy of U. Clanton, Johnson Space Center, Houston.

(i.e., an iron-sulfide liquid that separated immiscibly from the silicate magma).[9]

In metamorphosed breccia 66095,89, the troilite (in an assemblage with goethite, sphalerite, Zn- and Cl-rich phases, and a Pb-rich phase) shows different degrees of shock from simple cracking to severe crushing, undulose extinction, and stress twinning.[11] Irregularly shaped and intensely shocked grains of troilite in eutectic intergrowth with kamacite were found in an enstatite chondrite from soil sample 15602,29.[12]

*Optics*

Troilite is opaque, light gray to dark gray in polished section,[8] or a bright metallic yellow.[2] Troilite from fines samples, viewed with a binocular microscope, has a light to dark brassy color, often with iridescent patches on irregular surfaces.[13]

Fig. 2–3 Spongy mass of iron and troilite. Darker gray laths are ilmenite. From *EPSL* **12**, 8. Photograph courtesy of Elsevier Press and A. J. Gancarz, Division of Geological and Planetary Sciences, California Institute of Technology, Pasadena.

### *Chemical Composition*

Most of the analyses of troilite show it to be close to stoichiometric FeS. Minor Ti, Mn, Cr, Ni, and Co may be present. Troilite, coexisting with the oxides ilmenite and/or spinels, contains more Ti than troilite coexisting with the silicates such as pyroxene and plagioclase[7] (Table 2-1; [1], [2]). In some troilite the Ni content is as high as 0.9%.[4] In metamorphosed breccia 66095,89, the troilite that occurred together with sphalerite contained 0.03% Zn, but Zn was not detectable in the troilite from a mineral assemblage without sphalerite.

### *X-Ray Data*

X-Ray powder and single-crystal X-ray measurements of lunar troilite gave a unit cell that is in close agreement with that for stoichiometric FeS. The cell is hexagonal, with space group $P\bar{6}2C$, $a = 5.692 \pm 0.002$ Å, $c = 11.750 \pm 0.003$ Å, and a cell content of 12 FeS. The cell represents a superstructure of the high-temperature simple NiAs type.[2]

The X-ray powder pattern of a phase reported to be pyrrhotite revealed it to be actually troilite. The composition, determined by microprobe, was

**Table 2-1 Analyses of Sulfides**

| | [1] 12004,11 | [2] 12063,9 | [3] 68841 | [4] 12021 | [5] 12021,134 | [6] 12021,134 | [7] 68841 | [8] 66095,78 | [9] 66095,80 | [10] 15602,29 |
|---|---|---|---|---|---|---|---|---|---|---|
| Ti | 0.24 | 0.02 | — | 0.27 | — | — | — | — | — | 0.07 |
| Cr | — | <0.02 | — | — | — | — | — | — | — | 0.82 |
| Fe | 63.4 | 63.4 | 63.1 | 63.58 | 30.0 | 40.4 | 12.7 | 14.8 | 17.6 | 25.50 |
| Cu | — | — | — | — | 33.6 | 22.8 | 60.7 | — | — | — |
| Mg | — | — | — | 0.04 | — | — | — | — | — | 27.97 |
| Zn | — | — | — | — | — | — | — | 51.0 | 48.2 | — |
| Ni | 0.10 | <0.03 | 0.08 | 0.02 | — | — | 0.07 | — | — | 0.23 |
| Co | 0.12 | 0.08 | 0.09 | 0.11 | 0.85 | 0.87 | 0.2 | — | — | — |
| S | 36.4 | 35.9 | 35.8 | 35.66 | 35.2 | 35.7 | 26.2 | 33.0 | 33.7 | 36.88 |
| Total | 100.26 | 99.41* | 99.11* | 100.00* | 99.65 | 99.77 | 99.97* | 98.8 | 99.5 | 100.00* |

*References*

[1] Taylor et al, *Suppl.* **2,** 862. From a porphyritic basalt. Coexisting with ilmenite and spinel. Troilite.
[2] ———, *ibid.* From a porphyritic basalt. *Analysis includes Mn = 0.01%. Coexisting with plagioclase and pyroxene. Troilite.
[3] Carter and Padovani, *Suppl.* **4,** 324. From a fines sample. *Analysis includes P = 0.04%. Troilite.
[4] Weill et al, *Suppl.* **2,** 417. From an ophitic basalt. *Analysis includes Si = 0.16% and Ca = 0.16%. Troilite.
[5] Taylor and Williams, *Am. Mineral.* **58,** 953. From an ophitic basalt. Chalcopyrite.
[6] ———, *ibid.* From an ophitic basalt. Cubanite.
[7] Carter and Padovani, *Suppl.* **4,** 324. From a fines sample. *Analysis includes P = 0.1%. Bornite.
[8] El Goresy et al, *Suppl.* **4,** 744. From a metamorphosed breccia. Sphalerite.
[9] Taylor et al, *Suppl.* **4,** 835. From a metamorphosed breccia. Sphalerite.
[10] Haggerty, *Apollo 15 Lunar Sample*, 86. From an enstatite chondrite in a soil sample. *Analysis, which includes Mn = 7.53% and Ca = 1.00%, has been recalculated from original analysis reported as sulfides and with MgS content determined by difference. Niningerite.

ideally stoichiometric.[9] An Fe-S phase immediately adjacent to chalcopyrite in porphyritic basalt 12021,134 may be pyrrhotite. The material is less than 10 $\mu$ wide, and definite identification has not been made.[14]

*References*

1. Dence et al, *Suppl.* **1,** 324.
2. Evans, *Suppl.* **1,** 399–400.
3. Carter and MacGregor, *Suppl.* **1,** 247.
4. Mason et al, *Suppl.* **1,** 659–660.
5. Bailey et al, *Suppl.* **1,** 179.
6. Brown et al, *Suppl.* **1,** 205–206.
7. Taylor et al, *Suppl.* **2,** 862.
8. El Goresy et al, *Suppl.* **3,** 336–337.
9. Skinner, *Suppl.* **1,** 891–892, 894.
10. Haggerty et al, *Suppl.* **1,** 528.
11. El Goresy et al, *EPSL* **18,** 413–414.
12. Haggerty, *Apollo 15 Lunar Samples*, 86.
13. Frondel, J. W., personal observation.
14. Taylor and Williams, *Am. Mineral* **58,** 954.

## MACKINAWITE FeS, or $(Fe,Ni)_{1+x}S$

*Occurrence and Form*

A tentative identification of mackinawite was made of a strongly anisotropic phase present as fine spots and bands within a troilite–iron spherule. The spherule occurred in a chondrulelike body in fines sample 10085-4-14. The optical properites of the spots and bands were similar to those of mackinawite, but because the material is fine-grained no definite distinction could be made between it and graphite.[1] Subsequent descriptions appear to regard mackinawite as a valid lunar phase.

Mackinawite occurs within troilite in ophitic basalt 12038. In porphyritic basalts 12018 and 12063, the mackinawite is found together with troilite and chalcopyrite.[2] It is believed that the mackinawite in 12063 was formed at subsolidus temperatures below 135°C and that the phase could have formed as a breakdown product of troilite.[3] The presence of mackinawite, which has an upper stability temperature below 200°C, demonstrates that subsolidus reequilibration and exsolution processes were still active during cooling to temperatures not far above 25°C.[2]

*Optics*

In reflected light the mackinawite exhibits strong bireflectance and, with crossed nicols, strong birefringence.[2]

### *Chemical Composition*

Ideally, mackinawite is FeS. The grain size of the lunar material ($< 2\mu$) permitted only semiquantitative probe analysis. The formula may be written $(Fe,Ni)_{1+x}S$.[4]

### *References*

1. Simpson and Bowie, *Suppl.* **1,** 884.
2. El Goresy et al, *Suppl.* **2,** 221.
3. Taylor et al, *Suppl.* **2,** 868.
4. Taylor and Williams, *Am. Mineral.* **58,** 952.

## PENTLANDITE $(Fe,Ni)_9S_8$ (tentative)

### *Synonymy*

Heazlewoodite; Jedwab, *Suppl.* **4,** 866.
NiFe sulfide; El Goresy et al, *Meteorit.* **6,** 266.
Fe,Ni sulfide; El Goresy et al, *EPSL* **13,** 121.

### *Occurrence and Form*

An unidentified flamelike exsolution, observed within some troilite at its contact with native iron (in fine-grained basalt 10072-46) may be possible pentlandite.[1] In breccia 14315,9, a phase identified as pentlandite was observed replacing troilite.[2] Quite possibly the Ni-Fe sulfide spots, noted on glassy spheres in a number of Apollo 16 soil samples, are pentlandite. They have been identified tentatively as heazlewoodite,[3] which is an invalid mineral name for impure pentlandite.[4] The spots consist of aggregates of granular scales,[3] and their mode of occurrence suggests deposition from a vapor.[5]

### *Optics*

The above-mentioned spots are opaque, highly anisotropic, and of a striking bronze color reminiscent of graphite.[5]

### *Chemical Composition*

Probe analysis of the spots on the glass spheres yielded a Ni, Fe, S composition. The Ni content of 5 to 10% is too high for the phase to be troilite,[6] and the Fe content of less than 11% is too low for troilite.[3] Earlier identification of pentlandite from 14315,9, gave only qualitative determination of Ni, Fe, and S. It has been suggested that the formation of the pentlandite by reaction with the troilite and migrating Ni indicates a temperature below 610°C. A sulfidization reaction could take place on the lunar surface from the impact of a sulfur-bearing carbonaceous chondrite.[2]

Until better chemical analyses are obtained for this phase, the identification of pentlandite must be accepted as tentative.

### *References*

1. Simpson and Bowie, *Suppl.* **1,** 880.
2. Ramdohr, *EPSL* **15,** 113–115.
3. Jedwab, *Suppl.* **4,** 866.
4. Palache et al; *Dana VII, Vol. 1,* 243.
5. Jedwab, Preprint, *Lun. Sci. IV Conf.*
6. Jedwab, *Lun. Sci. IV, Abstr.,* 413–414.

## CHALCOPYRITE $CuFeS_2$

### *Occurrence and Form*

Originally described from Apollo 11 rocks as only an occasional speck of sulfide resembling chalcopyrite,[1] this phase was considered doubtful as a lunar mineral. Chalcopyrite was qualitatively identified as occurring on the outside of native copper rims around FeNi metal grains in basalt 15475. From porphyritic basalt 12021,134, chalcopyrite was definitely identified. It occurs with cubanite along cracks and grain boundaries with troilite (Figure 2–4).[2]

### *Optics*

The chalcopyrite is anisotropic and therefore cannot be the isotropic (and discredited) "chalcopyrrhotite."

### *Chemical Composition*

The composition of chalcopyrite, determined by microprobe, was further verified by X-ray scans (Table 2-1; [5]). Some minor Co was detected.[2]

### *References*

1. Agrell et al, *Suppl.* **1,** 97.
2. Taylor and Williams, *Am. Mineral.* **58,** 953.

## CUBANITE $CuFe_2S_3$

### *Occurrence and Form*

In porphyritic basalt 12021,134, cubanite has been found, together with chalcopyrite, along cracks and grain boundaries within troilite (Figure 2–5).[1]

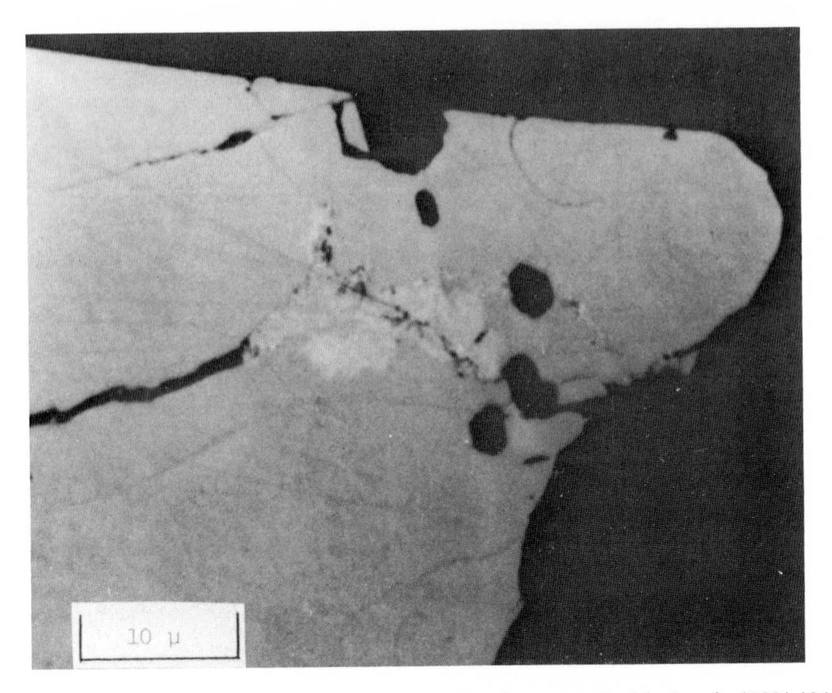


Fig. 2–4 Chalcopyrite (very light gray) in troilite from porphyritic basalt 12021,134. From *Am. Mineral.* **58,** 953. Photograph courtesy of the Mineralogical Society of America and L. A. Taylor, Department of Geology, University of Tennessee, Knoxville.

*Optics*

Cubanite is anisotropic, and therefore cannot be confused with the isotropic Cu, Fe sulfide "chalcopyrrhotite."[1]

*Chemical Composition*

Microprobe analysis of cubanite revealed some minor Co (Table 2-1; [6]). It is believed that the cobalt in the cubanite (as well as in the chalcopyrite) substitutes for the iron in the structure.[1]

*Reference*

1. Taylor and Williams, *Am. Mineral.* **58,** 953.

BORNITE $Cu_5FeS_4$

*Occurrence and Form*

In a metallic spherule (from fines sample 68841) consisting mainly of interlocking grains of schreibersite and metallic iron, bornite was found together

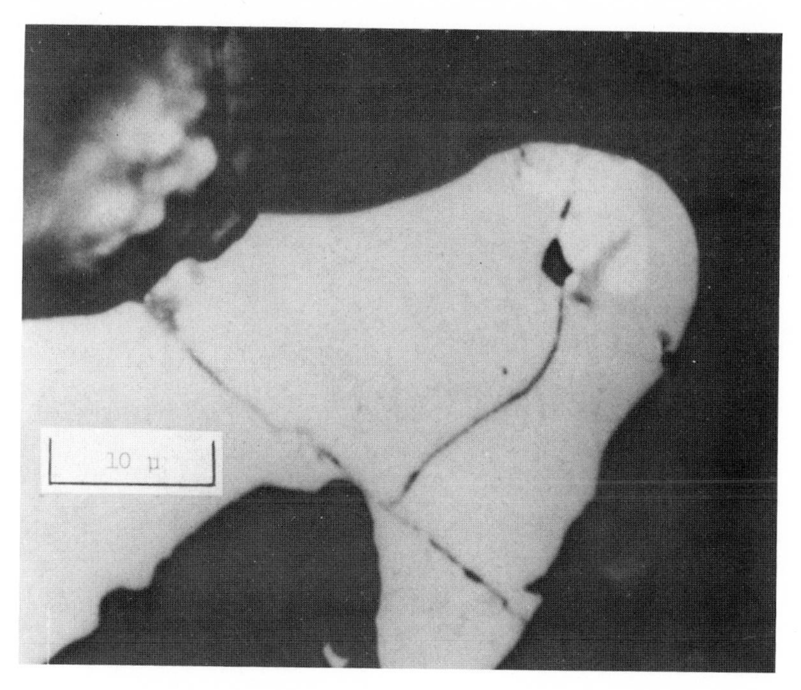


Fig. 2–5 Cubanite (very light gray) in troilite from porphyritic basalt 12021,134. From *Am. Mineral.* **58,** 953. Photograph courtesy of the Mineralogical Society of America and L. A. Taylor, Department of Geology, University of Tennessee, Knoxville.

with minor troilite (Figure 2–6). The association of bornite with these other phases implies a moderate temperature environment during its formation.[1]

*Chemical Composition*

The identification of bornite was made on the basis of its chemical analysis (Table 2-1; [7]). The apparent excess of Fe and S in the analysis could be due to admixed troilite.[1]

*Reference*

1. Carter and Padovani, *Suppl.* **4,** 324–325.

TALNAKHITE $Cu_9(Fe,Ni)_9S_{16}$ (tentative)

*Synonymy*

"Chalcopyrrhotite"; El Goresy et al, *Suppl.* **2,** 221.
Cubanite II; Taylor and Williams, *Am. Mineral.* **58,** 952.

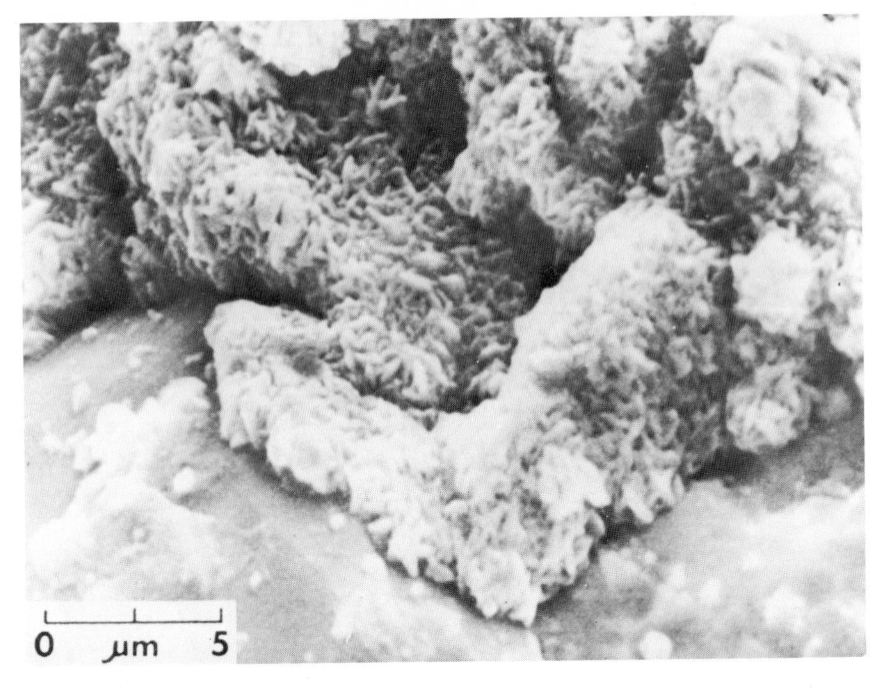


Fig. 2–6 Bornite on metallic substrate. From *Suppl.* **4**, 325. Photograph courtesy of J. L. Carter, University of Texas at Dallas.

*Occurrence and Form*

A late-formed phase found in porphyritic basalt 12018 together with mackinawite, as inclusions in troilite, was identified tentatively as "chalcopyrrhotite."[1] "Chalcopyrrhotite" with an "ideal" composition of $CuFe_4S_6$ has been discredited, and actually may be cubanite II[2], an isotropic form of cubanite (which see). It has been suggested that the lunar "chalcopyrrhotite" is talnakhite.[3] In a study of thermal stability of assemblages in the Cu-Fe-S system it was noted that on heating to about 80°C, talnakhite breaks down to tetragonal cubanite and bornite. On cooling, the transformation is reversed.[4] The occurrence of both lunar cubanite and lunar bornite supports the identification of talnakhite. Perhaps the cubanite and bornite resulted from an incomplete reversal of the transformation.

*Optics*

The phase is yellowish, chalcopyrite-colored, and isotropic.[3]

### *Chemical Composition*

In the original description of this lunar phase, its fine grain size prohibited complete probe analysis, but the presence of copper was determined qualitatively.[1] Better chemical analysis and/or X-ray study are needed to verify definitely lunar talnakhite. Terrestrial material has a formula of $Cu_9(Ni)_9S_{16}$.[5]

### *References*


1. El Goresy et al, *Suppl.* **2,** 221–222.
2. Yund and Kullerud, *J. Petrol.* **7,** 481.
3. Taylor and Williams, *Am. Mineral.* **58,** 952.
4. Cabri, *Econ. Geol.* **62,** 910–914.
5. Cabri and Harris, *Econ. Geol.* **66,** 673.


## CHALCOCITE $Cu_2S$ (doubtful)

Traces of chalcocite were reported from Apollo 12 rocks without supporting data. Its lunar occurrence is in doubt.[1] The material may be the abovementioned talnakhite.

### *Reference*


1. El Goresy et al, *Suppl.* **2,** 222.


## SPHALERITE ZnS

### *Synonym*

Zincblende; Ramdohr, *Fortschr. Mineral.* **48,** 50.

### *Occurrence and Form*

A tiny grain of sphalerite was discovered in porphyritic basalt 12018,49. No data were given, but identification was believed to be "probably certain."[1] Sphalerite has been identified definitely from metamorphosed breccia 66095. Here it occurs in a complex assemblage with troilite, goethite, two unidentified zinc- and chlorine-rich phases, and an unknown lead-rich phase. The sphalerite forms narrow reaction rims along the cracks penetrating troilite or surrounding troilite grains (Figure 2–7). Both sphalerite and troilite are surrounded by goethite and the unidentified phases. In enstatite chondrite meteorites, sphalerite is a rare mineral coexisting with titanian-chromian troilite, daubreelite, niningerite, or ferromagnesian alabandite. In iron meteorites sphalerite occurs mainly in troilite nodules

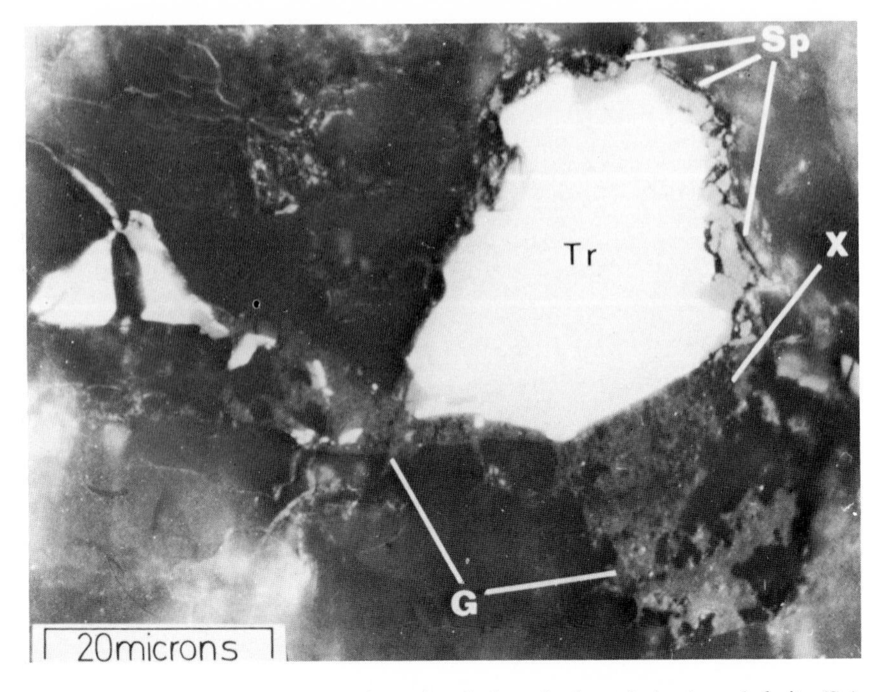


Fig. 2–7 Shocked troilite (Tr) grain replaced along its boundaries by sphalerite (Sp), goethite (G), and unidentified Zn-, Fe-, and Cl-bearing sulfates and phosphates (X). From *Suppl.* **4,** 744. Photograph courtesy of A. El Goresy, Max-Planck-Institut für Kernphysik, Heidelberg, W. Germany.

together with chromian troilite and daubreelite or zincian daubreelite. None of these minerals was observed with the sphalerite of 66095. It is suggested that sphalerite (with its volatile Zn) could hardly have survived an impact event and have been incorporated in the lunar rocks while all the other minerals disappeared. It is more likely that the presence of sphalerite in 66095 can be explained by a reaction between volatile-bearing vapors rich in zinc and chlorine, depositing sphalerite as reaction rims around troilite, and probably $FeCl_2$ around metallic FeNi. Reaction rims of sphalerite and goethite around troilite occur also in polymict breccia 67455,8.[2] In another investigation of 66095, the reaction rims between troilite and the zinc-bearing chloride-sulfate material were not observed. Instead, small ($< 10\ \mu$) individual grains (commonly one per troilite grain) were noted, sometimes within the troilite. This appears to rule out a troilite–sulfate reaction. Much of the troilite does not contain sphalerite, even in regions of the thin sections showing abundant evidence of oxidation, and the sphalerite could be possibly of lunar origin.[3]

### *Optics*

In thin section sphalerite appears gray and has high reflectivity compared with goethite.[4]

### *Chemical Composition*

Analyses of sphalerite from 66095 (Table 2-1; [8], [9]) show that it is an iron-rich variety.[2]

### *References*

1. Ramdohr, *Fortschr. Mineral.* **48,** 50.
2. El Goresy et al, *Suppl.* **4,** 744, 746–747.
3. Taylor et al, *Suppl.* **4,** 835.
4. El Goresy et al, *EPSL* **18,** 414

## NININGERITE MgS (or Nss)*

### *Occurrence and Form*

Niningerite, from an enstatite chondrite in soil sample 15602,29, occurs in discrete subrectangular or angular fragments with a maximum size of 250 × 100 $\mu$. Two of the 15 identified grains contain troilite lamellae (5 $\mu$ wide) in a Widmanstätten-like pattern.[1]

### *Chemical Composition*

The niningerite is not stoichiometric MgS but contains multiple sulfides in solid solution[1] (Table 2-1; [10]).

### *References*

1. Haggerty, *Apollo 15 Lunar Samples*, 85–86.

## UNKNOWN SULFIDE PHASE

Bordering the troilite lamellae in the niningerite (see above) an unidentified light gray transitional phase is rich in Mg, Mn, Ca, and S, but depleted in Fe relative to the niningerite host.[1]

### *References*

1. Haggerty, *Apollo 15 Lunar Samples*, 86.

* Nonstoichiometric solid solution.

MOLYBDENITE $MoS_2$ (doubtful)

In some lunar fines samples (in particular 15601,113[1]) bright bluish metallic flakes, with crystalline outlines and traces of cleavages, have been identified as molybdenite by probe analysis and X-ray diffraction. A lunar source is doubted, however, and the flakes are probably a terrestrial contaminant (e.g., from the lubrication of screw threads in the lunar sample cabinets).[2]

*References*

1. Jedwab, *Apollo 15 Lunar Samples*, 109.
2. ———, *Suppl.* **4,** 472.

## Phosphide: Schreibersite

SCHREIBERSITE $(Fe,Ni,Co)_3P$

*Synonym*

Fe-schreibersite; El Goresy et al, *EPSL* **13,** 129.

*Occurrence and Form*

Trace amounts of schreibersite have been found in the breccias and soils of Apollo 11 samples. In polished section it has been observed intergrown with cohenite.[1] A splinter of schreibersite has been seen together with idiomorphic kamacite crystals.[2] In polished section, a metallic sample from breccia 10048-18a exhibits a eutecticlike structure consisting of a very fine dispersion of schreibersite plus metal with lesser amounts of troilite and cohenite.[3] The eutectic structure is believed to have formed from the last liquid to solidify.[4] Eutectic intergrowths of schreibersite and metal have been observed in some shocked rocks (e.g., highly metamorphosed breccia 66095). Here the schreibersite could be either of meteoritic origin or formed by reaction between preexisting pyroxene and apatite.[5] In glass-coated highland basalt fragment 64455, schreibersite is present as discrete grains and bleblike inclusions in metal. Cores in metal spherules rimmed with troilite often are composed of a eutectic intergrowth of cobaltian nickel-iron and schreibersite. The texture suggests that melting and recrystallization of metal, sulfide, and phosphide have occurred.[6] In contrast, from norite 62295, somewhat rounded schreibersite (together with troilite in association with composite nickel-iron) shows no evidence of partial melting or of a phosphide–metal eutectic.[7]

In many metal particles it is believed that schreibersite, occurring as randomly oriented blebs within or at the edges of metal, was precipated from an Fe-Ni-P phase during cooling.[8] In breccia 66055, spherical particles

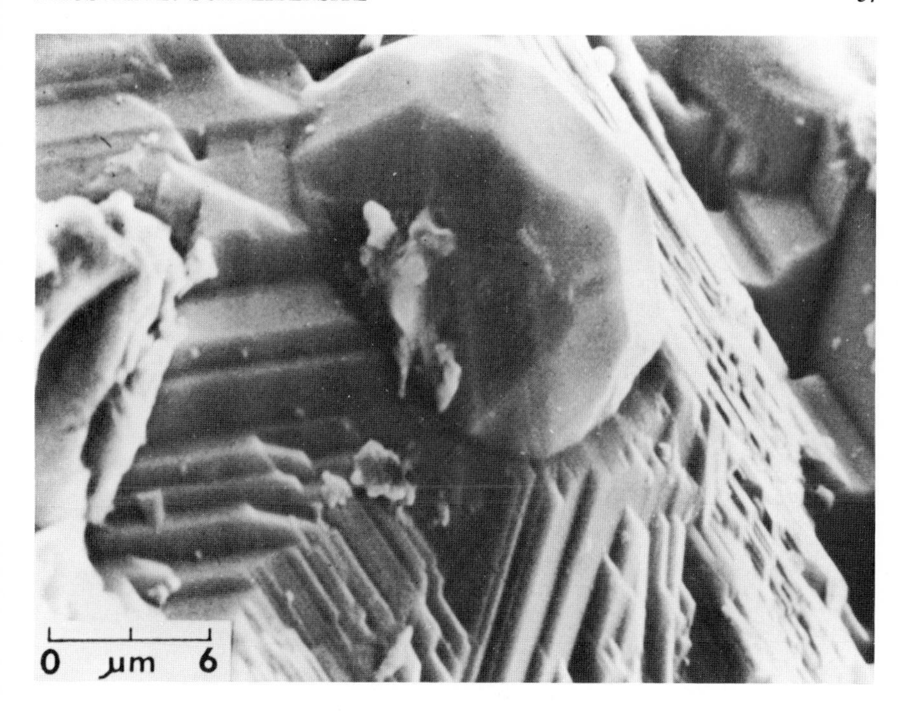


Fig. 2–8 Schreibersite crystal on metallic iron growth steps. From *Suppl.* **4,** 325. Photograph courtesy of J. L. Carter, University of Texas at Dallas.

of associated metal and schreibersite strongly suggest a liquid phase origin (Figure 2–8). This does not necessarily rule out a meteoritic origin. However, since the rocks in question are found on the lunar surface where there is ample evidence of widespread igneous activity and impact melting, and where the necessary reducing conditions are believed to prevail, origin by reduction of P and Fe from the silicate melt appears more likely.[9] It has been suggested that complex Fe-Ni-Co-P blebs are characteristic of Apollo 16 feldspathic rocks and that the schreibersite occurring in these highland crystalline rocks need not have been contributed by impacting meteorites.[10]

*Optics*

In polished section schreibersite appears light gray in contrast to dark gray troilite and black cohenite.[4]

*Chemical Composition*

Since the crystalline lunar rocks have an extremely low Ni content, the high-Ni content of the schreibersite (e.g., that which occurs in a eutectic

Table 2-2 Schreibersite Analyses

| | [1] 64585 | [2] Apollo 12 Particle F2 | [3] 66055 | [4] 66095,89 | [5] 63549 | [6] 61568 |
|---|---|---|---|---|---|---|
| Fe | 75.1 | 70.25* | 72.1 | 66.7 | 61.2 | 50.9 |
| Ni | 9.4 | 14.0 | 12.16 | 17.1 | 22.0 | 32.7 |
| P | 15.7 | 15.5 | 13.16 | 14.8 | 15.3 | 15.4 |
| Co | 0.4 | 0.25 | 0.13 | 0.09 | 0.5 | 0.1 |
| S | 0.1 | — | — | — | 0.1 | 0.06 |
| Total | 100.7 | 100.00 | 97.55 | 98.69 | 99.1 | 99.16 |

*References*

[1] Gooley et al, *Suppl.* **4,** 801. From a mesostasis-rich rock fragment in a rake sample.

[2] Goldstein and Yakawitz, *Suppl.* **2,** 184–185. From a soil sample. *Fe by difference.

[3] McKay et al, *Lun. Sci. IV, Abstr.*, **488**. From a microbreccia.

[4] El Goresy et al, *Suppl.* **4,** 742. From a metamorphosed breccia.

[5] Gooley et al, *Suppl.* **4,** 801. From a diabase rock fragment in a rake sample.

[6] ———, *ibid.* From a poikilitic rock fragment in a rake sample.

structure with kamacite in particle F2 from an Apollo 12 soil sample) has caused it to be considered as meteoritic in origin[11] (Table 2-2; [2]). Schreibersite that is considered indigenous to the Moon occurs in two chemical varieties in KREEP basalt 14310: Ni-free and stoichiometric (referred to also as Fe-schreibersite) and with a Ni content up to 28%.[12]

In a troctolite fragment from breccia 68815,148, an iron particle rimmed by a phosphide is P-deficient for schreibersite. It is probably a submicron intergrowth of iron and schreibersite.[10] In several metal grains from metamorphosed breccia 66095,89, inclusions of blebs or small idiomorphic crystals of schreibersite are very low in Co compared to the coexisting kamacite (Table 2-2; [4]).

*References*

1. Frondel et al, *Suppl.* **1,** 460.
2. Ramdohr and El Goresy, *Sci.* **167,** 618.
3. Adler et al, *Suppl.* **1,** 92.
4. Goldstein et al, *Suppl.* **1,** 506.
5. El Goresy et al, *Suppl.* **4,** 740, 742.

6. Grieve and Plant, *Lun. Sci. IV, Abstr.*, 317–318.
7. Agrell et al, *Lun. Sci. IV, Abstr.*, 15–16.
8. Gooley et al, *Lun. Sci. IV, Abstr.*, 304.
9. McKay et al, *Suppl.* **4,** 817.
10. Brown et al, *Suppl.* **4,** 516–518.
11. Goldstein and Yakowitz, *Suppl.* **2,** 182–187.
12. El Goresy et al, *Lun. Sci. Conf. '72, Abstr.*, 203.

## Carbides: Cohenite, Moissanite (Tentative), Ramdohr's Phase (2), and Aluminum Carbide

### COHENITE $(Fe,Ni)_3C$

*Synonym*

Carbide; Goldstein and Axon, *Suppl.* **4,** 773.

*Occurrence and Form*

In a polished section of a metallic fragment from an Apollo 11 fines sample, cohenite has been observed together with rims or inclusions of troilite, or intimately associated with schreibersite.[1] From soil sample 14003, metal particle 27.11 contains cohenite that shows signs of incomplete dissolution in the metal.[2] From soil sample 68501, particles of cohenite, 10 to 30 $\mu$ in size were observed at the edge of and within a 150-$\mu$ metal particle (Figure 2–9). In another particle small quantities of cohenite were found coexisting with large proportions of schreibersite.[3] In a complex metal grain from metamorphosed breccia 66095,78, cohenite (with schreibersite) occurs within a troilite mantle of the nickel-iron grain (Figure 2–10).[3] The cohenite is presumed to be of meteoritic origin[1] or at least formed by reaction between meteoritic carbon and metallic nickel-iron. The extremely low carbon content of the lunar material does not account for the formation of cohenite in the lunar rocks.[4]

*Optics*

Cohenite is black. Under crossed nicols it reveals a lamellar pronged structure interpreted as magnetic domains, which are known to occur in meteoritic cohenite.[4]

*Chemical Composition*

Cohenite in a eutecticlike metal particle from breccia 10046-18 contains 6.5% Ni and 8% P. The major elements in iron meteorites (i.e., Fe, S, Ni,

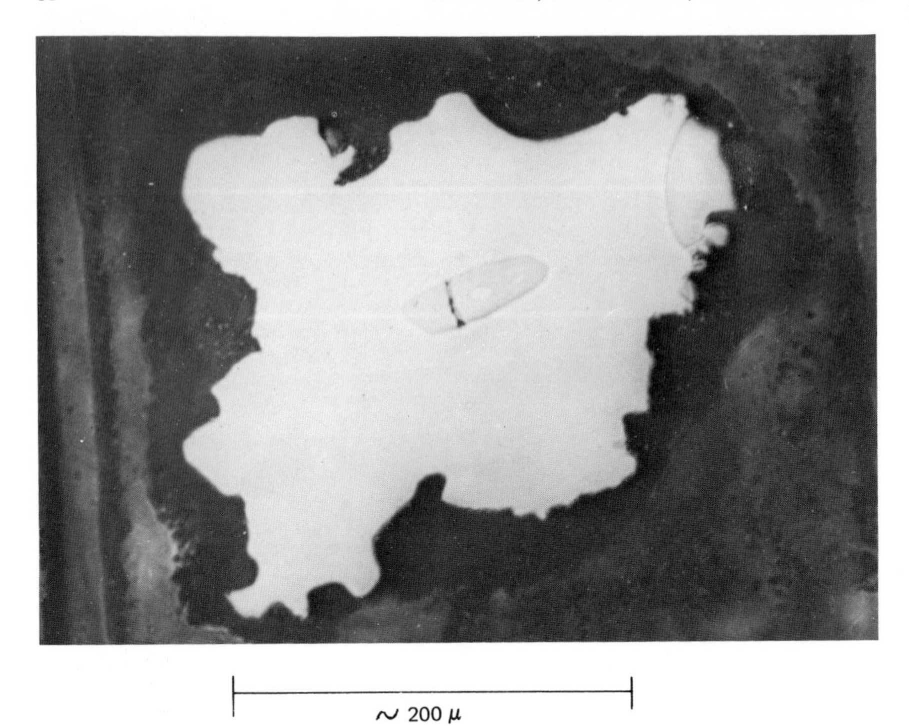


Fig. 2–9 Cracked, boat-shaped particle of meteoritic cohenite in kamacite. A portion of a larger cohenite particle is present at the edge of the metal. From *Suppl.* **3,** 1053. Photograph courtesy of J. I. Goldstein, Department of Metallurgy and Materials Science, Lehigh University, Bethlehem, Pa.

Co, P, and C) are present in the particle, further supporting the suggestion of a meteoritic origin for the cohenite.[5] Analyses are given in Table 2-3, [1] and [2].

*References*


1. Frondel et al, *Suppl.* **1,** 460, 467.
2. Goldstein et al, *Suppl.* **3,** 1040, 1050.
3. Goldstein and Axon, *Suppl.* **4,** 768.
4. El Goresy et al, *Suppl.* **4,** 740.
5. Goldstein et al, *Suppl.* **1,** 508.


## MOISSANITE SiC (tentative)

*Synonym*

Silicon carbide; Gay et al, *EPSL* **9,** 124.

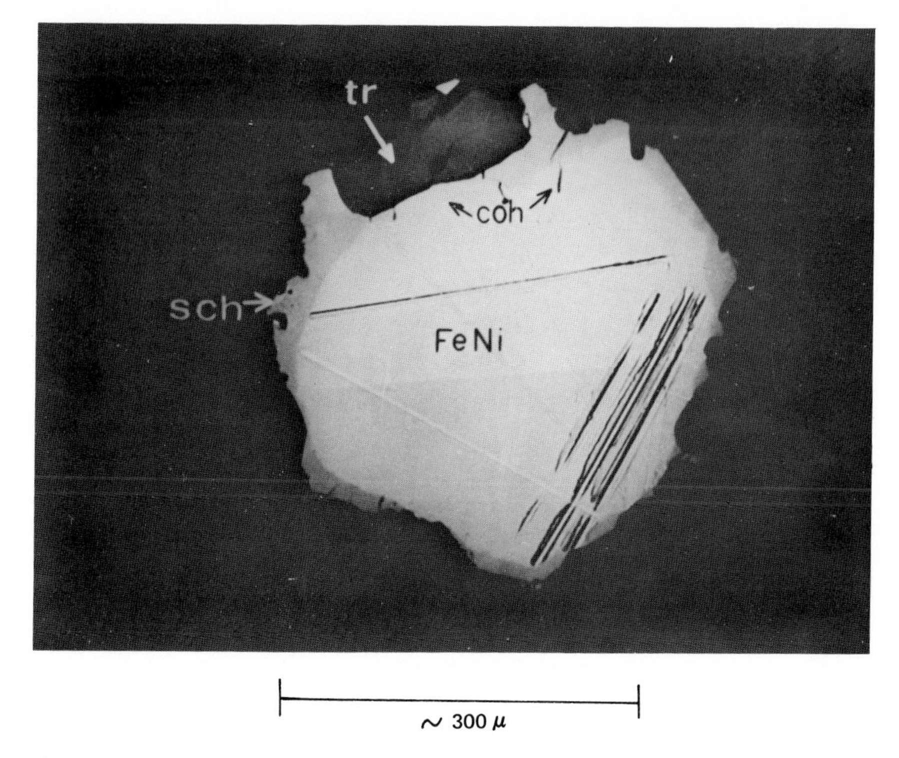


Fig. 2–10 Composite grain of metallic FeNi, cohenite (coh), schreibersite (sch), and troilite (tr). From *Suppl.* **4,** 741. Photographs courtesy of A. El Goresy, Max-Planck-Institut für Kernphysik, Heidelberg, W. Germany.

### *Occurrence and Form*

Originally two crystals of silicon carbide in breccia 12013,10 were thought to be a possible contaminant.[1] However, repeated observations of monocrystalline particles of SiC in a number of Apollo 16 soil samples lend credence to the suggestion that these particles are naturally occurring moissanite. It is argued that since the fines samples were not subjected to cutting and polishing, the possibility of the SiC phase being the terrestrial synthetic carborundum is considerably reduced.[2] None of the particles was attached to any lunar material.[3] The grains show no crystal faces, and the forms are the result of fracturing. A good cleavage has been noted.[2]

### *Optics*

The fragments of silicon carbide are pale blue and have a very high dispersion in reflected light. The fragments appear almost opaque in transmitted light, because of their extreme relief.[2]

**Table 2-3 Analyses of Cohenite and Aluminum Carbide**

| | [1] 66095,78 | [2] 66095,80 | [3] Luna 20 Thin Section 863 |
|---|---|---|---|
| Al | — | — | 74.0 |
| Fe | 88.15 | 90.6 | 0.4 |
| Ni | 0.81 | 1.28 | — |
| C | 10.92* | 7.79* | 20.0 |
| Co | 0.10 | 0.25 | — |
| P | 0.02 | 0.08 | — |
| Total | 100.00 | 100.00 | 99.6* |

*References*

[1] El Goresy et al, *Suppl.* **4,** 742. From a metamorphosed breccia. *The value for C obtained by difference. Cohenite.

[2] Taylor et al, *Suppl.* **4,** 831. From a metamorphosed breccia. *The value for C obtained by difference. Cohenite.

[3] Tarasov et al, *Suppl.* **4,** 347. From an olivine-plagioclase rock fragment.

*Analysis includes

Si = 0.2% K = 0.1%
Mg = 0.1% S = 0.1%
Ca = 0.3% O = 4.4%

Aluminum carbide (or aluminum oxycarbide).

### *Chemical Composition*

Microprobe analysis yielded very high Si, with traces of Ca and Fe. Determination of C and O is beyond the capabilities of the instrument.[2]

### *X-Ray Data*

X-Ray diffraction patterns confirm the identification of $\alpha$-SiC, but lunar origin of the phase moissanite awaits proof.[2]

### *References*

1. Gay et al, *EPSL* **9,** 124.
2. Jedwab, *Suppl.* **4,** 868–869.
3. ———, *Lun. Sci. IV, Abstr.,* 413.

## Ramdohr's Phase (2)

An unidentified phase, originally referred to as mineral D,[1] occurs as idiomorphic grains intergrown with ilmenite, rutile, and baddeleyite. The phase is opaque; it has some internal reflections but low reflectivity. Its chemical composition has not been determined, but it was noted that its hardness is greater than 9.[2] If this material is the SiC just discussed, the possibility of lunar origin is strengthened because of its relationship with the other lunar minerals.

*References*

1. Ramdohr and El Goresy, *Sci.* **167,** 617.
2. ———, *Naturwiss.* **57,** 102.

## Aluminum Carbide $Al_4C_3$

*Synonym*

Aluminum oxycarbide; Tarasov et al, *Suppl.* **4,** 346.

*Occurrence and Form*

In thin section 863 of a Luna 20 sample, a grain of aluminum carbide 40 $\mu$ long and 7 to 20 $\mu$ wide (believed to be more properly an aluminum oxycarbide) was observed in close contact with olivine and plagioclase. The grain is a platelike, irregularly shaped segregation and is deeply embedded in the thin section. After additional polishing of the thin section, the contacts of the carbide with olivine and plagioclase became more distinct. An olivine inclusion within the main carbide part was observed, and patches of intergrown carbide and olivine were exposed in the platy end of the grain. The end also was overlapped by olivine. It is believed that terrestrial contamination is excluded as a source of the aluminum carbide by its relationship with the other minerals.[1]

*Optics*

The $Al_4C_3$ grain has a high reflectivity, similar to that of metallic iron.[1]

*Chemical Composition*

An analysis of the aluminum carbide is given in Table 2-3 [3].

*References*

1. Tarasov et al, *Suppl.* **4,** 345–347.

# CHAPTER THREE | SIMPLE OXIDES AND OXIDES OF FERRIC IRON

## Simple Oxides: Baddeleyite, Rutile, Unidentified Titanium Phases, Ilmenite, Corundum, and (Tentative) Wüstite; Unidentified Oxides

### Baddeleyite $ZrO_2$

*Synonymy*

Ferroan baddeleyite; Simpson and Bowie, *Suppl.* **2,** 207.
Hafnian baddeleyite; Ramdohr and El Goresy, *Sci.* **167,** 615.
Phase C (Lovering); Lovering and Wark, *Suppl.* **2,** 155.
Zirconium oxide; Reid et al, *Suppl.* **1,** 753.
Zr-phase enriched with U; Lovering and Wark, *Suppl.* **2,** 155.

*Occurrence and Form*

Baddeleyite is found as large, untwinned isometric crystals or aggregates of grains.[1] Thick tabular untwinned grains up to 75 $\mu$ long have been observed associated with ilmenite.[2] In fine-grained basalts its grain size ranges from only 3 to 46 $\mu^2$, but in coarse-grained basalts some baddeleyite grains are as large as 128 $\mu^2$. Baddeleyite also has been observed intergrown with zirconolite or tranquillityite (see Figure 5–1).[3] Apparently a late differentiate in ophitic basalts 12036, 12039, and 12051, baddeleyite is associated with K-feldspar, $K_2O$–$SiO_2$-rich residual glass, fluor-apatite and yttrian-cerian whitlockite.[4] In KREEP besalt 14310, baddeleyite occurs most commonly with ilmenite as both rounded and bladed inclusions. Sometimes the baddeleyite is found alone or with troilite or ulvöspinel.[5]

*Optics*

Compared with ilmenite, baddeleyite has a lower reflectivity and a lilac tint. It is transparent and not metamict, or semi-opaque with high reflectance and white internal reflections.[2] Some micron-sized grains are reddish.[6] A pale greenish yellow or sometimes mottled red has been seen in the intergrowths with zirconolite.[3] In spinel-bearing breccia 14066, one yellow transparent baddeleyite grain was found.[7]

*Chemical Composition*

Almost all baddeleyite analyses report some hafnium content. With $HfO_2$ content 2% or greater, the phase has been termed hafnian baddeleyite (Table 3-1; [2]).[4] A ferroan variety, high also in $TiO_2$, has no reported $HfO_2$ (Table 3-1; [1]).[8] Some baddeleyite is low or completely lacking in U, Th, or REE.[9] Much baddeleyite, however, is uranium-enriched. In some rocks, areas of U concentration greater than 40 parts per million (ppm) are thought to be baddeleyite and appear to be the source of fission

**Table 3-1 Baddeleyite Analyses**

| | [1] | [2] | [3] | [4] | [5] |
|---|---|---|---|---|---|
| | | | | Luna 20 | |
| | 12038,67 | 12036,9 | | 22002,3 | 14310 |
| $SiO_2$ | <0.1 | 0.39 | 0.12 | 0.18 | — |
| $Al_2O_3$ | <0.1 | — | — | 0.54 | — |
| $TiO_2$ | 2.4 | 1.97 | 3.13 | 1.82 | — |
| FeO | 7.4 | 3.25 | 0.96 | 0.45 | — |
| MgO | — | <0.02 | 0.18 | 0.14 | 0.06 |
| MnO | — | <0.02 | <0.02 | 0.17 | — |
| $ZrO_2$ | 90.3 | 91.9 | 93.2 | 94.7 | 98.23 |
| $HfO_2$ | — | 3.23 | 1.60 | 1.65 | 1.70 |
| Total | 100.1* | 100.74* | 99.19* | 100.49* | 99.99 |

*References*

[1] Simpson and Bowie, *Suppl.* **2,** 212. From a coarse-grained basalt. *Analysis includes CaO = <0.1%. Ferroan baddeleyite.

[2] Keil et al, *Suppl.* **2,** 333. From an opthitic basalt. *Analysis includes $Na_2O$ = <0.02. Hafnian baddeleyite.

[3] ———, *ibid.* From an ophitic basalt. *Analysis includes $Na_2O$ = <0.02%. Baddeleyite.

[4] Brett et al, *Geochim. Cosmochim. Acta* **37,** 769. From a soil sample. *Analysis includes $Cr_2O_3$ = 0.13%, $V_2O_3$ = 0.06%, CaO = 0.16%, and $Nb_2O_5$ = 0.49%. Baddeleyite.

[5] El Goresy et al, *EPSL* **13,** 127. From a KREEP basalt. Baddeleyite.

tracks.[6] In fines sample 12028, the U in baddeleyite is $\geq$ 50 ppm. From Apollo 14 samples, especially KREEP basalt 14310, the baddeleyite has a U content of 80 to 400 ppm.[10] In 14310 the $ZrO_2$ content of the total rock (0.13%) is greater than in Apollo 11 and 12 samples, and this is attributed largely to the abundant presence of baddeleyite.[11]

*References*

1. Ramdohr and El Goresy, *Naturwiss.* **57,** 102.
2. Agrell et al, *Suppl.* **1,** 112.
3. Lovering and Wark, *Suppl.* **2,** 155–156.
4. Keil et al, *Suppl.* **2,** 333.
5. El Goresy et al, *EPSL* **13,** 127.
6. Lovering and Kleeman, *Suppl.* **1,** 628.
7. Christophe-Michel-Lévy et al, *Suppl.* **3,** 889.
8. Simpson and Bowie, *Suppl.* **2,** 212.
9. Brown et al, *Suppl.* **2,** 595.
10. Crozaz et al, *Lun. Sci. Conf.* '*72, Abstr.*, 149.
11. El Goresy et al, *Suppl.* **3,** 338.

## RUTILE $TiO_2$

*Synonymy*

Blue rutile; Jedwab, *Apollo* 15 *Lunar Samples*, 108.
Mg-Fe rutile; Haggerty, *Suppl.* **4,** 781.
Niobian rutile; Marvin, *EPSL* **11,** 7.
Zirconian rutile; Dowty et al, *Lun. Sci V, Abstr.*, 174.

*Occurrence and Form*

Both primary rutile and exsolved rutile are found in the lunar rocks, generally associated or intergrown with ilmenite. The rutile intergrowths that appear to be exsolution phenomena are thin lamellae ($< 1\ \mu$ across) oriented parallel to $\{01\bar{1}2\}$ of the ilmenite (Figure 3–1). Occurrences of primary rutile are distinctly different from the exsolved rutile in that they are complexly twinned and randomly oriented inclusions in the ilmenite. In fine-grained basalt 10003-37, both the ilmenite and the included rutile are hexagonal. The rutile shows well-developed sectorial twinning, probably about [101]. The rutile twin planes and the rutile–ilmenite contacts are sharp and well-defined. In microbreccia 10021-30, however, the outline of the rutile crystal is somewhat irregular. It embays and is embayed by ilmenite host; twinning is less well defined, and parts of the grain are polycrystalline. In both cases secondary exsolved rutile (together with

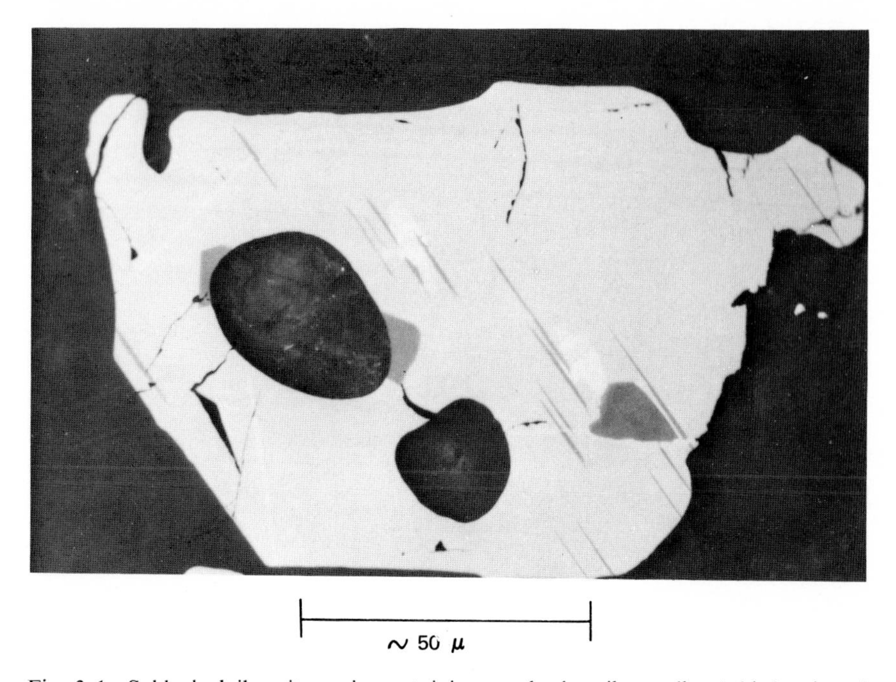


Fig. 3–1 Subhedral ilmenite grain containing exsolved rutile needles (white) oriented parallel to the $\{01\bar{1}2\}$ of ilmenite and lensoidal rods of spinel (dark gray) oriented parallel to $\{0001\}$ of ilmenite. Photograph courtesy of S. Haggerty, Department of Geology, University of Massachusetts, Amherst.

spinel) also is present in the ilmenite. The exsolved rutile lamellae terminate on contact with primary rutile, but the associated spinel tends to penetrate the primary rutile. A rutile core in 10003-37 is occupied by an intergrowth of iron and troilite, late-stage phases on which the sectorial twin planes in the rutile appear to have exercised some control. The coexistence of metallic iron and rutile might suggest the breakdown of ilmenite under highly reducing conditions, but the euhedral outline, the well-defined contacts, and the sector twinning all indicate a primary rather than a secondary origin.[1]

Some wide crosscutting lamellae or rare discoidal rutile grains are limited almost entirely to shocked samples.[2] In the mesostasis of fine-grained basalt 10017[3] and of ophitic basalt 10058, the rutile occurs as needles associated with geikielo-ilmenite.[4] Rutile is present in a Luna 20 sample both as oriented lamellae and irregular areas in ilmenite.[5] In a bronzite grain in microbreccia 14305,4, acicular crystals of rutile are aligned parallel to the *c*-axis of the bronzite.[6] In peridotite 15445,10, pyroxene grains are heavily charged with tiny inclusions of rutile.[7] A loose fragment (found in the sealed bag that contained vuggy basalt 15555,105) is approximately 0.5 mm in size and consists of an agglomerate of micron-sized irregular rutile grains.[8] Similar agglomerates (0.2–1 $\mu$ in size) of microcrystals of rutile have been found in a number of Apollo 16 soil samples. Microcrystals in the form of needles, short prisms, or rounded grains, occur on or in silicates and silica glass. This association and the variety of habit suggest formation by more than one process (e.g., possibly the condensation of a Ti-rich vapor generated by impact, and/or crystallization by impact).[9]

Niobian rutile from microbreccia 12070,35 is enclosed in ilmenite.[10] Niobian rutile occurs as rounded grains or lamellae within ilmenite grains also in a 0.5 $\times$ 2 mm KREEP basalt fragment from microbreccia 14162,16. All but one of the rutile grains are less than 2 $\mu$ in size.[11]

In microbreccia 14321,21, an armalcolite decomposition assemblage consists of magnesian ilmenite plus a Mg-Fe-rutile.[12]

*Optics*

In polished section exsolved rutile appears as a white, bright anisotropic phase.[13] Some lamellae are almost water-clear.[2] Needles of rutile are mostly white and anisotropic.[1] Rutile from basalt 15555,105 is deep ultramarine blue. In transmitted light it is opaque but transparent on edges. With reflected light it is transparent with very high dispersion.[8] Blue rutile has been found again in a number of Apollo 16 fines samples, together with grains that are pure white in reflected light. In transmitted light the same grains are dark brown or opaque.[9] The rutile from fines sample 10004[14] and fine-grained basalt 10017[3] has been described as reddish or brown on thin edges. In ophitic basalt 10058, the rutile needles are reddish brown.[4]

The niobian rutile from microbreccia 12070,35 is anisotropic, tawny yellow, uniaxial (+), with both indices of refraction well over 2.10. The $\omega$ is tawny yellow; $\epsilon$ is olive-green.[10]

*Chemical Composition*

Some lunar rutile is essentially pure $TiO_2$, and some contains several percent FeO and appreciable Al (Table 3-2; [4], [5]).[15] Many rutile analyses contain some Nb. The moderate Nb content of rutile blebs in the enstatite of peridotite 15445,10 is consistent with its derivation as an exsolution phenomenon (Table 3-2; [6]).[7]

In other rocks (e.g., breccia 12070,35 and a KREEP fragment from breccia 14162,16), the Nb content is high enough for the phase to be considered a niobian rutile (Table 3–2; [1], [3]). It is suggested that

**Table 3-2 Rutile Analyses**

| | [1] | [2] | [3] | [4] | [5] | [6] |
|---|---|---|---|---|---|---|
| | | | | | Luna 20 | |
| | 14162,16 | 14321,21 | 12070,35 | 10058 | 22003,1 | 15445,10 |
| $SiO_2$ | 0.61 | 0.93 | — | — | 0.13 | — |
| $TiO_2$ | 85.3 | 87.29 | 89.9 | 96.62 | 97.23 | 98.0 |
| $Al_2O_3$ | 0.82 | 0.13 | — | 1.91 | 0.02 | — |
| $Cr_2O_3$ | 2.65 | 0.56 | 3.2 | 0.30 | 0.48 | 0.03 |
| $V_2O_3$ | 0.22 | — | 0.4 | — | — | — |
| FeO | 0.61 | 7.45 | — | 0.22 | 2.34 | 0.1 |
| MnO | 0.12 | 0.14 | — | 0.03 | 0.01 | — |
| MgO | 0.03 | 3.20 | — | 0.05 | 0.04 | <0.1 |
| CaO | 0.52 | 0.28 | — | 0.37 | 0.10 | <0.1 |
| $Nb_2O_5$ | 7.1 | 0.55 | 6.4 | — | — | 1.6 |
| $ZrO_2$ | 0.70 | 0.07 | — | — | — | 0.1 |
| Total | 98.76 | 100.60 | 101.3* | 99.50 | 100.35 | 100.1* |

*References*

[1] Hlava et al, Preprint for *Niobian Rutile in an Apollo 14 KREEP fragment*, submitted to *Meteoritics*. From a KREEP fragment in a microbreccia. *Analysis includes CeO = 0.08%. Niobian rutile.

[2] Haggerty, *Suppl.* **4,** 781 From a microbreccia. Mg-Fe rutile.

[3] Marvin, *EPSL* **11,** 8. From a microbreccia. *Analysis includes $Ta_2O_5$ = 0.2%, $La_2O_3$ = 0.4%, and $Ce_2O_3$ = 0.8%. Nb may possibly be as $Nb^{4+}$. Niobian rutile.

[4] Agrell et al, *Suppl.* **1,** 111. From a basalt. Rutile.

[5] Haggerty, *Geochim. Cosmochim. Acta* **37,** 859. From a soil sample. Rutile.

[6] Anderson, *J. Geol.* **81,** 221. From a peridotite. *Analysis includes $Ta_2O_5$ = <0.02%, $HfO_2$ <0.05%, and La and Ce <0.1%. Rutile.

niobian rutile is most likely to be found in rocks of KREEP chemistry, since the same process that concentrates K, REE, P, U, Th, Zr, and other related elements also concentrates the Nb in the lunar environment. The oxidation state of the Nb in extraterrestrial niobian rutiles is still in question. One assumption is that Nb in meteorites and lunar rocks is present as $Nb^{4+}$ because of the low oxygen fugacities involved and because $NbO_2$ is isostructural with rutile. It is also possible that the Nb may be present as $Nb^{3+}$. The rutile structure, however, can accommodate up to 15 mol % $Nb_2O_5$ in solid solution, and, until a definite value is determined for the oxidation state of extraterrestrial Nb, the commonly accepted value of $Nb^{5+}$ is used.[11]

In fine-grained basalt 10017, several grains of ilmenite contain lamellae of both rutile and chromite. The rutile is almost stoichiometric, and the chromite lamellae abut against and sometimes are displaced by the rutile lamellae. The origin of the rutile is difficult to explain. One suggestion, that the rutile might have been formed by the reduction of ilmenite,

$$(\text{e.g., } 2FeTiO_3 \rightarrow 2TiO_2 + 2Fe + O_2)$$

is rejected here because no native iron has been observed adjacent to the lamellae or within the ilmenite. In fine-grained basalt 10020, small ($<5\ \mu$) areas in the centers of some ilmenite grains have a composition approaching $FeTi_2O_5$. These areas might have produced the rutile lamellae by the following decomposition on cooling:[16]

$$FeTi_2O_5 \rightarrow FeTiO_3 + TiO_2.$$

An armalcolite decomposition essemblage in microbreccia 14321,21 indicates that such a reaction has occurred. The armalcolite contains both magnesian ilmenite and a Mg-Fe rutile (Table 3-2; [2]). Repeated analyses of the rutile produced consistently coherent values and an approach to stoichiometry suggesting that part of the Ti is replaced by Fe and Mg.[12] From spinel troctolite 65785, a zirconian rutile has been reported to occur together with ilmenite, metallic Ni-Fe, troilite, whitlockite, and a Cr-Zr-REE armalcolite. The $ZrO_2$ content of the rutile is 3.8%. From melt rock 60615 the Zr-rutile has up to 6.0% $ZrO_2$.[17]

### *X-Ray Data*

An X-ray diffraction pattern of niobian rutile from 12070,35 has 16 lines, indexed on a tetragonal rutile lattice. Computed from these are:[10]

$$a = 4.600 \pm 0.001 \text{ Å}$$
$$b = 2.962 \pm 0.002 \text{ Å}$$

### *References*


1. Haggerty et al, *Suppl.* **1,** 518, 526, 527.
2. Ramdohr and El Goresy, *Naturwiss.* **57,** 102.
3. French et al, *Suppl.* **1,** 439–440.
4. Agrell et al, *Suppl.* **1,** 97.
5. Haggerty, *Geochim. Cosmochim. Acta* **37,** 865.
6. Klein and Drake, *Suppl.* **3,** 1103.
7. Anderson, *J. Geol.* **81,** 219.
8. Jedwab, *Apollo 15 Lunar Samples*, 108–109.

9. ———, *Suppl.* **4,** 863–864.
10. Marvin; *EPSL* **11,** 7–9.
11. Hlava et al, Preprint for "Niobian Rutile in an Apollo 14 KREEP Fragment"; submitted to *Meteoritics.*
12. Haggerty, *Suppl.* **4,** 781–783.
13. Simpson and Bowie, *Suppl.* **1,** 874.
14. Bailey et al, *Suppl.* **1,** 187.
15. Wood et al, *Spec. Rep.* **333,** 148.
16. Dence et al, *Suppl.* **1,** 321–323.
17. Dowty et al, *Lun. Sci. V, Abstr.*, 174.

## Unidentified Titanium Phases

### *$TiO_2$*

The formula $TiO_2$, used as a mineral name, was applied to small grains ($< 20\ \mu$ in diameter) in ilmenite from some Apollo 11 samples. The composition of the phase appeared to be pure $TiO_2$.[1] Possibly rutile is implied.

### *$TiO_2$ Polymorph*

Rare, small (up to 3 $\mu$) grains of a reddish-orange opaque phase have been observed in cracks in porphyritic basalts 12004 and 12022. Probe analysis yielded a composition of $TiO_2$, but the color of the grains is different from any of the known $TiO_2$ polymorphs.[2]

### *Titanium-Rich Phase*

A vesicle in an agglutinate fragment of Luna 20 soil sample 22002,3 yielded a brick-red opaque phase, about 3 $\mu$ in diameter. Its reflectivity is somewhat higher than that of the transition metal oxides. The phase is either isotropic or exhibits very weak reflection birefringence. X-Ray analysis (using SEM) shows the phase to be an oxide with Ti as the predominant cation. Subordinate Si, Al, Mg, Fe, and either P or Zr are present in approximate decreasing order of abundance; Si $< 5\%$. Optically and analytically the phase is identical with the $TiO_2$ polymorph just described. It is believed this phase is of lunar origin because it was found in material from different lunar missions and in thin sections prepared in different laboratories.[3]

### *Needlelike Exsolution*

Rare needles have been observed in titanian chromite from some fine-grained Apollo 11 basalts.[4] No other data are given, but possibly the needles are rutile or some $TiO_2$ polymorph.

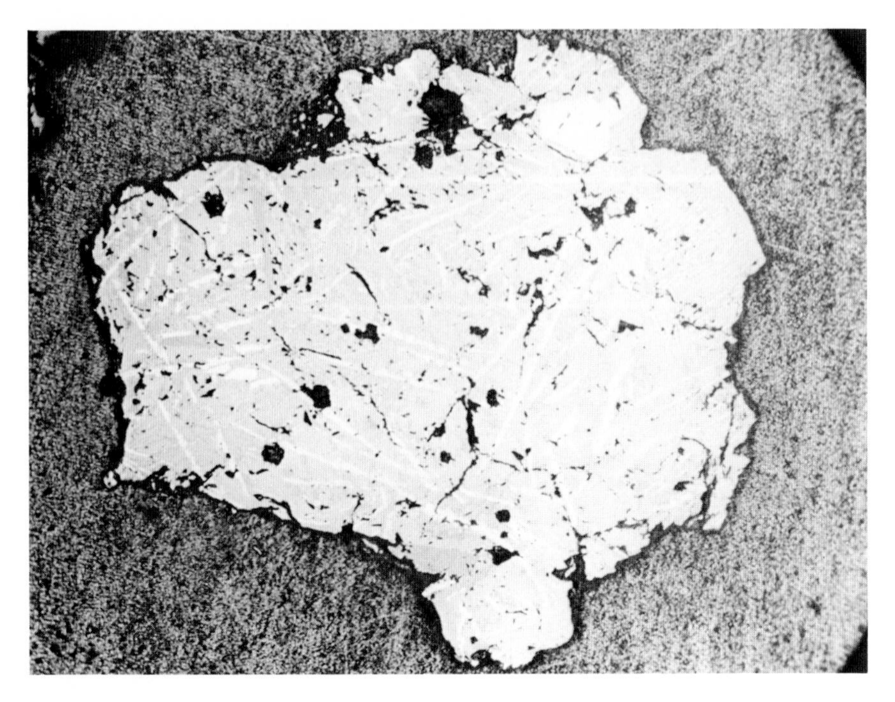

Fig. 3–2 Rock fragment (from soil sample) showing phenocrysts and skeletal crystals of ilmenite (light gray). X 135. Photograph courtesy of E. N. Cameron, Department of Geology and Geophysics, University of Wisconsin, Madison.

### *References*

1. Keil et al, *Suppl.* **1,** 585.
2. Brett et al, *Suppl.* **2,** 310.
3. Brett et al, *Geochim. Cosmochim. Acta* **37,** 768.
4. Simpson and Bowie, *Suppl.* **1,** 874.

## ILMENITE $FeTiO_3$

### *Synonymy*

Magnesian ilmenite; Agrell et al, *Suppl.* **1,** 112.
Geikielo-ilmenite; Agrell et al, *Suppl.* **1,** 111, 115.
Mg-ilmenite; Weill et al, *Suppl.* **1,** 941.
Zirconian ilmenite; Brown et al, *Suppl.* **3,** 149, 152.

### *Occurrence and Form*

Ilmenite is the most common lunar opaque mineral, sometimes making up to 20% of the volume of the rocks (Figure 3–2).[1] In Apollo 11 rocks, after

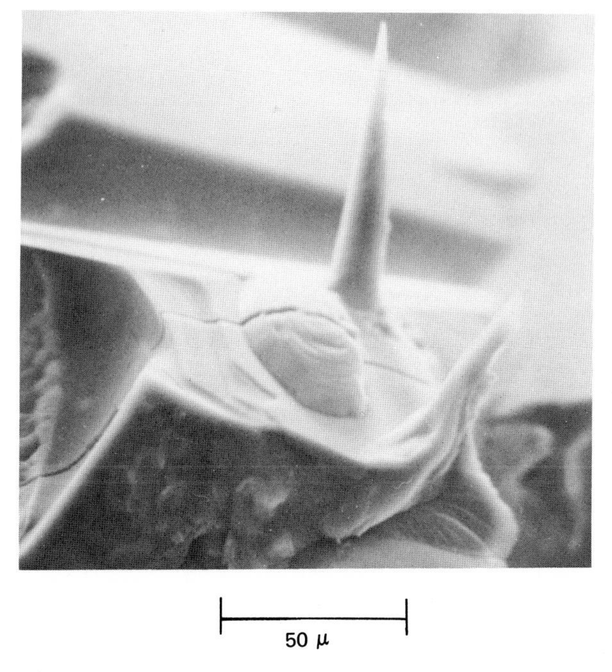


Fig. 3–3 Ilmenite (blocky crystal) associated with troilite bleb and silicate pillar. From *Suppl.* **2,** 929. Photograph courtesy of J. Jedwab, Université Libre de Bruxelles, Belgium.

clinopyroxene and plagioclase, ilmenite is the most prominent phase.[2] Its morphology is extremely varied, ranging from anhedral to subhedral to euhedral.

In the crystalline rocks ilmenite forms

1. Blocky euhedral to subhedral crystals (Figures 3–3 and 3–4).
2. Thin plates paralleling (0001) with rhombohedral modifications.
3. More rarely, coarse skeletal crystals with entrapped pyroxene, troilite, and metallic iron (Figures 3–5 and 3–6).[3]

Blocky grains sometimes have cores of armalcolite or chromian ulvöspinel and probably have formed by reaction of these included phases with liquid during cooling.[3] A suggestion has been made that the coarsely crystalline ilmenite from Luna 16 may be pseudomorphous after armalcolite,[4] but no evidence of ilmenite pseudomorphs after armalcolite was found in the Luna 20 samples.[5] Ilmenite overgrowths on armalcolite also contain lamellae of rutile.

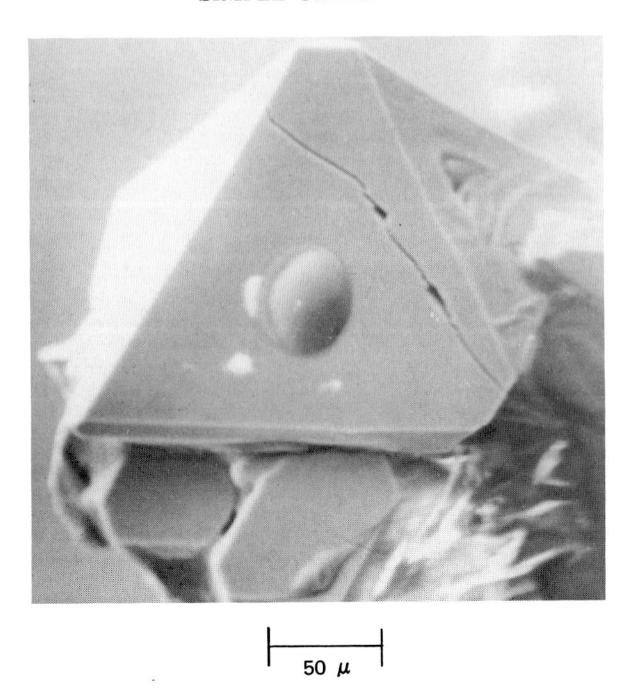


Fig. 3–4 Pseudo-octahedron of ilmenite. Bleb is silicate overgrowth. From *Suppl.* **2,** 928. Photograph courtesy of J. Jedwab, Université Libre de Bruxelles, Belgium.

Ilmenite occurs in the lunar dust as thin plates flattened on (0001), a number of them showing hexagonal outline. In granular fine-grained rocks, spherical cavities representing former gas bubbles are lined with ilmenite plates (Figure 3–7).[1] In the Luna 20 samples, ilmenite is present in five distinct modes:

1. As a primary crystalline phase.
2. As a product of subsolidus reduction of Cr-Al-ulvöspinel.
3. As an exsolution phase from chromian pleonaste.
4. As a phase containing chromite plus rutile.
5. As oriented lamellae apparently of exsolution origin in pyroxene.[6]

Ilmenite crystallized from early to almost final stages of crystallization of the igneous rocks.[2] In KREEP basalt 14310, for example, lath-shaped ilmenite began to crystallize at about the onset of pyroxene crystallization and continued to crystallize until, finally, in the mesostases, it formed as small laths and dendritic crystals.[7] Late-stage ilmenite results also from a

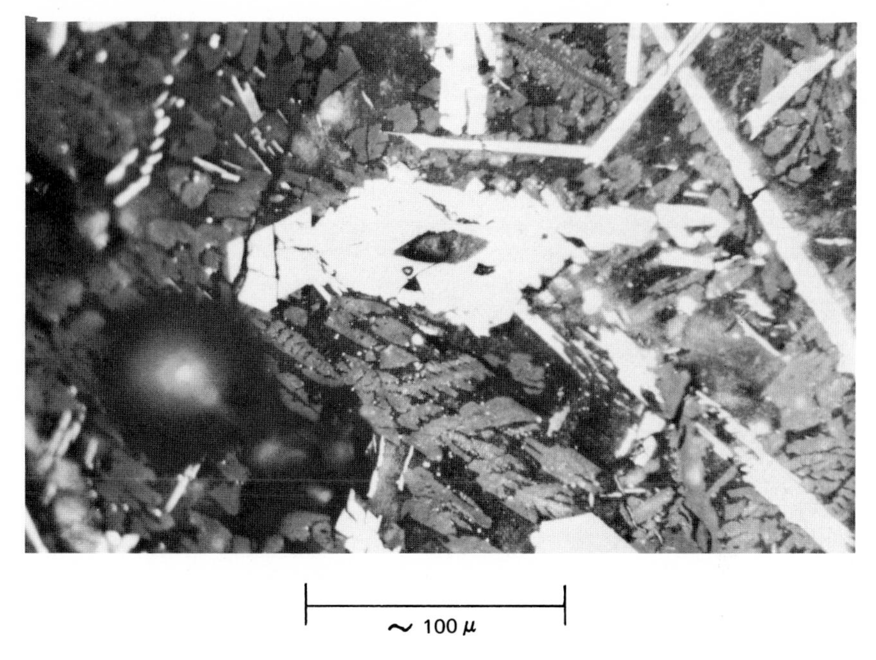


Fig. 3–5 Skeletal ilmenite crystal (white) in center of picture. Photograph courtesy of S. Haggerty, Department of Geology, University of Massachusetts, Amherst.

subsolidus reduction of chromian ulvöspinel to native iron plus ilmenite, and this process has been observed in rocks from all Apollo missions 11 through 15.[8] This breakdown is developed to a great extent in basaltic rocks 14053 and 14072, with all stages of reduction to complete breakdown.

*Optics*

Ilmenite is distinctly anisotropic and is deep reddish tan in oil immersion.[3] Ilmenite is usually opaque, but thin plates are transluscent, and some dark brown thin transparent plates have been observed.[9] Ilmenite high in magnesium exhibits considerable reflectance birefringence; reflection pleochroism is especially marked also, ranging from pinkish buff to brown gray.[5]

Ilmenite is uniaxial (−). Reflectivity for ilmenite from fine-grained basalt 10049-29 is given in Table 3-3.[10]

Twinning is rare in ilmenite,[9] but special studies on deformation in lunar ilmenites revealed twinning on $\{10\bar{1}1\}$ and, less prominently, on $\{0001\}$.[11] In a Luna 20 recrystallized troctolitic anorthosite fragment, fine twin lamellae occur in some ilmenite grains. In one grain two different orientations are represented; therefore, the morphology of the lamellae

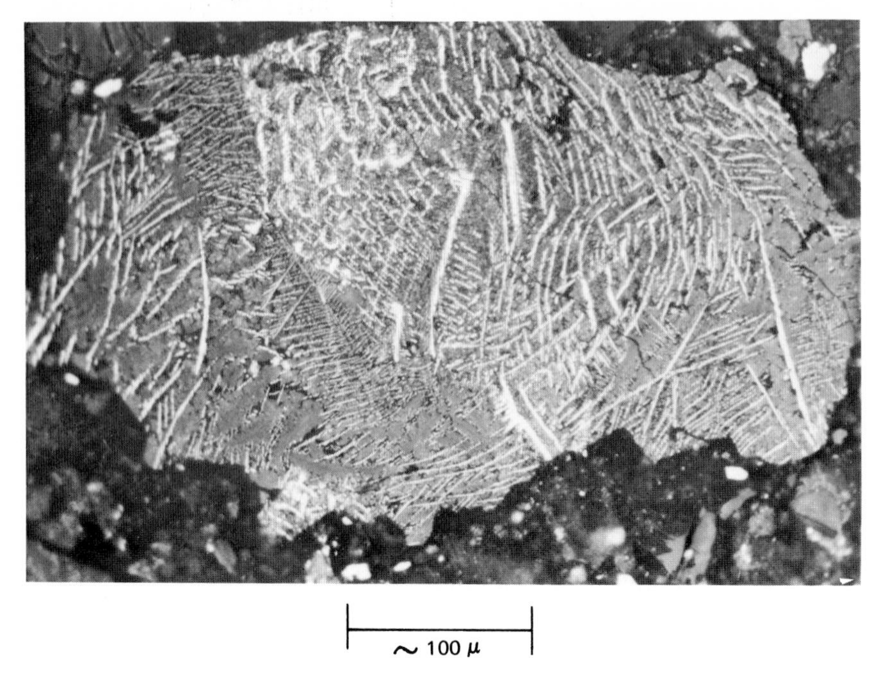


Fig. 3–6 Fernlike arrays and subskeletal laths of ilmenite. Photograph courtesy of S. Haggerty, University of Massachusetts, Amherst.

**Table 3-3 Reflectivity of Ilmenite from 10049-29***

| Wavelength (nm) | $R_\omega$ | $R_\epsilon$ |
|---|---|---|
| 450 | 18.9 | 15.8 |
| 470 | 18.6 | 15.6 |
| 500 | 18.5 | 15.4 |
| 520 | 18.5 | 15.4 |
| 546 | 18.7 | 15.7 |
| 589 | 18.9 | 16.1 |
| 620 | 19.2 | 16.6 |
| 640 | 19.5 | 17.0 |

*Fine-grained basalt. All values in percent. Accuracy ±0.02%. Cameron; *Suppl.* **1,** 225.

Fig. 3–7 Ilmenite platelet in vug in fine-grained basalt 10072,23. Photograph courtesy of D. Walker, Harvard University, Cambridge, Mass.

appears to differ from that of shock-induced twins in ilmenite. Both the lamellae and the host ilmenite exhibit strong reflection pleochroism, and electron microprobe analysis and optical examination suggest that the lamellae also are ilmenite.[5] Zoning in ilmenite has not been detected, either optically or with the microprobe.[10]

### *Chemical Composition*

In general, lunar ilmenite is close to stoichiometric, with minor amounts of Cr, Al, and Mn present in solid solution (Table 3-4; [1]–[3]).[12]

**Table 3-4 Analyses of Ilmenites**

| | [1] | [2] | [3] | [4] | [5] 10057–67 | [6] | [7] |
|---|---|---|---|---|---|---|---|
| | 12052 | Luna 20 | Luna 16 | 14310 | Type 2 | Type 1 | 10085 |
| $SiO_2$ | — | — | 0.45 | 0.39 | 0.09 | 0.09 | <0.01 |
| $Al_2O_3$ | — | — | 0.28 | 0.10 | <0.03 | <0.03 | 0.05 |
| $TiO_2$ | 52.4 | 53.1 | 50.79 | 52.66 | 52.33 | 52.22 | 52.6 |
| $Cr_2O_3$ | 0.1 | 0.16 | 0.27 | 0.38 | 0.53 | 0.53 | 0.78 |
| FeO | 46.4 | 45.7 | 48.14 | 44.62 | 44.84 | 44.38 | 45.3 |
| MnO | 0.4 | 0.40 | 0.45 | 0.32 | 0.15 | 0.15 | 0.33 |
| MgO | — | 0.15 | 0.15 | 0.83 | 1.14 | 1.39 | 1.23 |
| CaO | — | 0.04 | 0.34 | 0.12 | 0.21 | 0.21 | — |
| Total | 99.3 | 99.55 | 100.87 | 99.45* | 101.97* | 99.21* | 100.29 |

*References*


[1] Champness et al, *Suppl*, **2,** 364. From a porphyritic basalt.

[2] Brett et al, *Geochim. Cosmochim. Acta* **37,** 767. From a fines sample.

[3] Haggerty, *EPSL* **13,** 348. From a fines sample.

[4] Gancarz et al, *EPSL* **16,** 309. From a KREEP basalt. *Analysis includes $Na_2O$ = <0.01% and $K_2O$ = 0.03%.

[5] Lovering and Ware, *Suppl.* **1,** 635–636. From a fine-grained basalt. *This type of ilmenite has a high Ti abundance with low cation total; quadrivalent ions exceed divalent ions by 0.10 formula units. Analysis has been recalculated assuming $Ti_2O_3$ also is present and = 2.44%, so that $^{4+}$ ions = total $^{2+}$ ions. Analysis also includes $Na_2O$ = <0.04%, $K_2O$ = <0.02%, and $ZrO_2$ = 0.24%.

[6] Lovering and Ware; *Suppl.* **1,** 635–636. From a fine-grained basalt. In this type of ilmenite the Ti abundance is normal with equivalent quadrivalent and divalent ions; therefore the phase cannot contain significant $Ti^{3+}$ (or $Fe^{3+}$). *Analysis includes < $Na_2O$ = <0.04%, $K_2O$ = <0.02%, and $ZrO_2$ = 0.24%.

[7] Raymond and Wenk, *Contrib. Mineral. Petrol.* **30,** 136. From a microanorthositic fragment in a soil sample.

| | [8] | [9] | [10] | [11] | [12] | [13] | [14] |
|---|---|---|---|---|---|---|---|
| | 10072–49 | 14257,2 | 65015 | Luna 20 | 14258,28 (1419–7) | 10085–4–10 | Terrestrial Ilmenite |
| $SiO_2$ | 0.04 | — | 0.26 | 0.18 | 0.3 | — | <0.02 |
| $Al_2O_3$ | 0.17 | 0.60 | 0.13 | — | 0.19 | 1.64 | 0.2 |
| $TiO_2$ | 54.2 | 59.35 | 53.58 | 53.7 | 51.7 | 56.30 | 47.6 |
| $Cr_2O_3$ | 0.72 | 0.85 | 0.54 | 0.39 | 0.58 | 0.34 | 0.08 |
| FeO | 41.7 | 34.30 | 39.57 | 37.1 | 37.7 | 32.39 | 48.0 |
| MnO | 0.35 | — | 0.32 | 0.36 | 0.20 | 0.34 | 0.42 |
| MgO | 3.12 | 3.85 | 5.00 | 8.0 | 8.2 | 9.63 | 1.38 |
| CaO | 0.03 | — | — | 0.58 | 0.31 | 0.44 | — |
| Total | 100.33 | 99.36* | 99.45* | 100.31 | 99.18 | 101.08 | 98.40* |

*References*


[8] Kushiro and Nakamura, *Suppl.* **1,** 619. From a fine-grained olivine basalt.

[9] Klein and Drake, *Suppl.* **3,** 1103. From a microbreccia fragment in coarse fines. *Analysis includes NiO = 0.41%. Presence of $Ti^{3+}$ may be indicated, since there is an excess of titanium if it is all calculated as $Ti^{4+}$.

[10] Albee et al, *Lun. Sci. IV*, *Abstr.*, 26. From a polymict KREEP-rich rock with Ba, U, and Th. *Analysis includes $ZrO_2$ = 0.05%.

[11] Brett et al, *Geochim. Cosmochim. Acta* **37,** 767. From a fines sample.

[12] Powell and Weiblen, *Suppl.* **3,** 846. From an anorthositic fragment in a fines sample.

[13] Agrell et al; *Suppl.* **1,** 110–111. From a microanorthositic fragment.

[14] Raymond and Wenk; *Contrib. Mineral. Petrol.* **30,** 136. From a Pliocene basaltic lava flow in Omo Basin, Ethiopia. *Analysis includes $V_2O_3$ = 0.70%.

Minor Zr sometimes is present (Table 3-4; [5], [6]). It has been suggested that ilmenite is one of the main host minerals for Zr in the lunar rocks,[12] and there appears to be a partitioning of $ZrO_2$ in favor of the ilmenites.[8] Ilmenites associated with ulvöspinel and baddeleyite in KREEP basalt 14310 and basalt 14073, contain 0.17 to 0.57% $ZrO_2$. This may represent some degree of substitution of $Zr^{4+}$ for $Ti^{4+}$.[8] The name zirconian ilmenite has been used.[13] There is an appreciable geikielitic content in some ilmenite (Table 3-4; [8]–[13]). A magnesian ilmenite with MgO = 9.63% (Table 3-4; [13]) is closely associated with rutile in a manner that suggests simultaneous crystallization rather than exsolution.[2] High-magnesian ilmenite ($Fe_{0.92}Mg_{0.05}TiO_3$) is common in devitrified glasses of (mestastable?) breccias.[14] Some investigators have stated that there is no apparent relationship between the MgO content of the ilmenite and its grain size or mineral association.[15] Others have noted that ilmenite associated with armalcolite has a higher Mg/Fe ratio than ilmenite in rocks without armalcolite, and the relations of magnesian ilmenite with armalcolite suggest that the two minerals are high-temperature early phases.[10] Ilmenite from fine-grained olivine basalt 10072-49 has inclusions of magnesian armalcolite and rutile lamellae, and a higher Ti content than ilmenite without these inclusions (Table 3-4; [8]). It is possible that originally the host ilmenite contained excess Ti (over stoichiometric $FeTiO_3$) which, with lowering temperatures, was exsolved as rutile. In this case the Ti probably was trivalent. Alternatively, the ilmenite originally may have been magnesian armalcolite that was converted to magnesian ilmenite with lowering temperature, and the excess Ti was exsolved.[16] A slight excess of FeO in the ilmenite has been attributed to the possible presence of some $Fe_3O_4$.[17] Some excess of Fe + Mg, if real, may reflect possible $Fe^{3+}$ substitution for $Ti^{4+}$.[18] In a special study of U concentration in lunar minerals, ilmenite from porphyritic basalt 12021 showed $UO_2 = 72 \pm 22$ to $820 \pm 160$ ppb.[19]

*X-Ray Data*

Lunar and terrestrial ilmenite have virtually identical crystal structure parameters. This has been demonstrated in a special study comparing lunar and terrestrial ilmenites with chemical compositions fairly close to stoichiometric $FeTiO_3$ (Table 3-4; [7], [14]).[20] The cell parameters of these two specimens are in reasonable agreement with those for synthetic $FeTiO_3$[20] and with those of an average of measurements on seven Apollo ilmenites (Table 3-5; [1]–[4]).[21] The Fe-Ti order was high in both the lunar and terrestrial ilmenites, although the lunar ilmenite had slightly better order (less Ti on Fe 1 sites.) This was attributed either to the ordering mechanism, which goes faster at high Ti content, or to the possibility that

**Table 3-5 Cell Parameters of Ilmenites (Å)**

| | [1] 10085 | [2] Terrestrial Ilmenite | [3] Synthetic Ilmenite | [4] Apollo 11 Basalts | [5] 10047,13 | [6] Terrestrial Ilmenite | [7] 10047 | [8] Terrestrial Ilmenite |
|---|---|---|---|---|---|---|---|---|
| *a* | 5.085 ± 0.001 | 5.091 ± 0.001 | 5.087 | 5.087 ± 0.001 | 5.0886 ± 0.0005 | 5.092 ± 0.0009 | 5.085 ± 0.005 | 5.083 |
| *c* | 14.088 ± 0.004 | 14.056 ± 0.003 | 14.085 | 14.078 ± 0.005 | 14.082 ± 0.002 | 14.026 ± 0.005 | 14.04 ± 0.02 | 14.04 |

*References*

[1] Raymond and Wenk, *Contrib. Mineral Petrol.* **30,** 136. From a microanorthositic fragment in a fines sample.
[2] ———, *ibid.* From a Pliocene basaltic flow in Omo Basin, Ethiopia.
[3] ———, *ibid.*
[4] Stewart et al; *Suppl.* **1,** 929. Average of seven ilmenite analyses.
[5] Bayer et al, *EPSL* **16,** 273. From an ophitic ilmenite basalt.
[6] ———, *ibid.* From the Urals.
[7] Minkin and Chao, *Suppl.* **2,** 239. From an ophitic basalt. Unshocked ilmenite.
[8] Palache et al, *Dana VII,* **1,** 535. From Quincy, Mass.

the terrestrial crystal might have cooled more rapidly in the superficial lava flow. The measured disorder was barely resolved by X-ray diffraction.[20] Another investigation comparing a lunar ilmenite with a terrestrial ilmenite yielded cell parameters that seemed to show an increase in the *c* axis for lunar ilmenite (Table 3-5; [5]). This effect was considered to be due to the larger ionic radius of $Ti^{3+}$ as compared to the average radius of $Fe^{2+}$ and $Ti^{4+}$, assuming the following substitutional mechanisms:

| | | |
|---|---|---|
| lunar ilmenite | $2Ti^{3+}$ | for $Fe^{2+} + Ti^{4+}$ |
| terrestrial ilmenite | $2Fe^{3+}$ | for $Fe^{2+} + Ti^{4+}$ |

This assumption is reasonable if lunar ilmenite contains no $Fe^{3+}$ but an excess of $Ti^{3+}$.[22] However, the cell parameters of an unshocked ilmenite from 10047[10] are virtually identical to those for another terrestrial ilmenite[23] (Table 3-5; [7], [8]), although the X-ray data on these two samples are not accompanied by chemical analyses. It would appear that until we have many more X-ray and chemical analyses of both lunar and terrestrial ilmenites, the lunar ilmenite and its earthly counterpart can be assumed to be the same.

*References*

1. Bailey et al, *Suppl.* **1,** 170, 177–178.
2. Agrell et al, *Suppl.* **1,** 97, 111.
3. Haggerty et al, *Suppl.* **1,** 514–517.
4. Haggerty *EPSL* **13,** 348–349.
5. Brett et al, *Geochim. Cosmochim. Acta* **37,** 766–767, 769.
6. Haggerty et al, *Geochim. Cosmochim. Acta* **37,** 864.
7. Ridley et al, *Suppl.* **3,** 163.
8. El Goresy et al, *Suppl.* **3,** 338–339, 344.
9. Dence et al, *Suppl.* **1,** 322.
10. Cameron, *Suppl.* **1,** 225, 227–229.
11. Minkin and Chao, *Suppl.* **2,** 237, 239.
12. Frondel et al, *Suppl.* **1,** 463.
13. Brown et al, *Suppl.* **3,** 149, 152.
14. Essene et al, *Suppl.* **1,** 394.
15. El Goresy et al, *Suppl.* **2,** 223.
16. Kushiro and Nakamura, *Suppl.* **1,** 619–620.
17. Ramdohr and El Goresy, *Naturwiss.* **57,** 100.
18. Walter et al, *Suppl.* **2,** 351.
19. Thiel et al, *EPSL* **16,** 38.
20. Raymond and Wenk, *Contrib. Mineral. Petrol.* **30,** 135–136, 138.
21. Stewart et al, *Suppl.* **1,** 928–929.

22. Bayer et al, *EPSL* **16,** 273.
23. Palache et al, *Dana VII*, **1,** 535.

## CORUNDUM $\alpha$-$Al_2O_3$

### *Occurrence and Form*

Originally the occurrence of corundum in the lunar samples was thought to be due to terrestrial contamination.[1] It is now believed that the phase originated on the Moon, possibly as condensation from a vapor phase produced by impact.[2] The grains of corundum found in core sample 10004 are small (20–200 $\mu$).[1] From some Apollo 11 fines subrounded or elongated grains (40–100 $\mu$) may be agglomerates of still finer grains.[2] In contrast to shocked grains from Apollo 11 and 12 samples, an unshocked 200-$\mu$ corundum grain has been found in fines sample 14163. This crystal is twinned on {0001}, an unusual type of twinning for terrestrial corundum.[3]

### *Optics*

The corundum grains are red.[1]

### *Chemical Composition*

Microprobe analysis of the material from Apollo 11 fines revealed Al as the major element, with traces of Si, K, and Ca.[2] The corundum grain from 14163 is very pure alumina, without iron or other noticeable impurities.[3]

### *X-Ray Data*

X-Ray study of Apollo 11 corundum yielded the space-group[1] $R\bar{3}c$ and $d$ spacings of $\alpha$-$Al_2O_3$. Cell parameters of this material are:[2]

$$a = 4.767 \pm 0.003 \text{ Å}$$
$$c = 12.974 \pm 0.007 \text{ Å}.$$

The corundum from 14163 has cell parameters of[3]

$$a = 4.75 \text{ Å}$$
$$c = 12.97 \text{ Å}.$$

### *References*

1. Crozaz et al, *Sci.* **167,** 565.
2. Kleinman and Ramdohr, *EPSL* **13,** 19–22.
3. Christophe-Michel-Lévy et al, *Suppl.* **3,** 892.

WÜSTITE FeO (tentative)

X-Ray examination of inclusions in olivines from microbreccia 14321 revealed that one or two crystals may contain traces of a wüstite-like structure.[1] No other data are given.

*Reference*

1. Gay et al, *Suppl.* **3,** 359.

UNIDENTIFIED OXIDES

*ZnO Particles*

From breccia 12013,10 a few reddish crystallites have been observed, apparently contiguous with metallic brass particles. The crystallites are composed predominantly of a random aggregation of very fine ZnO particles.[1]

*References*

1. Gay et al, *EPSL* **9,** 124.

*Unknown Bluish Mineral*

In breccia 14315,9, a very thin crust (too thin for microprobe) or reaction rim of an unknown bluish mineral was observed between ilmenite and troilite grains.[1] Possibly the same phase was noted as an unidentified anisotropic phase of weaker reflectivity and darker "lilac" hue in a 12-$\mu$ rounded area in kamacite from norite 62295. The area is a composite of schreibersite and the unknown phase.[2]

*References*

1. Ramdohr, *EPSL* **15,** 114.
2. Agrell et al, *Lun. Sci. IV*, *Abstr.*, 16.

## Oxides of Ferric Iron: Goethite, Hematite, and Magnetite

Hydrated oxides of iron occur mainly as reddish stains around and between cracked metal grains; however, their presence in a number of lunar rocks (especially in a group of Apollo 16 rocks, termed "Rusty Rocks"[1]) is of interest because of the ferric iron and water contents. Both $Fe^{3+}$ and water are extremely rare on the Moon. The total amount of these oxides (relative to the volume of the rocks) is insignificant, but the phases may be indicators

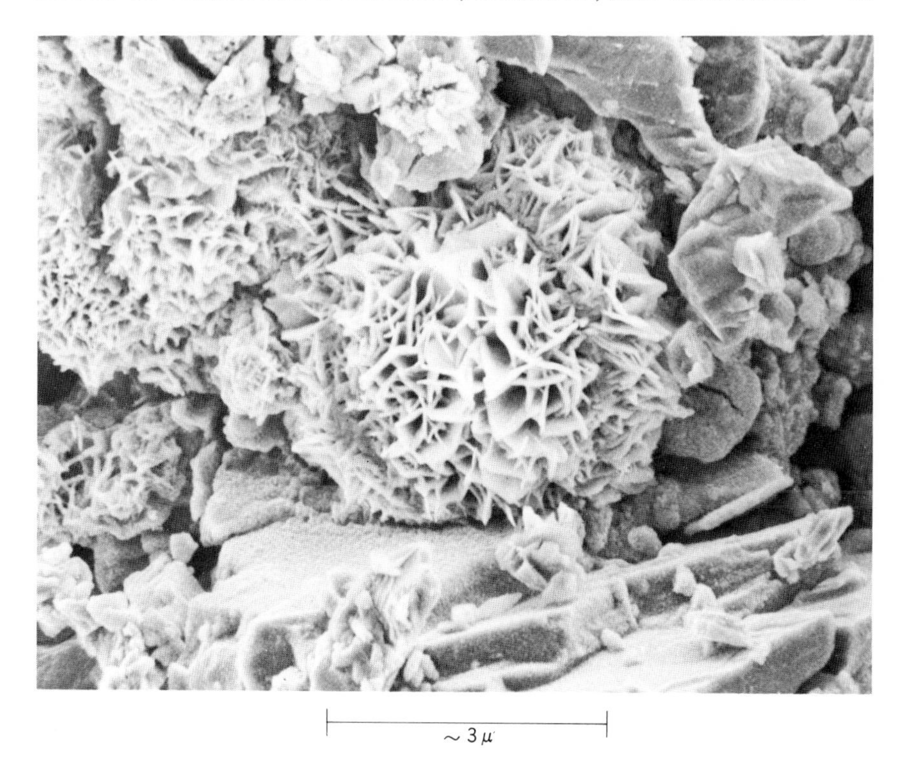


Fig. 3–8 SEM photograph of iron "rosettes," possibly lepidocrocite crystals. NASA photograph S-73-17705, courtesy of U. Clanton, Johnson Space Center, Houston.

of events or processes that are lunar or extralunar. The Fe(OH) may be goethite [$\alpha$-FeO(OH)], akaganeite [$\beta$-FeO(OH)], or lepidocrocite [$\lambda$-FeO(OH)] (Figure 3–8). It is even possible that all three polymorphs exist as lunar phases, but this has not been demonstrated. In general, the FeO(OH) phases have been referred to as goethite, or "goethite-like," and this practice will be observed in the following discussion.

Various suggestions for the origin of the goethite have been made and are summed up as follows:

1. The small amount of water required to form the goethite could have been trapped during condensation and aggregation of an impact-produced cloud. The impacting body may have been a comet or a carbonaceous chondrite. It is suggested that although meteoritic components of lunar soils and microbreccias have been observed frequently, the rarity of goethite in microbreccias and its sporadic distribution over a couple of square centimeters of examined sample do not favor such an origin.

2. If the source of the water is endogenous, it may have been present in the original rock from which the impact ignimbrite was derived. Frozen into the glassy constituent, the water could have reacted with the iron during subsequent cooling or thermal annealing. It is possible also that the small amount of volatiles in an impact ignimbrite may be redistributed or concentrated in a manner similar to that observed in terrestrial ignimbrite sheets. Comparable recrystallization and vapor transport may have taken place in lunar analogs.

3. Fumarolic or pneumatolitic processes might have occurred. Gas escape is evident in the vesicular lunar basalts. If a basaltic intrusion happened to occur within an impact ignimbrite sheet, both thermal recrystallization and vapor transport by gas escaping from the magma could have affected the enclosing rocks, and the metallic iron would have acted as a water scavenger.[2]

Once formed, goethite would be stable under thermal conditions at the lunar surface, provided it was sealed within the rocks.[2] In glass-coated basalt 64455, which appears to have melted during impact, goethite occurs only within the crystalline portion of the sample. This suggests that the alteration of the metal grains to goethite took place in a lunar environment, probably prior to the formation of the glass coating.[3] If the goethite was near enough to the surface to have access to the lunar atmosphere, it should eventually break down to hematite and water vapor.[2] Both hematite and magnetite have been reported from lunar samples,[4] but their identification is tentative.

Still another explanation for the occurrence of goethite is that after the lunar samples were brought to Earth, the lawrencite present in the rocks absorbed water from the terrestrial atmosphere and oxidized the iron in the rocks to the ferric state.[5] The occurrence of lawrencite ($FeCl_2$) on the Moon has been presumed repeatedly (e.g., El Goresy,[6] Christophe-Michel-Lévy,[7] Taylor[1]). The actual presence of a lunar $FeCl_2$ phase has not been verified, although the existence of such a phase is suggested by the following circumstances:

1. In some vugs of highly metamorphosed Apollo 14 breccias, a chlorine peak in the EDX* spectrum of an iron mass in one vug indicates the possible presence of metal chlorides.[8]

2. The observed occurrence of a birefringent yellow-brown mineral with a peculiar gridlike structure. The material contained Fe and probably Cl, but it decomposed under the electron beam.[9]

* An energy-dispersive X-ray system attached to the SEM.

Possibly also genetically connected with goethite are two unidentified Zn- and Cl-bearing phases that are chemically distinct. With these phases there is an unknown Pb-rich phase.[6]

*References*

1. Taylor et al, *Suppl.* **4,** 830.
2. Agrell et al, *Lun. Sci. III, Rev. Abstr.*, 8.
3. Grieve and Plant, *Lun. Sci. IV, Abstr.*, 317–318.
4. Taylor et al, *EOS* **54,** 356.
5. Lunar Sample Analysis Planning Team, *Sci.* **181,** 620.
6. El Goresy et al, *EPSL* **18,** 414, 416.
7. Christophe-Michel-Lévy et al, *Bull. Soc. Mineral.* **96,** 364.
8. McKay et al, *Suppl.* **3,** 739, 748.
9. Frondel et al, *Suppl.* **1,** 469.

## GOETHITE $\alpha$-FeO(OH)

### *Synonymy*

Akaganeite; Taylor et al, *Suppl.* **4,** 831.
Lepidocrocite; Carter and Padovani, *Suppl.* **4,** 326.
Limonite; *Lunar Sample Information Catalog, Apollo 16*, 269.
Fe-Cl-Ni oxyhydrate; Bell and Mao, *Lun. Sci. V, Abstr.*, 50.

### *Occurrence and Form*

Goethite was reported but questioned from fines sample 10084,96.[1] From microbreccia 14301/9 and 14301/19, goethite is found as rusty haloes (2–3 $\mu$ wide) around or replacing kamacite.[2] In breccia samples 66095,87 and 66095,78, goethite occurs in two distinct assemblages: one with metallic FeNi and the other with troilite and sphalerite. In the first assemblage the size of the goethite reaction rims around metal blebs varies widely from a few microns to a fraction of a micron. Cracks radiating from the metal blebs contain submicroscopic goethite vein fillings that penetrate deep into the silicate matrix. Rust rims are found around every metal bleb in the shocked but non-fused silicate matrix. The metallic spherules in the shocked fused silicate glass veins, however, do not show such goethite reactions. In the goethite-troilite-sphalerite assemblages there are also two Zn- and Cl-rich phases and a Pb-rich phase[3] (see below). Shocked breccia 61016 also contains many kamacite grains with rusty rims of "goethite" composition (Figure 3–9). Some of the rimmed kamacite grains are completely enclosed in maskelynite. It is believed that the rust could not have been introduced into the rock after the shock that transformed the plagioclase

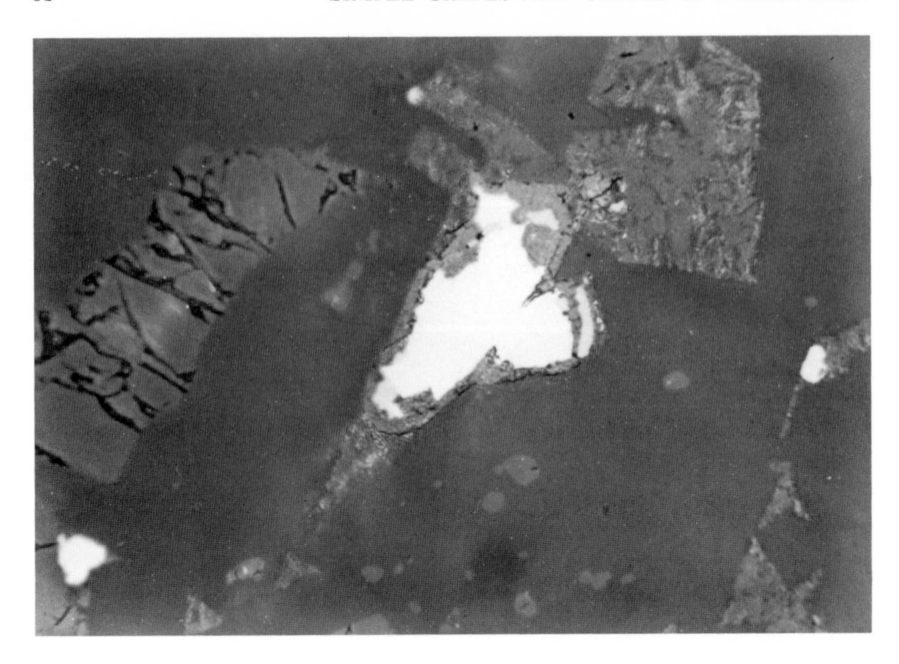

Fig. 3–9 A red "goethite" border (light gray) surrounding a kamacite grain (white). Polished section viewed in reflected light. X 1500. From *Bull. Soc. Mineral.* **96,** 364. Photograph courtesy of M. Christophe-Michel-Lévy, Université de Paris, France.

into maskelynite.[4] Goethite-like minerals were observed in soil samples 66081,5 and 69941,13.[5] In 69941 the phase occurs as well-developed, hollow, boxlike twinned crystals found commonly in the contact between the silicate minerals (especially olivine) and metallic iron. It has been suggested that here the hydrated iron oxide is lepidocrocite [$\lambda$-FeO(OH)] that has formed through the hydration of lawrencite. The hydrated material appears to have formed later than the surfaces of the iron particles in the samples, since impact pits, present in small numbers on the surfaces of some of the metallic grains, are not observed on the hydrated material.[6] In another sample, 66095,80, the majority of the free metal phases, as well as some troilite, have been oxidized to various degrees and commonly contain rims of what appears to be goethite.[7] A recent investigation of this material (by means of X-ray diffraction and crystal field spectra techniques) has indicated that it is probably akaganeite [$\beta$-FeO(OH)].[8]

*Optics*

Under the microscope goethite is a gray reflecting phase with reddish internal reflections.[2] The material suggested to be lepidocrocite is described as deep ruby red to orange-red and crackled like a dessicated gel,[6] or crystalline ruby red and transparent.[9] From breccia 66095, the rusty areas viewed under a binocular microscope appeared to be brown, red-brown, red, yellow-brown, and orange films and spots. One spot in the center of a glass zap pit was called limonite.[10] This is just another name for cryptocrystalline goethite with adsorbed water.[11]

### *Chemical Composition*

Quantitative electron microprobe analyses of a number of goethite areas from 66095 yielded variable amounts of Ni, Cl, Ca, S, and P, in addition to major Fe (Table 3-6; [2]). The NiO content is quite variable—from

**Table 3-6 Analyses of Goethite and Lepidocrocite (?)**

| | [1] 14301,19 | [2] 66095,78 | [3] 66095,80 | [4] 68841 | [5] FeO(OH) |
|---|---|---|---|---|---|
| $SiO_2$ | 0.37 | — | 0.22 | — | — |
| $Fe_2O_3$ | 80.20 | 81.8† | 82.1 | 81.0 | 89.9 |
| NiO | 4.84 | 1.64 | 5.42 | 4.0 | — |
| CaO | <0.1 | 0.5 | 0.69 | tr. | — |
| $P_2O_5$ | 0.30 | 0.37 | — | — | — |
| $SO_3$ | 2.76 | 0.33 | — | 1.0 | — |
| Cl | 0.65 | 4.18 | 2.81 | 1.5 | — |
| OH } | — | — | } 7.98 | — | 10.1 |
| $SO_4$ } | — | — | | — | — |
| Total | 91.41* | 88.82* | 100.00* | 87.5* | 100.0 |

*References*

[1] Agrell et al, *Lun. Sci. III*, *Rev. Abstr.*, 9. From a microbreccia. *Analysis includes $Na_2O$ = 1.43%, $K_2O$ = 0.86%. $TiO_2$, MnO, and MgO all < 0.1%, and a presumed $H_2O$ content of about 8%. Goethite.

[2] El Goresy et al, *Suppl.* **4,** 743. From a breccia. †Fe expressed as FeO. *$H_2O$ presence (~ 10%) presumed from low total. Goethite.

[3] Taylor et al, *Suppl.* **4,** 836. From a breccia. *Analysis includes MgO = 0.42%, CoO = 0.36%, and OH,$SO_4$ calculated by difference. Akaganeite (?).

[4] Carter and Padovani, *Suppl.* **4,** 325. From a fines sample. *Value for $Fe_2O_3$ approximate; Values for NiO, $SO_3$, and Cl average for range of analyses; difference between total and 100% presumed to be due to presence of $H_2O$ and/or other volatiles. Lepidocrocite (?).

[5] Theoretical goethite.

1.64% to as much as 5.9%. The Cl content also is significantly high (up to 5.25%). The variable composition and totals may indicate that the analyzed material is a submicroscopic mixture of goethite and other compounds of uncertain nature. The chlorine may be present as an impurity in the goethite structure or in the form of minor lawrencite associated with FeO(OH). In addition, sulfur may be present as a sulfate and not as a sulfide molecule in the goethite structure.[3] A goethite-like mineral from soil sample 66081 had only 57.7% Fe and deteriorated rapidly under the electron beam.[5] The possible lepidocrocite from fines sample 69941 yielded only an approximate analysis (Table 3-6; [4]), partly because of deterioration under the electron beam. The rate of increase of Fe, observed to be 5% per minute with 0.2 $\mu$A of beam current, suggests the presence of volatiles and water. Water in the Moon may occur because of the reduction of iron-bearing materials to metallic iron by remobilized meteoritic carbon and trapped solar wind components in the lunar soil.[6] The akaganeite from 66095,80 has a composition that diverges somewhat from FeO(OH) (Table 3-6 [3]).[7]

As mentioned previously, the chlorine content of the goethite is variable (from 0.81 to 5.25%[3,7]), and electron microprobe X-ray scans of highly oxidized FeNi metal in breccia 67455 reveal the intimate association of the chlorine with the metal grains and the surrounding cracks. These observations strongly suggest the existence of lawrencite ($FeCl_2$), a phase that rapidly deliquesces and oxidizes in the presence of moisture. Oxidation of lawrencite in humid air includes the formation of ferrous hydroxide and a highly oxidizing acidic solution. The process is extremely rapid, and a polished section of a meteorite containing lawrencite begins to alter within seconds of exposure to Earth's atmosphere.[7] Lawrencite could be indigenous to the Moon, but in 66095 the association with schreibersite, kamacite, and taenite strongly suggests a meteoritic origin. In this case it is believed that the most plausible explanation for the hydration is that it occurred on Earth or during return to Earth aboard the spacecraft.[12] In breccia 61076, however, an observed grain of kamacite with an oxidized rim completely enclosed by maskelynite indicates that some of the oxidation process could have taken place on the Moon.[4]

*References*

1. Jedwab et al, *Sci.* **167,** 618.
2. Agrell et al, *Lun. Sci. III, Rev. Abstr.*, 7–9.
3. El Goresy et al, *Suppl.* **4,** 742–744.
4. Christophe-Michel-Lévy et al, *Bull. Soc. Mineral.* **96,** 363–364.
5. Taylor and Carter, *Suppl.* **4,** 302, 304.
6. Carter and Padovani, *Suppl.* **4,** 325–326.

7. Taylor et al, *Suppl.* **4,** 830–831, 836.
8. Taylor et al, *Geology* **2,** 430–432.
9. Padovani and Carter, *EOS* **54,** 356.
10. *Lunar Sample Information Catalog, Apollo 16*, 269.
11. Palache et al, *Dana VII*, **1,** 685.
12. Taylor, *Lun. Sci. IV, Abstr.*, 715.

## HEMATITE $Fe_2O_3$ (tentative)

### *Occurrence and Form*

Hematite was reported from fines sample 10084,96, but no data supporting the identification were given.[1] Again believed to be hematite, the phase was observed in fines sample 15301,116, where it occurred as isolated crystals with a scaly crystalline habit.[2] In a detailed study of more soil samples from Apollo 16 and 17 missions, the following have been observed: tiny smooth spheres (some $< 1\ \mu$), rounded irregular fragments, pellets, and individual or fragile clusters of microcrystals. Some of the pellets are attached to oxidizing metallic particles. Some of the spheres are associated with rutile and included in or deposited on silicates. One glassy fragment has many crystals both on its surface and enclosed in it, suggesting that the formation of the iron oxide predates the contact of the crystals with terrestrial atmosphere. A highly penetrative oxidation of lawrencite, however, cannot be ruled out.[3]

### *Optics*

In transmitted light the hematite is light to ruby red (nearly opaque), brownish-red, orange, and yellow. In reflected light it is ruby red to orange-red. The spheres are chiefly ruby red. The pellets are orange-red or yellow, and it has been suggested that the yellow pellets are oxidized outgrowths that have become detached from the iron particles in the samples.[3]

### *Chemical Composition*

All the above-described particles consist primarily of Fe with only traces of Si and Ca and very rare Ni.[3] This composition, together with their appearance, suggests that the particles are hematite, but proof will rest on eventual X-ray study.

### *References*

1. Jedwab et al, *Sci.* **167,** 618.
2. Jedwab, *Apollo 15 Lunar Samples*, 109.
3. ———, *Suppl.* **4,** 868.

## MAGNETITE $Fe_3O_4$, or $Fe^{2+}Fe_2^{3+}O_4$ (tentative)

### *Occurrence and Form*

From fines sample 10084,96, small spheres, 1 to 2 μ in diameter, were tentatively identified as magnetite. Their concentration in the total powder and concentrates is less than $1 \times 10^{-6}$.[1] Magnetite was believed to have been recognized also in porphyritic basalts 12018/49 and 12063/41.[2] Some of the goethite grains in breccia 66095,89 were reported to be altering to magnetite.[3]

### *Optics*

The spheres in 10084,96 are opaque, with a steel-gray shine and a reticulated surface.[1]

### *Chemical Composition*

Mössbauer spectra analyses of a number of Apollo 11 fines samples revealed some peaks that were assigned to a magnetic iron spinel (i.e., magnetite). These peaks deserve special comment, since the existence of magnetite in the lunar samples is still in question and the possibility that the peaks can be attributed to oxidation on handling has not been ruled out. Thus far the evidence for the presence of magnetite is only statistical and cannot be regarded as proof. If the mineral is present, it is not in a form that can be separated easily from the sample nor is it typical of magnetite habits in carbonaceous meteorites.[4]

### *References*

1. Jedwab et al, *Sci.* **167,** 619.
2. Ramdohr, *Fortschr. Mineral.* **48,** 51.
3. El Goresy et al, *Lun. Sci. IV*, 222–223.
4. Gay et al, *Suppl.* **1,** 493.

## UNIDENTIFIED PHASES $X_1$ AND $X_2$, AND A PB-BEARING PHASE

### *Occurrence and Form*

From breccia 66095, together with a grain of troilite being replaced along its boundaries by sphalerite and surrounded by goethite, there are two Zn- and Cl-rich phases and a Pb-rich phase. The first two phases have been designated as $X_1$ and $X_2$. Reaction between these phases and troilite may have produced the sphalerite and goethite.[1]

### *Optics*

It is not known whether phases $X_1$ and $X_2$ are opaque or transparent. Both are dark gray and (in reflected light) identical in color and in reflectivity which is only slightly higher than that of fayalite.[1] The low reflectivity makes it unlikely that the materials are sulfides.[2]

### *Chemical Composition*

Both phases $X_1$ and $X_2$ broke down slowly under the electron beam and only semiquantitative analyses could be obtained. Phase $X_1$ contains Fe, Ti, Zn, Ni, Cl, and S. This may represent a Cl-bearing sulfate of Fe, Ti, Zn, and Ni. It was estimated that Zn = 8.5% and Cl = 2.4%. Phase $X_2$ contains major Fe, P, S, Cl, Zn, and Ni with minor K, Ca, and Si. The presence of silicon may be due partly to excitation of neighboring silicates. This phase may be a mixture of a Cl-bearing sulfate and a phosphate of Fe, Ni, Zn, K, and Ca.

The third phase is submicroscopic and was detected by the probe. The phase contains 0.3 to 0.4% Pb, and its nature is unknown.

### *References*

1. El Goresy et al, *EPSL* **18,** 414–416.
2. ———, *Suppl.* **4,** 745.

# CHAPTER FOUR | MULTIPLE OXIDES: SPINEL GROUP

Detailed study of the lunar spinels has revealed a number of compositional complexities that have given rise to many synonyms and descriptive phrases. Some graphic representation of the components involved, as well as the nomenclature, will reduce confusion.

A diagram (Figure 4–1) modified from the spinel prism of Haggerty[1] indicates that the main components of the lunar spinels are chromite ($FeCr_2O_4$), ulvöspinel ($Fe_2TiO_4$), hercynite ($FeAl_2O_4$), and spinel ($MgAl_2O_4$). (The last component also may be termed "spinel proper.") Both picrochromite ($MgCr_2O_4$) and theoretical $Mg_2TiO_4$ enter into the composition of the lunar spinels to a lesser extent. The two sections of Figure 4–1 represent the two groups into which the lunar spinels may be divided broadly, that is, the chromite-ulvöspinel-hercynite series (section A) and the spinel-hercynite series[1] (section B). Some members of the first series have been variously described as chromian ulvöspinel, titanian chromite, and aluminian chromite. Members of the second series containing some $Cr_2O_3$ enrichment have been termed chromian pleonastes. Additional terms indicating compositional variation also appear in Figure 4–2. Figure 4–3 presents the generalized compositional trends in the spinels from the various lunar missions.

Ulvöspinel is structurally inverse; that is, divalent ions are in both the A and B sites. Hercynite, chromite, picrochromite, and spinel are structurally normal; that is, the divalent ions are in the A sites and the trivalent ions are in the B sites.[1] Hence the chromite-ulvöspinel-hercynite series may be regarded as a normal-inverse series and the spinel-hercynite series may be regarded as a normal series.[2]

Chromite-rich members of the normal-inverse series crystallize early, together with olivine; ulvöspinel-rich members crystallize later, either simultaneously with or after pyroxene.[3] Members of the normal-inverse

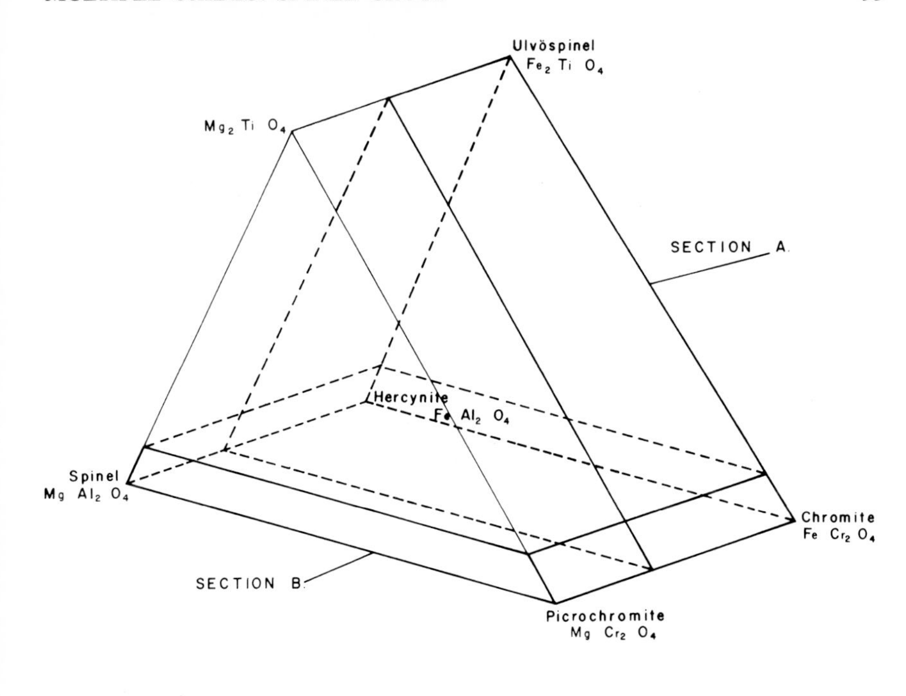


Fig. 4–1 Spinel prism showing main components of lunar spinels. Diagram is modified from the spinel prism of S. Haggerty, in *EPSL* **13,** 329.

series, which are typical of mare basalts, have been found in all lunar landing sites. However, Apollo 11 spinels are restricted to the intermediate members of the series, whereas Apollo 12 spinels generally show a discontinuity in the series—the "Apollo 12 miscibility gap." In the Apollo 14 spinels, however, an even more extensive discontinuity is displaced, with respect to the Apollo 12 gap, toward $Fe_2TiO_4$, and compositions comparable to those of Apollo 11 fill the Apollo 12 gap. Apollo 15 basalts contain some spinels that span the entire normal-inverse series, but there are not many spinels of intermediate compositions.[2] The compositional gap between chromite and ulvöspinel seems to be widest, and their compositions are closest to the extreme end members.[4] The range of compositions of the Apollo 15 spinels, specifically the degree of solid solubility and the relative Mg and Al enrichment and crystallization trends, are in close agreement with spinels in basalts from the Apollo 12 site.[5] Among the opaque minerals making up to about 2.5% of the volume of some Apollo 15 mare basalts (15074,1, 15474,1, 15514,3, 15534,1 and 2, 15564,7, and

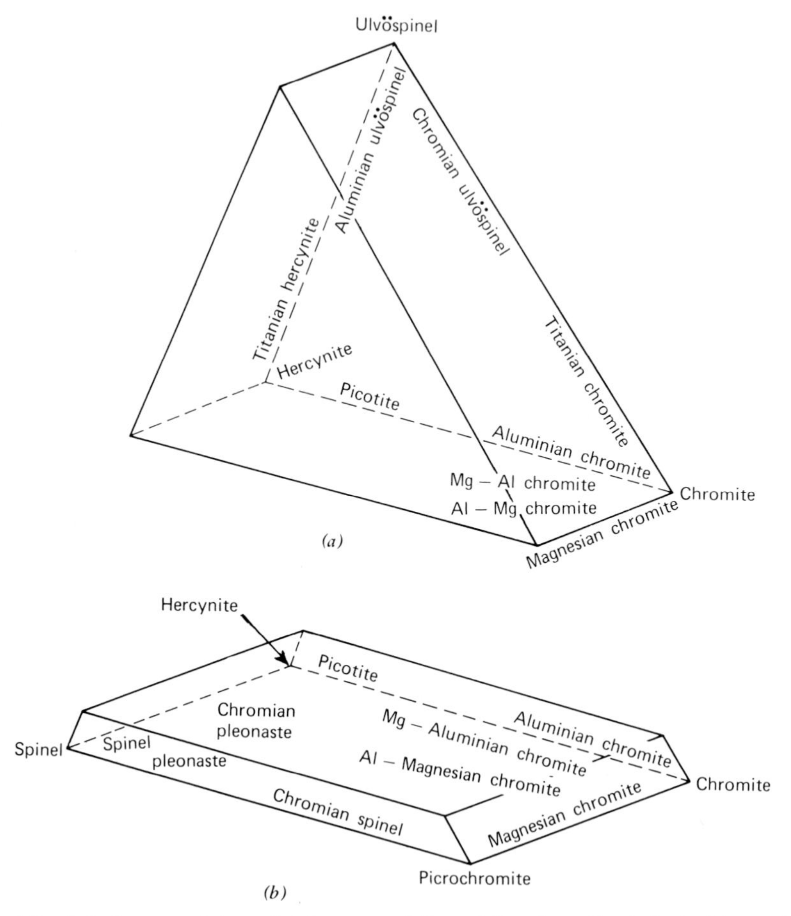


Fig. 4–2 Spinel nomenclature. (*a*) Lunar varieties of the chromite-ulvöspinel-hercynite series. (*b*) Lunar varieties of the spinel-hercynite series.

15604,3, 4, and 5) ulvöspinel plus chromite are more abundant than ilmenite.[6] Apollo 16 spinels are largely chromites, aluminian chromites, or titanian chromites. The last occur both as primary titanian chromite and as a titanian chromite formed by intense subsolidus reduction and reequilibration of former chromian ulvöspinels.[7] Trace amounts of chromite and chromian ulvöspinels occur in almost all the Apollo 17 basalts[8] and soil samples.[9] The overall fractionation trend observed for the Luna 16 spinels, from aluminian chromite to chromian ulvöspinel, generally conforms to the Apollo spinel trends except that the degree of ionic substitution in this series as well as the degree of substitution in the Mg-rich aluminian chro-

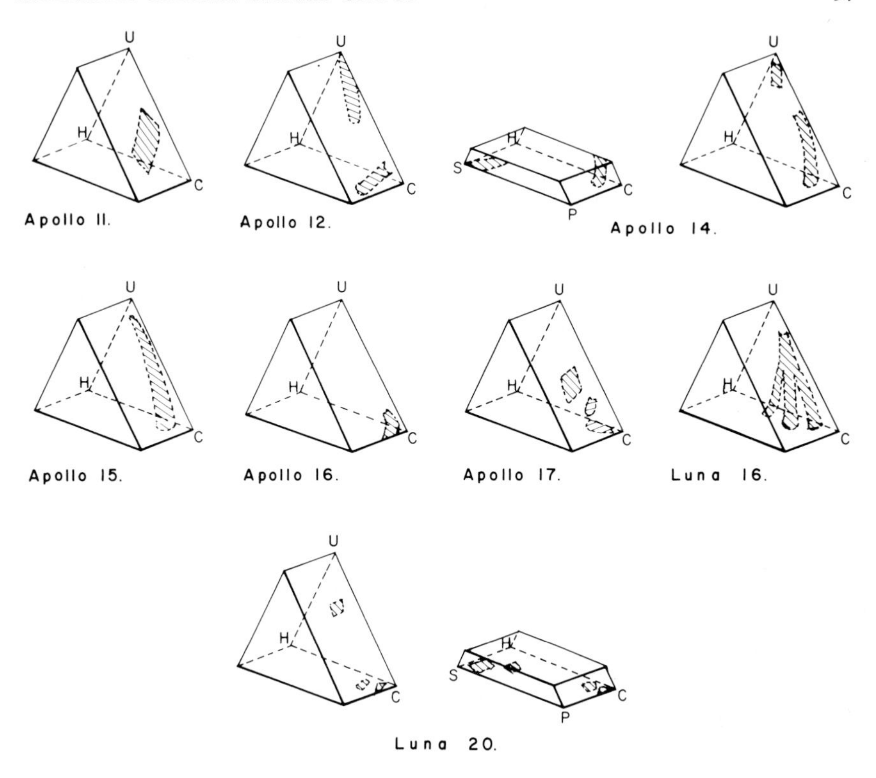


Fig. 4–3 Generalized compositional trends in the spinels from various lunar missions. Adapted from many diagrams from S. Haggerty and some other investigators.

mites is more variable than in the Apollo spinels.[1] Chromite, titanian chromite, and chromian ulvöspinel have been identified from Luna 20 samples.[10]

In the normal series from spinel to hercynite the substitution of Al by Cr extends the binary $MgAl_2O_4$-$FeAl_2O_4$ to include $FeCr_2O_4$-$MgCr_2O_4$. Minor Mn and Ti also are present. Members of this series, particularly the pleonastes and chromian pleonastes, have been identified in typical nonmare material. A single grain of spinel proper occurred in an Apollo 11 sample. Rare discrete grains of spinel have been identified from the Luna 16 samples, and spinels are fairly abundant in the Luna 20 samples. Some spinel pleonastes and chromian pleonastes have been found in Luna 20 samples, but these phases are most abundant in the Apollo 14 samples and have not yet been identified in the Apollo 12 material.[10] The paucity of pleonastes in lunar samples other than the Apollo 14 breccias suggests that a special portion of the lunar anatomy has been excavated at the Apollo 14 site.[11]

*References*

1. Haggerty, *EPSL* **13,** 329–330, 339.
2. ———, *Suppl.* **3,** 306–307.
3. Busche et al, *Am. Mineral.* **57,** 1729.
4. Dowty et al, *Suppl.* **4,** 437
5. Haggerty, *Meteorit.* **7,** 356.
6. Powell et al, *Lun. Sci. IV, Abstr.,* 597.
7. El Goresy et al, *Suppl.* **4,** 734, 736, 749.
8. Apollo 17 LSPET, *Sci.* **182,** 661.
9. Taylor et al, *EOS* **54,** 616.
10. Haggerty, *Geochim. Cosmochim. Acta* **37,** 859–862.
11. Roedder and Weiblen, *EPSL* **15,** 398.

## Normal Inverse Series: Chromite ($FeCr_2O_4$)–Ulvöspinel ($Fe_2TiO_4$)–Hercynite ($FeAl_2O_4$)

*Chromite Synonymy*

Al-chromite; Simpson and Bowie, *Suppl.* **2,** 207
Al-Mg chromite; Reid et al, *Geochim. Cosmochim Acta* **37,** 1025.
Al-Ti chromite; Haggerty, *Suppl.* **3,** 311.
Al-titanian chromite; Haggerty, *EPSL* **13,** 329.
Aluminian chromite; Cameron, *Suppl.* **2,** 194.
Aluminian-magnesian chromite; Haggerty, *EPSL* **13,** 330.
Aluminian-titanian chromite; Haggerty, *EPSL* **13,** 330.
Aluminous chromite; James and Jackson, *J. Geophys. Res.* **75,** 5802.
Aluminous chromite-rich spinel; Busche et al, *Am. Mineral.* **57,** 1740.
Aluminous-rich titanium-bearing chromite; Haggerty, *EPSL* **13,** 344.
Chrome-spinel (in part); Boyd et al, *Lun. Sci. Conf. '71, Abstr.,* 149.
Chromian spinel; Boyd et al, *Lun. Sci. Conf. '71, Abstr.,* 149.
Chromite-rich spinel; Skinner and Winchell, *Lun. Sci. Conf. '72, Abstr.,* 626.
Cr-rich spinel; Reid et al, *Geochim. Cosmochim. Acta* **37,** 1025.
Cr-spinel; Haggerty et al, *Suppl.* **1,** 524.
Cr-Ti-Fe oxide; Brown et al, *Suppl.* **2,** 590.
Cr-Ti spinel; Kushiro et al, *Suppl.* **2,** 487.
Fe-Ti-Cr spinel (in part); Brown et al, *Suppl.* **2,** 591.
Magnesian-aluminian chromite; Haggerty, *EPSL* **13,** 330.
Magnesian-chromite; Haggerty, *EPSL* **13,** 330.
Mg-aluminian chromite; Haggerty, *Suppl.* **3,** 311.
Mineral C (Ramdohr); Ramdohr and El Goresy; *Sci.* **167,** 617.
Phase B (Douglas); Douglas et al, *Sci.* **167,** 595.

Phase C (Brown); Brown et al, *Suppl.* **1,** 196.
Ramdohr's phase (1); Ramdohr and El Goresy, *Naturwiss.* **57,** 102.
Ti-Al-Cr spinel; Jakeš et al, *Lun. Sci. Conf. '72, Abstr.*, 384.
Ti-chromite; El Goresy et al, *Suppl.* **2,** 229.
Ti-poor Cr-spinel; von Engelhardt et al, *Lun. Sci. III, Rev. Abstr.*, 233.
Ti-rich chromite; Haggerty et al, *Suppl.* **1,** 526.
Titanian chromite; Agrell et al, *Suppl.* **1,** 83.
Titaniferous chrome-spinel; Brown et al, *Suppl.* **1,** 195.
Titaniferous chromite; Wood et al, *Suppl.* **1,** 970.
Titanium-aluminium chromite; El Goresy and Ramdohr, *Lun. Sci. Conf. '72, Abstr.*, 204.
Titanium chromite; Taylor et al, *Suppl.* **2,** 856.
Titanochrome spinel; Brown et al, *Suppl.* **1,** 205.
Titanochromite; Cameron, *Sci.* **167,** 623.

*Ulvöspinel synonymy*

Al-bearing chromian ulvöspinel; Ramdohr and El Goresy, *Sci.* **167,** 615.
Aluminian ulvöspinel; Haggerty, *EPSL* **13,** 334.
Chrome-titanium spinel; McKay et al, *Suppl.* **1,** 686.
Chrome-titano-spinel; Agrell et al, *Suppl.* **1,** 85.
Chrome ulvöspinel; Taylor et al, *Suppl.* **2,** 858.
Chromian ulvöspinel; El Goresy et al, *Suppl.* **3,** 340.
Chromite-ulvöspinel; Steele et al, *EPSL* **13,** 327.
Chromium ulvöspinel; El Goresy et al, *Suppl.* **3,** 339.
Cr-Al ulvöspinel; Haggerty, *Suppl.* **3,** 320.
Cr-rich titanomagnetite; Haggerty et al, *Suppl.* **1,** 526.
Cr-rich ulvöspinel; El Goresy et al, *Suppl.* **2,** 225.
Cr-ulvöspinel; El Goresy et al, *Suppl.* **2,** 229.
Fe-Ti-Cr oxide; Taylor et al, *Suppl.* **2,** 856.
Fe-Ti-Cr spinel (in part); Brown et al, *Suppl.* **2,** 591.
Ti-rich spinel; Reid et al, *EPSL* **10,** 352.
Titanomagnetite; Haggerty and Meyer, *Carnegie Inst. Yearb., '69*, 231.
Ulvite; El Goresy et al, *Suppl.* **2,** 228.

*Occurrence and Form*

Chromite occurs in all rock types except spinel troctolite.[1] Euhedral to subhedral grains have been noted as inclusions in troilite,[2] olivine, and pyroxene crystals.[3] In fine-grained basalts, chromite cores are separated from ulvöspinel rims by a compositional discontinuity, but porphyries with glassy to very fine matrices contain only chrome-rich spinels with very thin rims higher in Ti.[4] A strongly zoned chromite, low in Ti and distinct from

Fig. 4–4 Cluster of aluminian chromite crystals mantled by chromian-ulvöspinel. From *Suppl.* **2,** 195. Photograph courtesy of E. N. Cameron, Department of Geology and Geophysics, University of Wisconsin, Madison.

chromian ulvöspinel, was observed in microbreccia 10061-28.[5] In Apollo 15 olivine-microgabbros, chromite and ulvöspinel are found in separate rather than mantled grains.[6] Chromite with no ulvöspinel overgrowths occurs enclosed within early silicate phenocrysts.[7] Chromite inclusions in olivine and pyroxene are the earliest crystallization products in Apollo 12 rocks,[8] but in Luna 20 samples chromite is present both as a primary crystallization product and as exsolution from ilmenite. The widespread occurrence of the bimodal assemblage chromite plus rutile in ilmenite is

Fig. 4–5 Euhedral spinel grain (in center of field) has an aluminian chromite core with a rim of chromian ulvöspinel. White scale bar is 30 $\mu$. From *Suppl.* **2,** 209. Photograph courtesy of P. R. Simpson, Institute of Geological Sciences, London.

strongly suggestive of decomposition of a Cr-enriched $FeTiO_3$-$MgTiO_3$ solid solution. A chromite from a Luna 20 sample is exsolved from ilmenite and closely associated with oriented lamellae of rutile. The chromite is present in two textural modes: (1) as tapered lamellae transgressing the entire ilmenite crystal, and (2) as irregular lobate inclusions. The chromite–ilmenite contacts are optically and compositionally well defined.[9]

Aluminian chromite, titanian chromite, and chromian ulvöspinel occur in the three major types of Apollo 11 samples: in olivine and rock fragments of the loose surface fines, in microbreccias, and in basaltic rocks. The grains are usually euhedral, with the largest dimension 100 to 200 $\mu$. The crystals frequently are mantled by ilmenite, but mantling is not observed where the grains are inclusions in olivine and pyroxene.[10] Aluminian chromite is found as phenocrysts of sharply euhedral crystals or small crystal clusters (Figure 4–4);[11] as euhedral cores rimmed with ulvöspinel (Figure 4–5); as well-formed though slightly rounded octahedra (up to 75 $\mu$ across) included in olivine and, in rare instances, rimmed by ulvöspinel;[12] and also in groundmass of the rock with plagioclase and acicular pyroxene.[11] In Luna 20 material, two grains of magnesian-aluminian chromite are

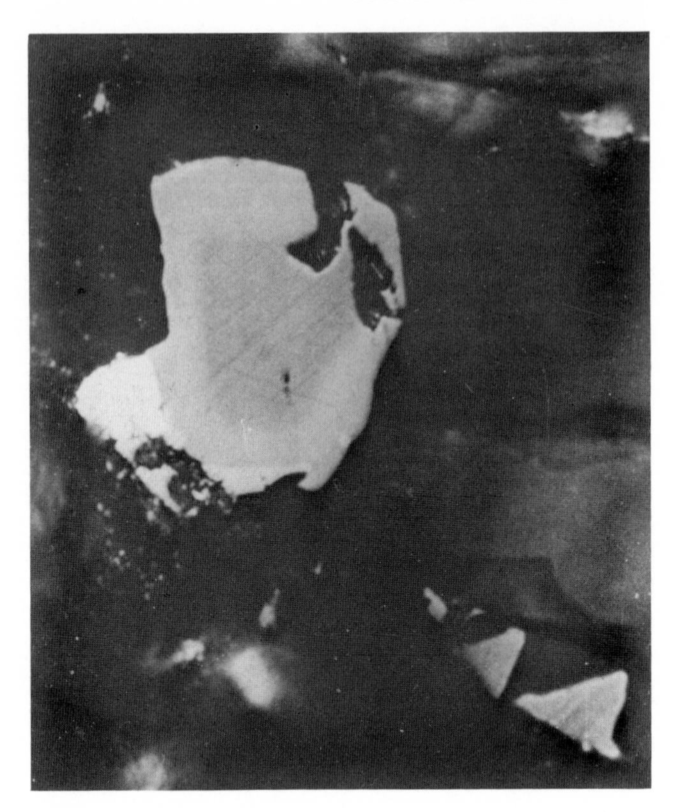

Fig. 4–6 Zoned titanian chromite. Oil immersion. X 1335. From *Suppl.* **1,** 232. Photograph courtesy of E. Cameron, Department of Geology and Geophysics, University of Wisconsin, Madison.

primary crystallization products. One occurs in maskelynite of thin section 512-2 from sample 22003,1. The other is in a glassy breccia particle 517-7 from the same sample. Both grains are anhedral and about 25 $\mu$ in diameter.[9] Titanian chromite forms rare brownish grains (Figure 4–6), in some instances enclosed in ilmenite[13] or associated with magnesian ilmenite, pyroxene, olivine, and plagioclase.[14] Grains not included in olivine have narrow rims of ilmenite or occur as small, partly resorbed cores in the ilmenite phenocrysts. This relationship suggests that the spinel reacted with the crystallizing liquid to form ilmenite.[15] A few grains contain rare rutile and ilmenite lamellae (Figure 4–7).[16] Textural relations in some Apollo 12 rocks suggest that the crystallization sequence started with titanian chromite.[17] Originally designated as mineral C and observed in breccias, it was noted that the form of titanian chromite was similar to titanomagnetite.[18]

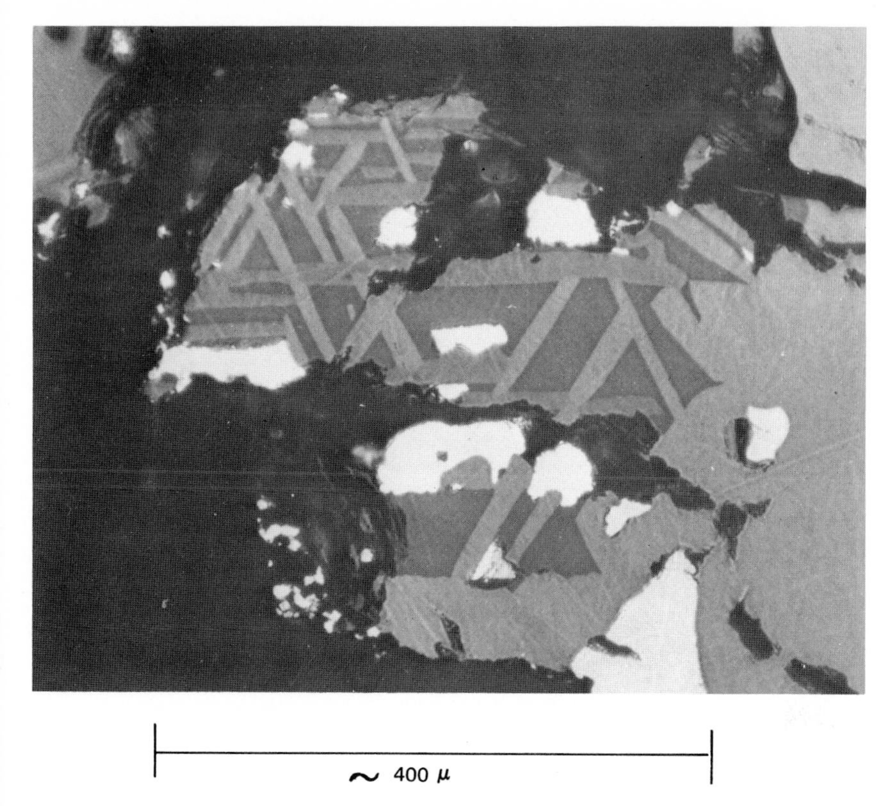


Fig. 4–7 Aluminian-titanian chromite (dark gray), plus oriented reduction lamellae of ilmenite (medium gray) along {111} spinel planes, plus iron (white). From *Suppl.* **3**, 309. Photograph courtesy of S. Haggerty, Department of Geology, University of Massachusetts, Amherst.

Two kinds of titanian chromites were found in brecciated anorthosite 67075,45. One was an early crystallized blue to gray spinel, occurring as partially crushed grains in the plagioclase groundmass or inclusions in olivine. This material is a primary titanian chromite without associated ilmenite or metal grains. The other, a blue spinel occurring as inclusions in pyroxene, was formed by intense subsolidus reduction and reequilibration of earlier crystallized ulvöspinel to chromite plus ilmenite and metal.[19] In nonmare basalt 14053 this intense reduction of a chromian-aluminian ulvöspinel is accompanied by the decomposition of fayalitic olivine to iron plus cristobalite.[20] In basalts 75081 and 74241 titanian chromite is not associated with the chromian ulvöspinel but is present as exsolution lamellae up to 10 $\mu$ wide within ilmenite.[21] Chromian ulvöspinel grains usually are

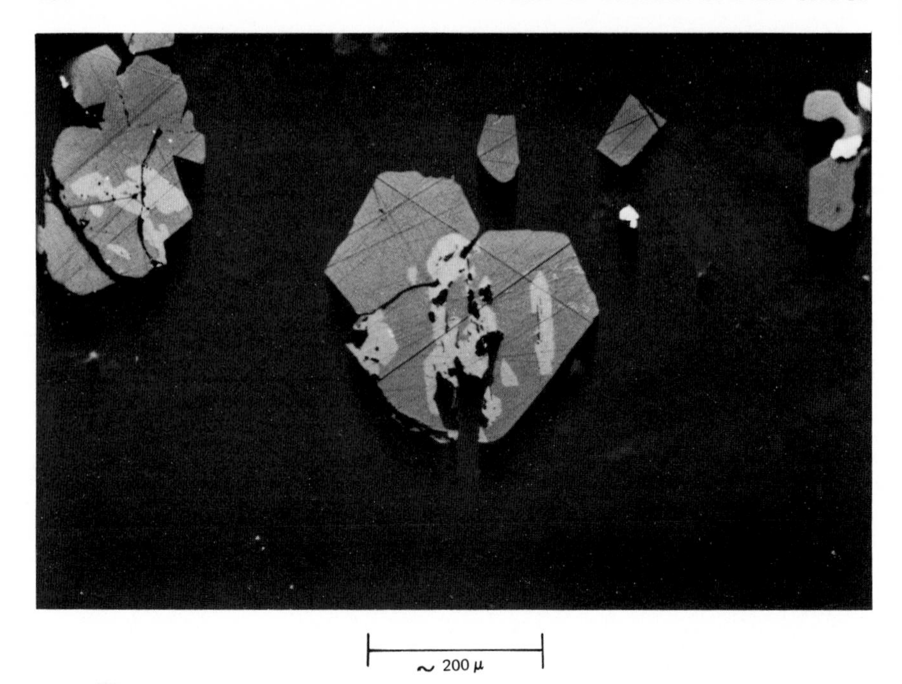


Fig. 4–8 Chromian-aluminian ulvöspinel (medium gray) with internal and external composite ilmenite (light gray). From *Suppl.* **3,** 309. Photograph courtesy of S. Haggerty, Department of Geology, University of Massachusetts, Amherst.

euhedral, from 100 to 200 $\mu$ in size; in many cases they are mantled by ilmenite (Figure 4–8).[5] Some grains are found as inclusions in K-feldspar (e.g., in breccia 12013).[22] Chromian ulvöspinel also forms rims on chromite, and this may be a reaction between chromite and the liquid.[23] In Apollo 17 samples, primary chromian-ulvöspinels comparable to those in Apollo 11 rocks show some subsolidus reduction to titanian chromite plus ilmenite and iron.[24] These small ($< 15\ \mu$) grains occur in all Apollo 17 samples.[21] In a few instances, chromian-ulvöspinel forms individual hollow crystals up to 5 $\mu$ in diameter in the glassy orange soil sample 74220.[25]

Ulvöspinel is a sparse accessory found mostly in coarse-grained rocks. It is rare in nonmare fragments and more abundant in mare basalts.[1] It occurs as tiny anhedral to subhedral grains in pyroxene or often intergrown with ilmenite (Figure 4–9).[26] It has been observed in late-stage interstitial vermicular patches (Figure 4–10).[13] Minute 5 to 10 $\mu$ ulvöspinel crystals, in fine-grained basalt 10020-40, display a cruciform habit parallel to axes [100], with arrowhead-shaped crystallites along {111} developing at the ends of the intersecting cross-arms. These features are typical of

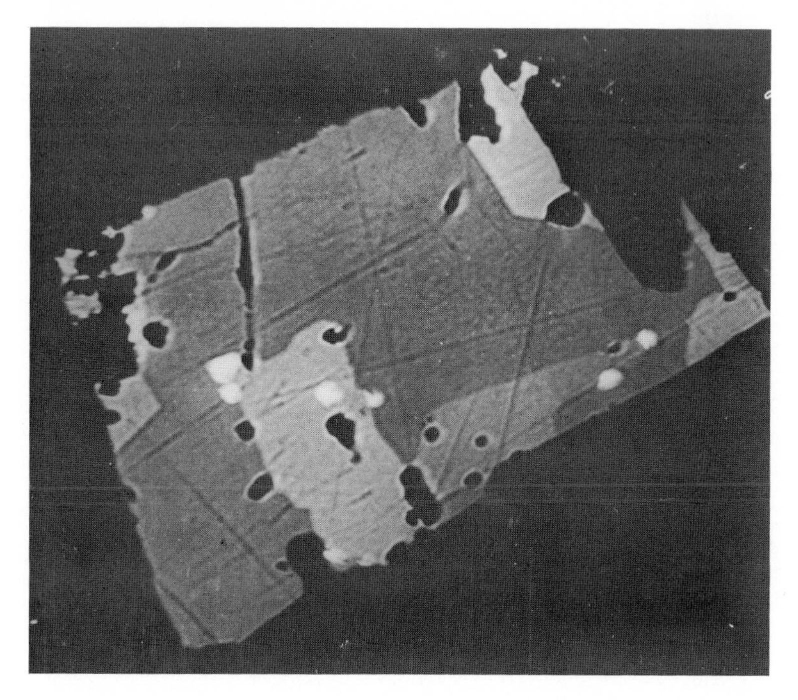

Fig. 4–9 Ulvöspinel (dark gray) intergrown with diversely oriented ilmenite (lighter gray). Oil immersion. X 700. From *Suppl.* **1,** 230. Photograph courtesy of E. N. Cameron, Department of Geology and Geophysics, University of Wisconsin, Madison.

spinels.[5] The morphology of the ulvöspinel and its interstitial texture with feldspar and pyroxene suggest that it was one of the last minerals to crystallize.[5] In Apollo 17 basalts, the ulvöspinel commonly shows evidence of subsolidus reduction to ilmenite plus metallic iron (Figures 4–11 and 4–12).[27] The polishing hardness of ulvöspinel is softer than that of ilmenite;[26] this property is an aid in identification is polished sections.

*Optics*

Chromite cores are a light pearl gray, rimmed by pinkish-grayish-tan chromian and aluminian ulvöspinels.[28] Chromites are also blue or gray,[8] isotropic, and zoned with respect to what probably are the octahedral faces.[5] From a basalt fragment from soil sample 14258, fine platelets of gray isotropic chromite were inclusions in discrete pyroxene crystals.[29]

Aluminian chromite is dark gray, isotropic, and usually overgrown by an isotropic lighter pinkish chromian ulvöspinel.[30] Its color may vary from gray to blue-gray, and in some instances it is weakly anisotropic in oil

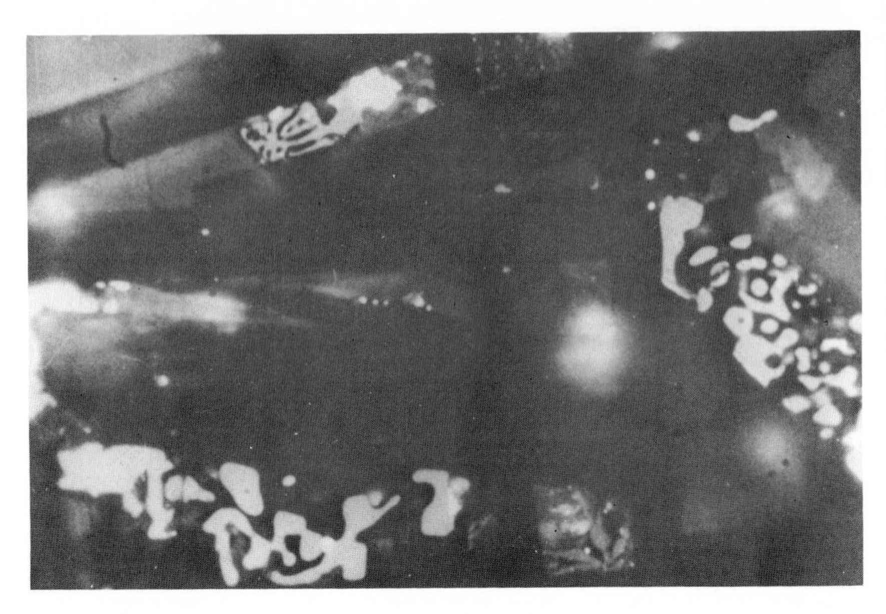

Fig. 4–10 Late-stage vermicular ulvöspinel. X 1350. From *Suppl.* **1,** 209. Photograph courtesy of M. Brown, Department of Geological Sciences, Durham University, England.

immersion.[11] It is opaque and usually optically homogeneous. One highly zoned grain, however, has a gray harder core enclosed in a tan softer rim of chromian ulvöspinel.[10]

Aluminian-titanian chromite usually is isotropic and gray (e.g., as in basalt 14053); however, some aluminian-titanian chromites are distinctly purple.[29]

Titanian chromites, compositionally close to chromite, are blue or gray and isotropic. With higher Ti content, they are tan, pink, or khaki, and anisotropic.[31] Titanian chromite observed in breccias has a reflectivity much lower than titanomagnetite. In oil immersion it has a dirty-gray reflectance color.[18] This is, no doubt, Ramdohr's phase (1), which has been given the same optical description[32] and has been called also mineral C (Ramdohr).[18] It is a cubic mineral with hexagonal exsolution lamellae.

Chromian ulvöspinels are opaque and usually optically homogeneous. They are brown, tan, pink, khaki, and tan with purple tint. Some are weakly anisotropic.[10] It has been suggested that the anisotropic spinels are not true spinels but a new group of nonisometric phases along the $Fe_2TiO_4$–$FeCr_2O_4$ join.[33] For example, a "brown spinel," referred to as an Fe-Ti-Cr oxide and distinctly anisotropic in reflected light, was thought not to be a true cubic phase. This phase, however, usually is associated with chromite

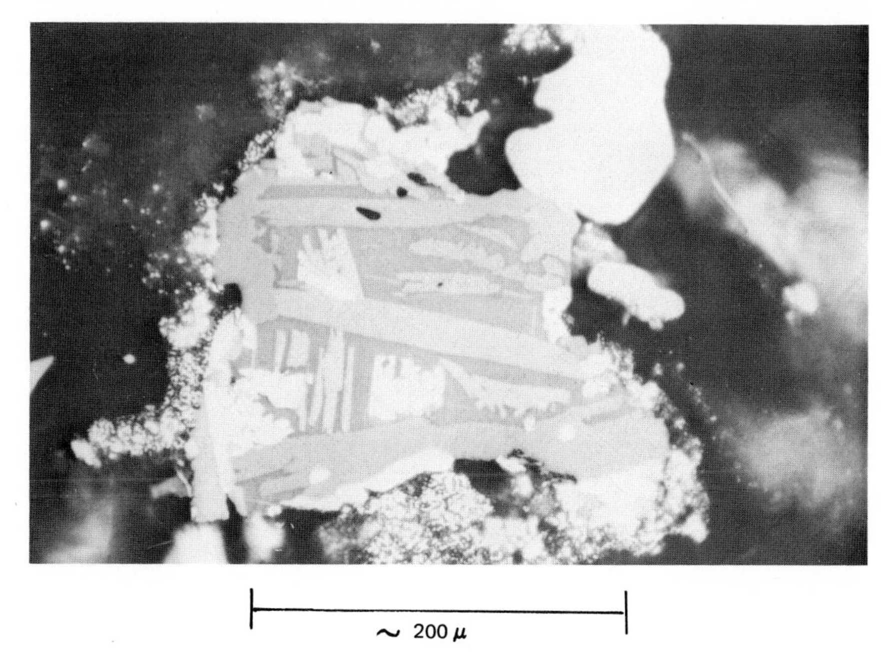


Fig. 4–11 From coarse-grained basalt 70035, a chromian-ulvöspinel grain (medium gray) shows reduction to ilmenite (light gray) plus iron (white). From *EOS* **54,** 593. Photograph Courtesy of S. Haggerty, Department of Geology, University of Massachusetts, Amherst.

and its composition closely fits the $AB_2O_4$ formula. It is, therefore, considered a chromian ulvöspinel.[17]

Ulvöspinel is isotropic, brownish-gray in air[26] or brown-reddish-gray in immersion oil, with reflectivity very similar to that for ilmenite.[32]

Reflectivities for some titanian chromite, aluminian chromite, and ulvöspinel are given in Table 4-1.

*Chemical Composition*

In general, the titanium contents of the lunar chromites are considerably higher than those for comparable terrestrial and meteoritic material.[8] Although some chromites are homogeneous, more commonly they are zoned, with Cr-, Mg-rich centers and Ti-, Fe-rich exteriors (Table 4-2; [3], [4]).[4]

Aluminian-magnesian chromites range compositionally to aluminian-titanian chromites (Table 4-2; [8], [9], [11]). Aluminian-titanian chromite, first noted in Luna 16 samples, had not been identified previously in lunar

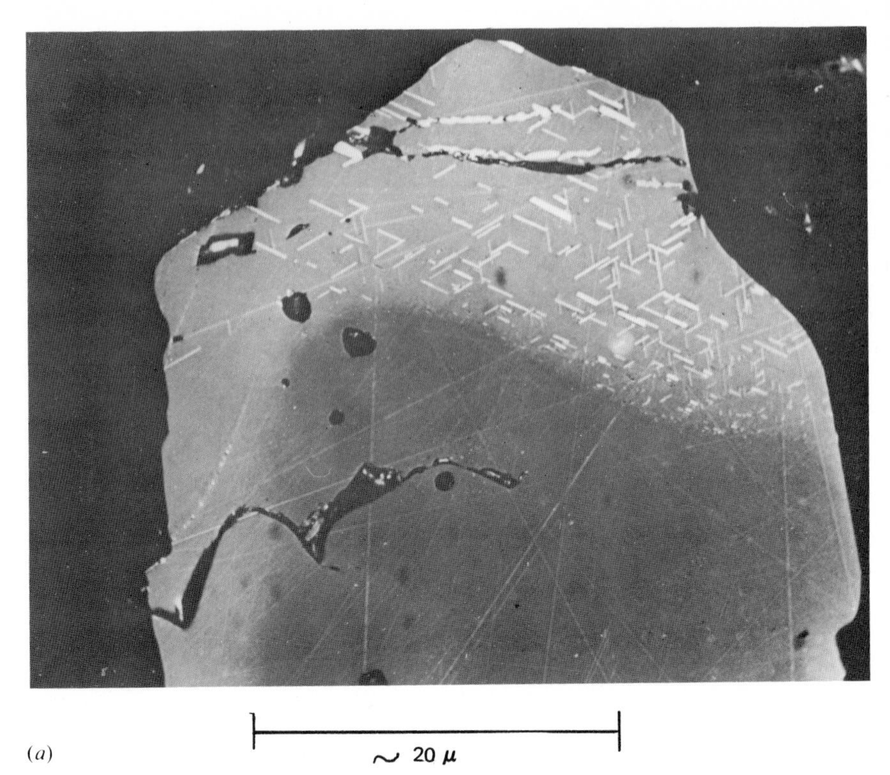


(*a*)

Fig. 4–12 (*a*) Breakdown in ulvöspinel (but not in dark gray chromite) to ilmenite plus native Fe (white) along {111} planes. (*b*) More advanced stage of ulvöspinel (dark gray) breakdown. In addition to ilmenite (light gray) plus native Fe (white), thin rims ($<1$ μ) of chromite occur around the ilmenite blades. (*c*) Complete breakdown of ulvöspinel (no longer present) to ilmenite (light gray) plus native Fe (white) plus chromite (dark gray). From *Suppl.* **3,** 344. Photographs courtesy of A. El Goresy, Max-Planck- Institut für Kernphysik, Heidelberg, W. Germany.

material or in terrestrial rocks. The high Al/Cr ratio of these spinels either reflects the composition of the parent liquid or results from a high-temperature metamorphic recrystallization in an Al-rich environment. These spinels grade into chromian ulvöspinels (Table 4.2; [14], [15]).[34]

Analyses of the anisotropic spinels along the chromite–ulvöspinel join show major amounts of $Al_2O_3$ and MgO. This unusual combination constitutes a mineral phase previously unrecorded.[35] It has been given the varietal name, titanian chromite, by the IMA Commission on Mineral Names,[10] instead of the proposed name *titanochromite*.[14] Many titanian chromite analyses show excess $TiO_2$ and a deficiency of other cations when

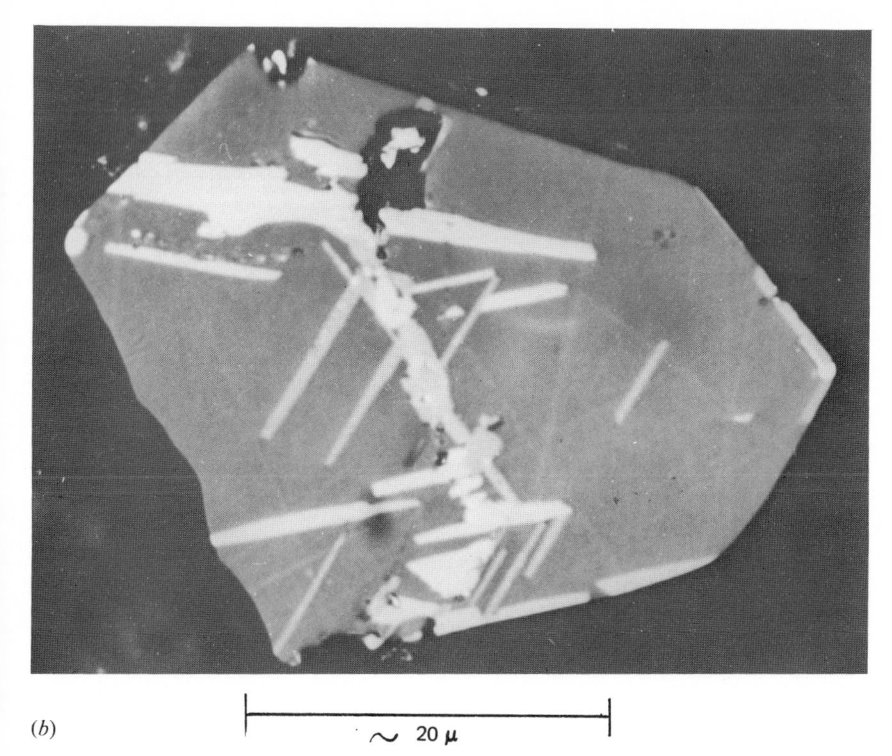


(*b*)

Fig. 4–12 (*continued*)

**Table 4-1 Reflectivities for Some Spinels**

| Wavelengths (nm) | Titanian Chromite* 1 | Titanian Chromite* 2 | Aluminian Chromite† | Ulvöspinel* |
|---|---|---|---|---|
| 450 | 14.0 | 14.8 | — | 15.1 |
| 470 | 13.8 | 14.8 | — | 14.7 |
| 500 | 13.6 | 14.8 | — | 14.6 |
| 520 | 13.4 | 14.8 | — | 14.8 |
| 546 | 13.6 | 15.0 | 14.2 ± 0.3% | 15.1 |
| 589 | 13.7 | 15.1 | — | 15.6 |
| 620 | 13.9 | 15.2 | — | 16.1 |
| 640 | 14.0 | — | — | 16.4 |

* Cameron, *Sci.* **167,** 624.

† Cameron, *Suppl.* **2,** 196.

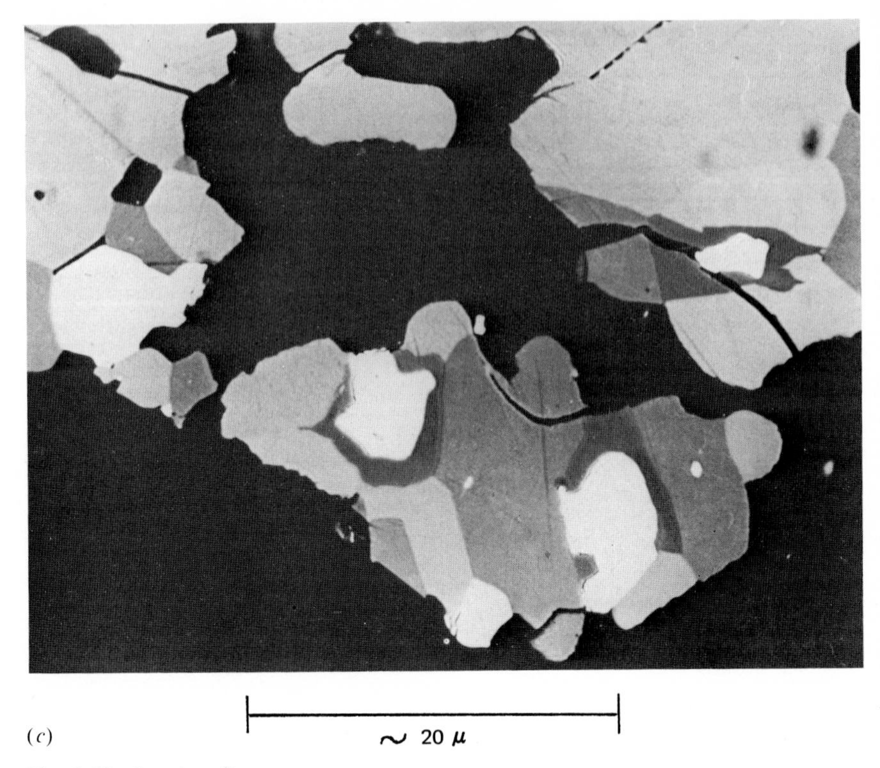


Fig. 4–12 (*continued*)

the structural formula is calculated on the assumption that all titanium is $Ti^{4+}$. This suggests that some of the titanium exists as $Ti^{3+}$. A titanian chromite analysis (Table 4-2; [13]) has been recalculated by distributing the titanium between $Ti^{4+}$ and $Ti^{3+}$ so that for 32 oxygens the cations total 24. The formula can be written as $(Mg_{8-x}Ti^{3+}{}_{x})$ $(Cr^{3+}{}_{16-x}, Ti^{4+}{}_{x}, Al)O_{32}$.[36] This phase with approximately 20% $TiO_2$ also has been called chromian ulvöspinel; with 22% $Cr_2O_3$, the Cr/Ti ratio is 1:1 (Table 4-2; [12]).[10]

The composition of the chromian ulvöspinels is such that they have been regarded either as Cr-rich titanomagnetites (e.g., with $Cr_2O_3 \sim 22–23\%$) or Ti-rich chromites.[5] The term titanomagnetite, however, has been used merely for purposes of comparison with terrestrial and meteoritic analogs of lunar titanian chromites or chromian ulvöspinels.[33] Chromian ulvöspinel associated with baddeleyite contains $< 0.02$ to 0.25% $ZrO_2$, and some substitution of $Zr^{4+}$ for $Ti^{4+}$ may be indicated.[37]

Ulvöspinel compositions closely approach the ideal formula, with only minor Al, Cr, Mn, and Mg content (Table 4-2; [19], [20]).[38] From Apollo 12 rocks, the anisotropic tan, pink, and khaki spinels, containing 60 to 90% $Fe_2TiO_4$, are the most $Fe_2TiO_4$-rich minerals yet found in nature (Table 4-2; [15]–[20]).[8] Ulvöspinels in all Apollo 17 samples contain only small amounts of Zr (usually $< 0.1\%$).[39] In a special study of U concentration in lunar materials, in ulvöspinel of porphyritic basalt 12021, the $UO_2 = 1030$ to $407 \pm 55$ ppb. The large uncertainty is due to great variation of U within a single crystal.[40]

Despite the Apollo 12 "miscibility gap," analyses from Apollo 12 spinels cover nearly the entire compositional range from Ti-free aluminian chromites to Cr-free ulvöspinels, and the analyses show that for a given increase in the atomic percent of Ti, the decrease in Cr + Al is twice as great. This is strong evidence for a complete and simple solid solution between ulvöspinel and chromite.[4]

### *X-Ray Data*

X-Ray study of a titanian chromite, enclosed in olivine from fines sample 10085,14, yielded a cell edge that was intermediate between those of ulvöspinel and chromite (Table 4-3; [1]).[10]

### *References*

1. Reid et al, *Geochim. Cosmochim. Acta* **37,** 764.
2. Goldstein et al, *Suppl.* **1,** 502.
3. Sellers et al, *Suppl.* **2,** 672.
4. Reid, *EPSL* **10,** 351–352.
5. Haggerty et al, *Suppl.* **1,** 524, 526–527.
6. Dowty et al, *Suppl.* **4,** 437.
7. Housely et al, *Lun. Sci. Conf. '71, Abstr.*, 189.
8. Haggerty and Meyer, *EPSL* **9,** 380, 382.
9. Haggerty, *Geochim. Cosmochim. Acta* **37,** 861.
10. Agrell et al, *Suppl.* **1,** 83–85, 111.
11. Cameron, *Suppl.* **2,** 194–197.
12. Champness et al, *Suppl.* **2,** 366–367.
13. Brown et al, *Suppl.* **1,** 205.
14. Cameron, *Sci.* **167,** 624.
15. James and Jackson, *J. Geophys. Res.* **75,** 5799.
16. Douglas et al, *Sci.* **167,** 595.
17. Taylor et al, *Suppl.* **2,** 856, 858.
18. Ramdohr and El Goresy, *Sci.* **167,** 617.
19. El Goresy et al, *Suppl.* **4,** 734.

**Table 4-2 Chromite–Ulvöspinel Analyses**

| | [1] | [2] | [3] | [4] | [5] | [6] | [7] | [8] | [9] | [10] |
|---|---|---|---|---|---|---|---|---|---|---|
| | | Luna 20 | 12020,8 | | | Luna 20 | | 14002, | | |
| | 15085/14 | 22002,3 | Core | Rim | Luna 16 | 22001,16 | Luna 16 | TE–1–8 | 12021 | 15475/125 |
| $SiO_2$ | — | — | — | — | 0.29 | — | 0.34 | — | 0.36 | — |
| $Cr_2O_3$ | 53.30 | 51.6 | 50.0 | 45.4 | 43.54 | 47.83 | 48.85 | 52.3 | 41.61 | 36.09 |
| $TiO_2$ | 2.46 | 4.5 | 3.60 | 5.02 | 5.25 | 1.85 | 0.95 | 0.9 | 7.90 | 11.68 |
| $Al_2O_3$ | 10.84 | 7.3 | 11.9 | 11.9 | 12.16 | 14.92 | 17.05 | 15.4 | 12.17 | 7.23 |
| FeO | 26.38 | 30.6 | 28.3 | 34.4 | 36.97 | 28.45 | 24.01 | 22.5 | 35.98 | 44.08 |
| MgO | 6.15 | 4.6 | 5.80 | 2.11 | 1.76 | 5.18 | 7.48 | 9.1 | 1.26 | 0.70 |
| MnO | 0.41 | — | 0.26 | 0.29 | 0.42 | 0.44 | 0.53 | 0.25 | — | 0.49 |
| CaO | — | — | — | — | 0.16 | — | 0.20 | — | 0.15 | — |
| Total | 99.54 | 99.44* | 99.86 | 99.12 | 100.55 | 99.46* | 99.41 | 100.45 | 99.52* | 100.27 |

*References*


[1] Brown et al, *Suppl.* **3,** 145. From a basalt. Chromite.
[2] Brett et al, *Geochim. Cosmochim. Acta* **37,** 765. From a fines sample. *Analysis includes $V_2O_3$ = 0.84%. Chromite.
[3] Kushiro et al, *Suppl.* **2,** 483. From a porphyritic basalt. Chromite.
[4] ———, *ibid.* From a porphyritic basalt. Aluminian chromite.
[5] Haggerty, *EPSL* **13,** 334. From a fines sample. Aluminian chromite.
[6] Reid et al, *Geochim. Cosmochim. Acta* **37,** 1015. From a fines sample. *Analysis includes $V_2O_3$ = 0.79%. Aluminian chromite.
[7] Haggerty, *EPSL* **13,** 334. From a fines sample. Magnesian-aluminian chromite.
[8] Steele and Smith, *Nature* **240,** 5. From a fines sample. Aluminian-magnesian chromite.
[9] Weill et al, *Suppl.* **2,** 416. From a porphyritic basalt. *Analysis includes $K_2O$ = 0.09%: Titanian-aluminian chromite.
[10] Brown et al, *Suppl.* **3,** 145. From a basalt. Titanian chromite.

| | [11] | [12] | [13] | [14] | [15] | [16] | [17] | [18] | [19] | [20] |
|---|---|---|---|---|---|---|---|---|---|---|
| | | | | | | 12064,6 | | | | |
| | 61156,31 | Luna 20 532–6 | 10085-4–14 | 12051/59 No. 19 A | Pink | Tan | Khaki | 12051 | 10044 | 14161/20 |
| $SiO_2$ | 0.91 | 0.54 | 0.40 | — | 0.19 | 0.33 | 0.17 | — | — | — |
| $Cr_2O_3$ | 31.12 | 22.43 | 21.98 | 17.3 | 9.34 | 3.64 | 1.50 | 0.03 | 0.22 | 0.18 |
| $TiO_2$ | 22.38 | 20.68 | 23.58* | 23.7 | 28.0 | 31.1 | 32.2 | 32.8 | 33.8 | 37.01 |
| $Al_2O_3$ | 8.63 | 9.79 | 6.25 | 5.2 | 2.64 | 2.66 | 2.32 | 2.48 | 1.8 | 0.02 |
| FeO | 27.06 | 39.80 | 42.30 | 53.2 | 58.6 | 61.8 | 62.9 | 62.2 | 62.1 | 61.69 |
| MgO | 8.94 | 5.12 | 5.75 | 0.46 | 0.21 | 0.17 | 0.13 | 0.21 | 0.14 | 0.75 |
| MnO | 0.52 | 0.33 | 0.34 | 0.24 | 0.27 | 0.16 | 0.29 | 0.25 | 0.50 | 0.40 |
| CaO | — | 0.29 | 0.01 | 0.39 | <0.01 | <0.01 | <0.01 | 0.47 | — | — |
| Total | 99.91* | 98.98 | 100.61 | 100.97* | 99.25 | 99.86 | 99.51 | 98.66* | 98.56 | 100.06 |

*References*


[11] Albee et al, *Suppl.* **4,** 578. From an annealed breccia. *Analysis includes $V_2O_3$ = 0.33% and $ZrO_2$ = 0.02%. Aluminian-titanian chromite.

[12] Haggerty, *Geochim. Cosmochim. Acta* **37,** 859. From a crystalline particle. Titanian chromite.

[13] Agrell et al, *Suppl.* **1,** 111. From a fines sample. Average of three points on same grain. *With redistribution of titanium between $Ti^{3+}$ and $Ti^{4+}$ to give a total of 24 cations for 32 oxygens, $TiO_2$ = 19.69%, $Ti_2O_3$ = 3.51%, and total = 100.23% Titanian chromite.

[14] Keil et al, *Suppl.* **2,** 329. From an ophitic basalt. *Analysis includes $V_2O_3$ = 0.48%. Chromian ulvöspinel.

[15] Haggerty and Meyer, *EPSL* **9,** 380. From a microgabbro. Chromian ulvöspinel.

[16] ———, *ibid.* From a microgabbro. Al-bearing chromian ulvöspinel.

[17] ———, *ibid.* From a microgabbro. Aluminian ulvöspinel.

[18] Busche et al, *Am. Mineral.* **57,** 1733.From a fine-grained basalt. *Analysis includes $V_2O_3$ = 0.22%. Aluminian ulvöspinel.

[19] Smith et al, *Suppl.* **1,** 904. From a microgabbro. Ulvöspinel.

[20] Brown et al, *Suppl.* **3,** 145. From a fines sample. Ulvöspinel.

**Table 4-3 X-Ray Data for Some Spinels**

| | [1] | [2] | [3] | [4] |
|---|---|---|---|---|
| | | Luna 20 | | |
| Sample Number | 10085,14 | (22002,2,1) | 14319,11b | 14319,11a |
| $a_0$ | 8.415 ±0.005 Å | 8.18 Å | 8.15 Å (approx.) | 8.143 ±0.01 Å |
| Space Group | | $F_{d3m}$ | | |

X-Ray Powder Pattern Data:

| hkl | I | d (Å) Measured | d (Å) Calculated |
|---|---|---|---|
| 022 | 1 | 2.974 | 2.975 |
| 113 | 10 | 2.536 | 2.537 |
| 004 | 2 | 2.102 | 2.104 |
| 024 | 05 | 1.719 | 1.718 |
| 115, 333 | 3 | 1.621 | 1.620 |
| 044 | 4 | 1.488 | 1.488 |
| 355, 137 | 1 | 1.095 | 1.095 |

*References*

[1] Agrell et al, *Suppl.* **1,** 86. From a fines sample. Titanian chromite.

[2] Cameron et al, *Geochim. Cosmochim. Acta* **37,** 788. Handpicked crystal from a fines sample. Chrome spinel.

[3] Roedder and Weiblen; *EPSL* **15,** 385. From a breccia. Low-Mg–high-Cr pleonaste.

[4] ———, *ibid.* From a breccia. High-Mg–low-Cr pleonaste.


20. Haggerty, *Nature* **234,** 117.
21. Taylor et al, *EOS* **54,** 616.
22. Drake et al, *EPSL* **9,** 113.
23. Brett et al, *Suppl.* **2,** 310.
24. Haggerty, *EOS* **54,** 593.
25. Carter et al, *EOS* **54,** 583.
26. Cameron, *Suppl.* **1,** 229, 231, 240.
27. Apollo 17 LSPET, *Sci.* **182,** 661.
28. Wood et al, *Spec. Rep.* **333,** 37.
29. Haggerty, *Suppl.* **3,** 312, 316, 318.
30. Simpson and Bowie, *Suppl.* **2,** 208.
31. Haggerty and Meyer, *EOS* **51,** 583.
32. Ramdohr and El Goresy, *Naturwiss.* **57,** 102.

33. Haggerty and Meyer, *Carnegie Inst. Yearb. '69*, 231–232.
34. Haggerty, *EPSL* **13,** 330, 334–335.
35. Simpson and Bowie, *Suppl.* **1,** 888.
36. Frondel, C., Private communication.
37. El Goresy et al, *Suppl.* **3,** 338–339.
38. Keil et al, *Suppl.* **1,** 586.
39. Taylor and Williams, *Lun. Sci. V, Abstr.*, 785.
40. Theil et al, *EPSL* **16,** 38.

## Normal Series: Hercynite ($FeAl_2O_4$)–Spinel ($MgAl_2O_4$)

### *Hercynite Synonymy*

Aluminous picotite; Drever and Johnston, *Nature* **235,** 30.
Chromian hercynitic spinel; Roedder and Weiblen, *EPSL* **15,** 376.
Chromian pleonaste; Wood et al, *Spec. Rep.* **333,** 271.
Chromiferous pleonaste; Christophe-Michel-Lévy and Lévy, *Suppl.* **3,** 887.
Chromium pleonaste; Haggerty, *Suppl.* **3,** 307.
Cr-pleonaste; Haggerty, *Geochim. Cosmochim. Acta* **37,** 859.
Hercynitic pleonaste; Roedder and Weiblen, *EPSL* **15,** 394.
Hercynitic spinel; Walter et al, *Lun. Sci. Conf. '72, Abstr.*, 685.
Magnesian spinel (in part); Christophe-Michel-Lévy and Lévy, *Suppl.* **3,** 887.
Orange spinel; Reid et al, *Geochim. Cosmochim. Acta* **37,** 1025.
Picotite; Agrell et al, *Suppl.* **1,** 98.
Pleonaste; Wood et al, *Spec. Rep.* **333,** 271.
Pleonaste-type spinel; Roedder and Weiblen, *EPSL* **15,** 376.
Reddish-purple spinel; Roedder and Weiblen, *Lun. Sci. Conf. '72, Abstr.*, 578.
Spinel-pleonaste; Haggerty, *Geochim. Cosmochim. Acta* **37,** 860.
Ti-free Al-Cr-Mg-Fe spinel; von Engelhardt et al, *Suppl.* **3,** 758.
Titanian hercynite; Haggerty, *EPSL* **13,** 329.
Titanian picotite; Roedder and Weiblen, *EPSL* **15,** 394.

### *Spinel Synonymy*

Aluminous spinel; Haggerty, *EPSL* **13,** 336.
Chrome spinel (in part); Anderson, *J. Geol.* **81,** 219.
High Mg spinel; Reid et al, *Geochim. Cosmochim. Acta* **37,** 1015.
Magnesian spinel (in part); Reid et al, *Geochim. Cosmochim. Acta* **37,** 1022.
Red spinel; Steele and Smith, *Lun. Sci. Conf. '72, Abstr.*, 636.

*Occurrence and Form*

Picotite, a variety of spinel in the series between hercynite and aluminian chromite,[1] was first identified from fine-grained basalt 10045. It occurred as tiny octahedra enclosed in olivine.[2] It was later found in a Luna 20 soil sample.[3]

A rare hercynite spinel (75% $FeAl_2O_3$), from feldspathic peridotite 12036,9, was found in an early $Al_2O_3$-rich pyroxene melt inclusion. This material may be partly responsible for the aluminous nature of the spinel.[4]

From Apollo 14 breccias, spinels intermediate between pleonaste and picotite occur mostly as monomineralic clasts, with only a few fragments embedded in shocked plagioclase. The grains (up to 0.2 mm long) are irregular or, in some cases, straight-sided, but never euhedral crystals.[5]

Most Luna 20 spinels occurring in spinel troctolite fragments are approximately spinel-hercynite solid solutions.[6] These form euhedral to subhedral crystals.[3] Chrome spinels (i.e., spinel-pleonastes) from peridotite 15445,10 occur with interlocking grains of pyroxene, olivine, plagioclase, and rutile (?), making up about 15% of the rock's volume.[7] In Apollo 17 basalts, lamellae of a Cr-rich spinel phase occur as inclusions parallel to the basal plane in ilmenite.[8] From Luna 20 sample 515-23, a strongly fractured spinel-pleonaste contains cyrstallographically oriented lamellae of rutile along {111} planes (Figure 4–13).[9] From breccias 14319 and 14303, from microbreccia 14321, and from soil sample 14162, the chromian pleonastes occur mainly as single-crystal fragments, some up to 0.2 mm long.[5] Chromian pleonaste, found frequently as individual grains (many $> 10\ \mu$), was observed in a thin section of breccia 14063-20 and in Apollo 14 rock debris. It has been referred to also as a magnesian spinel.[10]

Although rare in Apollo 11 samples, almost pure spinel was found in a lithic fragment of microbreccia 10019-22 (Table 4-4; [13]) together with plagioclase, plagioclase-glass, olivine, and pyroxene, all highly shocked.[11] In thin section of a clast in breccia 76315,11, some bands of crushed minerals contain several equant spinels.[8] Spinels with broad radial anorthite reaction rims (and with or without plates of armalcolite) occur in both clasts and the surrounding matrix of breccia 76055,10.[12] In Luna 20 samples, spinel that is essentially $MgAl_2O_4$ with a small degree of solid solution towards hercynite occurs in good euhedral crystals that are early crystallization products.[9] These have been found only in association with alumina-rich rocks.[13] From spinel troctolite 62295, spinel close to $MgAl_2O_4$ is abundant throughout the sample but most abundant in variolitic intergrowths. The grains, as long as 0.04 mm, are euhedral.[14] The spinel also occurs as minute octahedra included in plagioclase and, to a lesser extent, in olivine phenocrysts intergrown with feldspar. The larger crystals are

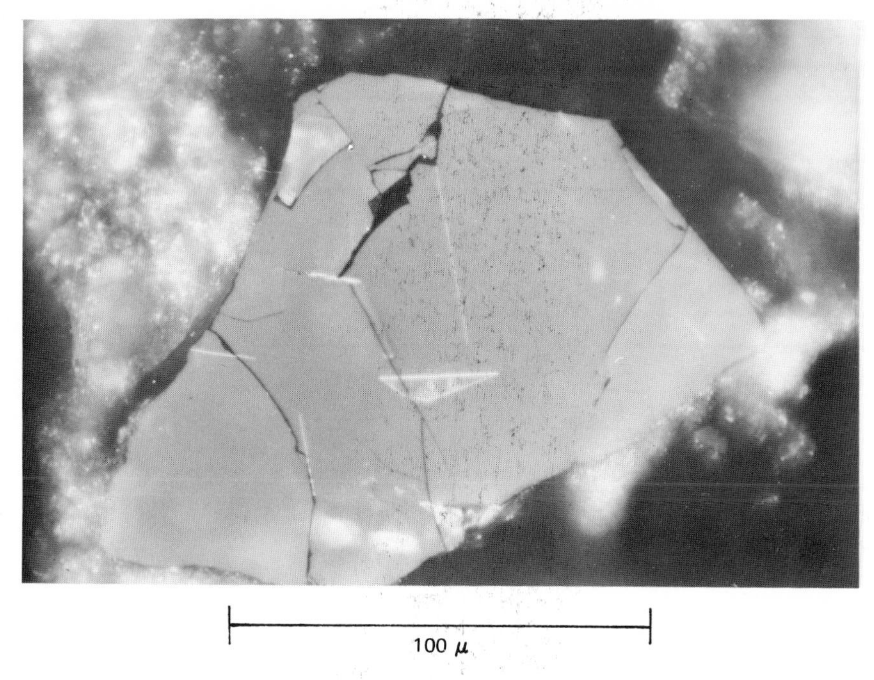


Fig. 4–13 From a Luna 20 breccia particle 515-23, white ilmenite lamellae are oriented parallel to the {111} planes of a spinel-pleonaste grain. From *Geochim. Cosmochim. Acta* **37,** 858. Photograph courtesy of S. Haggerty, Department of Geology, University of Massachusetts, Amherst.

found between feldspar laths.[15] A somewhat more Cr-rich spinel from the same sample occurs in rounded xenocrysts with well-developed very fine-grained reaction rims.[14]

From coarse-grained ilmenite basalt 70035, a phase intermediate between chromian pleonaste and aluminian-magnesian chromite has been identified.[16]

*Optics*

Picotite from fine-grained olivine basalt 10045 is coffee-brown.[2] Two grains, essentially picotite from a Luna 20 sample, are orange and distinct from the pink or red pleonastes in composition as well as color.[3] Aluminous picotite from soil and breccias of Apollo 14 is reddish-purple.[17]

A spinel-pleonaste, from thin section 515-23 of Luna 20 sample 22003,1, was burgundy in transmitted light and pale brown in reflected light. The grain, which occurs in a glass-rich breccia particle, is strongly fractured and contains lamellae of ilmenite crystallographically oriented along {111}

**Table 4-4 Hercynite Spinel Analyses**

| | [1] | [2] | [3] | [4] | [5] | [6] |
|---|---|---|---|---|---|---|
| | Luna 20 22001,16 | 14053 | 14055,7 | 12036,9 | 14319,11b | Luna 20 515–23 |
| $SiO_2$ | — | — | 0.75 | — | 0.44 | 0.10 |
| $Cr_2O_3$ | 25.27 | 18.80 | 5.51 | 4.6 | 8.69 | 10.06 |
| $TiO_2$ | 0.26 | 4.28 | 0.53 | 1.18 | 1.18 | 1.10 |
| $Al_2O_3$ | 45.26 | 38.23 | 57.80 | 55.0 | 53.0 | 54.43 |
| FeO | 12.27 | 24.94 | 22.50 | 33.4 | 22.4 | 15.93 |
| MgO | 17.38 | 11.32 | 11.00 | 5.6 | 12.6 | 17.31 |
| MnO | 0.17 | 0.08 | — | 0.16 | 0.18 | 0.13 |
| CaO | — | — | 0.31 | 0.46 | 0.19 | 0.02 |
| Total | 100.61 | 99.87* | 98.40 | 100.46* | 98.68 | 99.08 |

*References*

[1] Reid et al, *Geochim. Cosmochim. Acta* **37,** 1015. From a soil sample. Picotite.

[2] El Goresy et al, *EPSL* **13,** 125. From a medium-grained basalt. *Analysis includes $V_2O_3$ = 0.22%. Titanian picotite.

[3] Drever and Johnston, *Nature* **235,** 30. From a microbreccia. Aluminous picotite.

[4] Busche et al, *Am. Mineral.* **57,** 1733. From a feldspathic peridotite. *Analysis includes $V_2O_3$ = 0.06%. Hercynite.

[5] Roedder and Weiblen, *EPSL* **15,** 382. Average of three analyses from a breccia. Low-Mg–high-Cr pleonaste.

[6] Haggerty, *Geochim. Cosmochim. Acta* **37,** 859. From a glass-rich breccia particle. Spinel-pleonaste.

planes. It is comparable optically and texturally to some of the Apollo 14 chromian pleonastes. An optically similar single crystal of chromian-pleonaste is present in a feldspathic breccia particle 505-2 from the same Luna 20 sample. The crystal has an irregular bleached mantle from less than 1 to 15 $\mu$ wide. Although the mantle and core have a conspicuous optical contrast, they have only slight compositional differences.[9] The chromian pleonastes from breccias 14063, 14321, and 14318 and from fines samples 14003 and 14163 are light to bright pink, but from breccia 14066 the spinel grains are red. Their indices of refraction range from 1.75 to 1.84.[18] The spinels from Apollo 14 breccias are generally pink or red, but

| | [7] | [8] | [9] | [10] | [11] | [12] | [13] |
|---|---|---|---|---|---|---|---|
| | 14066 | | 14319, | 15445, | Luna 20 | Luna 20 | 10019– |
| | Core | Rim | 11a | 10 | 22002,2,1 | 532–3 | 22 |
| $SiO_2$ | 0.17 | 0.13 | 0.47 | — | — | 0.27 | — |
| $Cr_2O_3$ | 12.56 | 18.82 | 1.58 | 14.0 | 6.38 | 1.38 | 2.16 |
| $TiO_2$ | 0.19 | 0.56 | 0.52 | 0.03 | 0.24 | 0.01 | — |
| $Al_2O_3$ | 53.44 | 46.34 | 61.7 | 57.3 | 62.1 | 67.13 | 68.0 |
| FeO | 17.82 | 21.22 | 18.2 | 9.3 | 10.6 | 5.91 | 3.33 |
| MgO | 15.57 | 12.39 | 15.7 | 20.4 | 20.3 | 24.73 | 25.9 |
| MnO | 0.15 | 0.27 | 0.04 | — | 0.10 | 0.01 | — |
| CaO | 0.06 | 0.07 | 0.10 | — | — | 0.02 | — |
| Total | 99.96 | 99.80 | 98.31 | 101.03 | 99.72 | 99.46 | 99.39 |

*References*

[7] ———, *Suppl.* **3,** 311. From a microbreccia. Chromian pleonaste. (Low-Ti–chromian pleonaste).

[8] ———, *ibid.* From a microbreccia. Chromian pleonaste. (Low-Ti–chromian pleonaste).

[9] Roedder and Weiblen, *EPSL* **15,** 379. Average of three analyses from a breccia. High-Mg–Low-Cr pleonaste.

[10] Anderson, *J. Geol.* **81,** 221. From a peridotite clast. Chrome spinel.

[11] Cameron et al, *Geochim. Cosmochim. Acta* **37,** 783. Handpicked crystal from a soil sample. Chrome spinel.

[12] Haggerty, *Geochim. Cosmochim. Acta* **37,** 859. From a breccia fragment. Spinel.

[13] Keil et al, *Suppl.* **1,** 564. From a microbreccia. Spinel.

they vary in hue from pale purplish pink to almost opaque dark purplish red in transmitted light. Their color is very different from that of spinels from earlier lunar missions and may be responsible for some of the tentative "garnet or spinel" or "rutile" identifications of Apollo 14 LSPET.[5] Even though their reddish color has been ascribed to Cr,[19] only a moderate correlation between color and Cr content has been noted with $Cr_2O_3 < 5\%$, and the correlation is even less good with $Cr_2O_3 > 5\%$.

From spinel troctolite 62295 the more Cr-rich spinels (e.g., with 9–16% $FeCr_2O_4$ content) are pink, but true spinels ($MgAl_2O_4$ = 92–97%) occur as pale yellow to colorless grains[14] or minute octahedra.[15]

### *Chemical Composition*

Distinctive orange spinel from Luna 20 soil may be considered a picotite from its composition (Table 4-4; [1]). The picotites grade into hercynitic spinels. Small grains (10–40 $\mu$) included in pyroxene crystals of feldspathic peridotite 12036 have an $Al_2O_3$ content of 50 to 56% and approach the hercynite end member (Table 4-4, [4]).[20]

The pleonastes are those spinels along the $MgAl_2O_4$–$FeAl_2O_4$ join for which the Mg/Fe ratio ranges from 3 to 1. In general the lunar pleonastes contain some substitution of Cr for Al and more properly could be called chromian pleonastes.[21] They also are termed spinel-pleonastes (Table 4-4; [5], [6]). In breccia 14063, the low-Ti–chromian pleonastes have a composition between 0.5 $MgAl_2O_4$–$FeAl_2O_4$ and are unique to Apollo 14 material (Table 4-4; [7], [8]).[1]

The major spinel phase in Luna 20 samples approaches $MgAl_2O_4$ with some variable Cr and Fe (Table 4-4; [19]).[3] Red spinel proper from 14063, with only minor chromite and hercynite components, is believed to have formed by primary crystallization from a high-Al, low-Fe magma.[22] From peridotite clast 15445,10, the Mg/Fe ratio of the spinel (as well as of the accompanying olivine, pyroxene, and armalcolite) is exceptionally high compared with that from mare basalts, KREEP rocks, most anorthositic rocks, and most fragments in lunar breccias. By analogy with terrestrial rocks this suggests a comparatively primitive stage of differentiation.[7]

### *X-Ray Data*

A handpicked single crystal of chrome spinel from Luna 20 soil sample 22002,2,1 has a chemical analysis (Table 4-4; [11]) accompanying its X-ray data (Table 4-3; [2]). Two pleonastes from breccia 14319 were analyzed chemically and by X-ray. One is a low-Mg, high-Cr pleonaste (Table 4-4; [5]; Table 4-3; [3]). The other is an unshocked high-Mg, low-Cr pleonaste (Table 4-4; [9]; Table 4-3; [4]).

### *References*


1. Haggerty, *EPSL* **13,** 329, 339.
2. Agrell et al, *Suppl.* **1,** 98.
3. Reid et al, *Geochim. Cosmochim. Acta* **37,** 1022, 1024–1025.
4. Busche et al, *Am. Mineral.* **57,** 1739.
5. Roedder and Weiblen; *EPSL* **15,** 376–378, 390.
6. Brett et al, *Geochim. Cosmochim. Acta* **37,** 762, 764.
7. Anderson, *J. Geol.* **81,** 219, 221.
8. Apollo 17 LSPET, *Sci.* **182,** 661–663.
9. Haggerty, *Geochim. Cosmochim. Acta* **37,** 859, 860.

10. Christophe-Michel-Lévy et al, *Lun. Sci. Conf. '72, Abstr.*, 126.
11. Mason et al, *Lun. Sci. III, Rev. Abstr.*, 512.
12. Chao, *EOS* **54,** 584.
13. Meyer, *Geochim. Cosmochim. Acta* **37,** 947–948.
14. Hodges and Kushiro, *Suppl.* **4,** 1037.
15. Agrell et al, *Lun. Sci. IV, Abstr.*, **15.**
16. Weigand, *EOS* **54,** 621.
17. Roedder and Weiblen, *Lun. Sci. Conf. '72, Abstr.*, 578.
18. Christophe-Michel-Lévy and Lévy, *Suppl.* **3,** 887–888.
19. Drever and Johnston, *Nature* **235,** 31.
20. Busche et all *Spec. Publ. #3*, 50.
21. Wood et al, *Spec. Rep.* **333,** 90, 92.
22. Steele and Smith, *Lun. Sci. Conf. '72, Abstr.*, 636.

## Green Spinel (?)

In a thin section from breccia 14321, fine droplets of a green spinel were seen in the orthopyroxene crystals. The droplets are believed to have been incorporated during growth. No other identifying data are given.[1]

*References*

1. Gay and Brown, *Lun. Sci. Conf., '72, Abstr.*, 260.

# CHAPTER FIVE | MULTIPLE OXIDES: ZIRCONOLITE AND ARMALCOLITE

**Zirconolite** (Ca, Fe) (Zr, Ce) $(Ti, Nb)_2O_7$ (essentially)

*Synonymy*

"Dysanalyte"; Ramdohr and El Goresy, *Sci.* **167,** 617.
Perovskite (in part); Wenk et al, *Lun. Sci. III, Rev. Abstr.*, 797.
Phase $\beta$; Haines et al, *EPSL* **12,** 145.
Phase B; Lovering and Wark, *Suppl.* **2,** 154.
Phase Y; Brown et al, *Lun. Sci. III, Rev. Abstr.*, 97.
REE-bearing, high-Zr mineral; Ramdohr, *Fortschr. Mineral.* **48,** 48.
Red-brown mineral; Lovering et al, *Lun. Sci. III, Rev. Abstr.*, 494.
U-bearing zirkelite; Prinz et al, *Meteorit.* **6,** 302.
Zirkelite; Busche et al, *EPSL* **14,** 313.
Zr-Ti-Ca-rich phase; Lovering and Wark, *Suppl.* **2,** 151.
Zr-Ti phase; Albee et al, *Lun. Sci. Conf. '71, Abstr.* 57.
Zr-Ti-REE mineral; Roedder and Weiblen, *Geochim. Cosmochim. Acta* **37,** 1036.

It was suggested originally that a lunar Zr-Ti-rich phase[1] (also termed phase B[2]) was "dysanalyte," a variety of perovskite ($CaTiO_3$).[1] The lunar material, however, contains much Zr, not characteristic of terrestrial dysanalyte,[3] and its composition does not fit the perovskite formula type ($ABO_3$). The question of whether the lunar phase is zirkelite or zirconolite parallels the problem attending the terrestrial material.[4] The difficulties in nomenclature have been treated in detail by Busche et al.[5] It is probable that zirkelite and zirconolite are the same phase. However it has been claimed by some investigators that the composition of the lunar material fits into the zirkelite formula type ($AB_2O_5$),[5] whereas others state that the formula type of zirconolite ($AB_3O_7$) is more appropriate.[4] Certainly many of the analyses of the lunar material may be calculated as zirconolite, and in this discussion the material is referred to as zirconolite.*

*References*

1. Ramdohr and El Goresy, *Sci.* **167,** 617.
2. Lovering and Wark, *Suppl.* **2,** 154–155.
3. Ramdohr, *Fortschr. Mineral.* **48,** 48.
4. Wark et al, *Lun. Sci. IV, Abstr.*, 764.
5. Busche et al, *EPSL* **14,** 313–315, 318–319.

*Occurrence and Form*

Zirconolite occurs in very small isometric grains intergrown with troilite and bordered by ilmenite, pyroxene, and olivine.[1] One grain from ophitic basalt 12036,9 is only 9 × 10 μ. Zirconolite is relatively abundant in two lithic fragments from fines samples 14163,39 and 14257,3. The grains (the largest of which is 15 μ long) are surrounded by plagioclase and are not in contact with the ilmenite and baddeleyite.[2] In other samples zirconolite has been observed within baddeleyite aggregates (Figure 5–1).[3] In the groundmass of metamorphosed KREEP-rich breccia 65015 zirconolite occurs in about 20-μ anhedral grains coexisting with baddeleyite.[4] Irregular grains up to 40 μ across have been seen in coarse-grained basalt 10040-20.[5] In breccia 12013, tiny irregular grains, the largest 5 × 10 μ, occur mostly as spots but also as needles. They are more abundant in the light lithological portion of this sample. Because their composition is somewhat different from zirconolite (see under *Chemical Composition*) they have been referred to as phase β.[6] Zirconolite is a late phase that crystallized from the last U-enriched liquids interstitial in the lunar basalts.[5]

*References*

1. Ramdohr and El Goresy, *Naturwiss.* **57,** 102.
2. Busche et al, *EPSL* **14,** 315–317.
3. Ramdohr, *Fortschr. Mineral.* **48,** 47.
4. El Goresy et al, *Lun. Sci. IV, Abstr.*, 222–223.
5. Lovering and Wark, *Suppl.* **2,** 151, 154–155.
6. Haines et al, *EPSL* **12,** 145–146.

* A translation of *Study of Brazilian Zirkelite* by Pudovkina et al (Institute of Mineralogy, Geochemistry, and Crystal Chemistry of Rare Elements, Moscow) has just been completed and kindly sent to me by Dr. Michael Fleischer of the U.S. Geological Survey. The study, made on some of the originally described zirkelite material from Brazil, further substantiates the identity of the two phases, zirkelite and zirconolite, represented by formula type $AB_3O_7$. The study also suggests that the name *zirkelite* has priority and that the name zirconolite should be discredited.

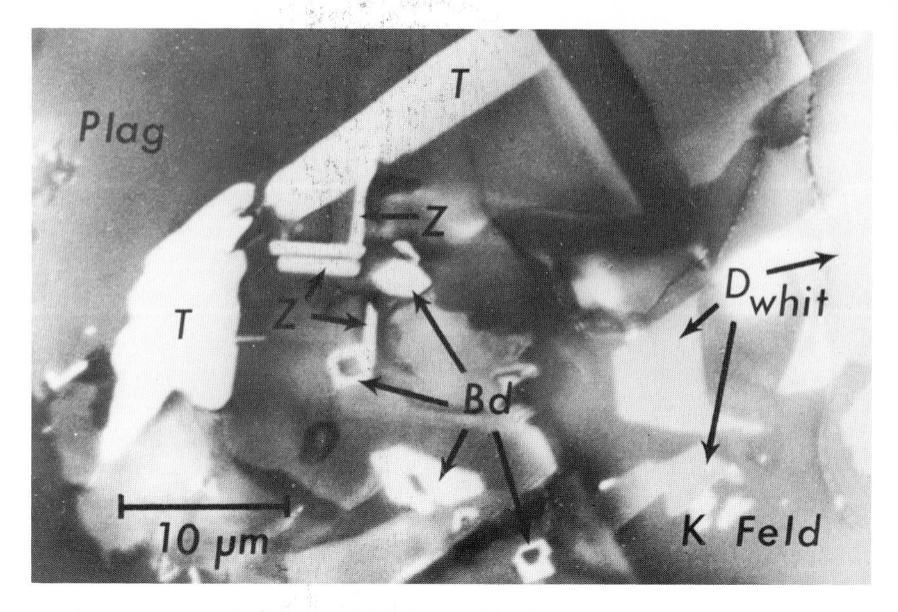


Fig. 5–1 In mesostasis of basalt 10047, 30, lathlike zirconolite crystals (Z) occur together with baddeleyite (Bd), tranquillityite (T), whitlockite (Whit), and potash feldspar (K Feld). D = phosphate-phases (i.e., apatite and/or whitlockite). From *Suppl.* **2,** 156. Photograph courtesy of J. F. Lovering, School of Geology, University of Melbourne, Parkville, Victoria, Australia.

### *Optics*

In reflected light zirconolite is gray[1] or somewhat bluish, with an adamantine luster similar to that of sphalerite.[2] In reflected light some zirconolite has a reflectivity slightly lower than that of ilmenite.[1] Still other zirconolite has been described as resembling, in reflected light, tranquillityite and baddeleyite. The zirconolite is either isotropic or weakly anisotropic.[3]

In transmitted light zirconolite is strongly reddish brown[3] and has beautiful deep-red internal reflections.[2] In transmitted light, the zirconolite from basalt 15555 is dark brown rather than foxy red, like tranquillityite.[4] In a troctolite clast from breccia fragment 516-1 of a Luna 20 soil sample, two crystals of zirconolite are yellowish brown rather than deep reddish brown. It is suggested that the color difference may be due to the extreme thinness of the grains in the sample.[5]

### *References*

1. Busche et al, *EPSL* **14,** 317.
2. Ramdohr and El Goresy, *Naturwiss.* **57,** 102.
3. Lovering and Wark, *Suppl.* **2,** 154.

4. Brown et al, *Suppl.* **3,** 149.
5. Roedder and Weiblen, *Geochim. Cosmochim. Acta* **37,** 1042.

### *Chemical Composition*

Lunar zirconolite has a generally constant composition with major Zr and Ti; Fe, Ca, Y, and REE are lesser but significant constituents. The phase differs from the silicate tranquillityite, another Zr, Ti, Fe-bearing mineral. Zirconolite has little or no Si, lower Fe, and higher Ti and Zr contents than tranquillityite.

Through studies of synthesized zirconolite structure phases and natural terrestrial phases of the same composition, it has been determined that they all have similar monoclinic structures. Their compositions conform to a generalized formula of $(M_{I})^{2+}$ $(M_{II})^{4+}$ $(M_{III})^{4+}O_7$, in which

$M_{I}{}^{2+}$ = Fe, Ca, and other minor divalent cations

$M_{II}{}^{4+}$ = $Zr^{4+}$, with minor $U^{4+}$, $Th^{4+}$, and $Pb^{2+}$

$M_{III}{}^{4+}$ = $Ti^{4+}$, with minor $Nb^{5+}$, $Ta^{5+}$, $Al^{3+}$, and $Cr^{3+}$.

The substantial amounts of $Y^{3+}$ and $REE^{3+}$ present are distributed equally between the $M_{I}{}^{2+}$ and $M_{II}{}^{4+}$ sites. (Calculations for a number of zirconolite analyses are given in Table 5-1.) As the temperature of formation increases, additional $Zr^{4+}$ can go into the $M_{III}{}^{4+}$ sites together with $Ti^{4+}$. Synthetic zirconolites formed at 1300°C have Ti/Zr = 2.75, and those formed at 1500°C have Ti/Zr = 1.31. The Ti/Zr ratio in the lunar zirconolites ranges from 1.07 to 1.52, suggesting temperatures of formation between 1400 and 1450°C.[1]

Although most of the lunar phases of composition similar to zirconolite can be calculated reasonably as formula type $AB_3O_7$,[1] the analyses of phase $\beta$ does not permit such a calculation (Table 5-1; [9]). Phase $\beta$ is still considered a mystery because of the low summation of the analysis, for which no great accuracy has been claimed. (Phase $\beta$ may or may not be zirconolite). The main importance of phase $\beta$ is that it is a major carrier of U, Th, and Pb. It should be emphasized here that terrestrial zirconolite is also high in U and Th.[2] The U content of zirconolite in Apollo 14 mare basalts is about 170 ppm. In Apollo 14 nonmare basalts zirconolite contains up to 400 ppm U.[3] Zirconolite from basalt 75035 is greatly enriched in Y, with $Y_2O_3$ = 12.8%.[4]

### *X-Ray Data*

Data from an X-ray investigation of a terrestrial zirconolite are given in Table 5-2. X-Ray study of a 20-$\mu$ zirconolite grain from mare basalt 10047,68 is in progress.[1]

Table 5-1 Analyses of Zirconolite

| | [1] Luna 20–516–1 | | [2] 12036,9 | | [3] 14257,3 | | [4] 10046,20 | | [5] 10047,20 | | |
|---|---|---|---|---|---|---|---|---|---|---|---|
| | a | b | a | b | a | b | a | b | a | b | |
| CaO | 10.7 | 0.700 | 4.6 | 0.320 | 8.6 | 0.558 | 7.31 | 0.483 | 2.63 | 0.196 | |
| MgO | 0.36 | 0.033 | 0.58 | 0.055 | 0.72 | 0.066 | 0.65 | 0.060 | 0.01 | 0.000 | A |
| FeO | 4.3 | 0.220 | 6.5 | 0.352 | 5.9 | 0.299 | 5.95 | 0.307 | 9.06 | 0.527 | |
| MnO | 0.03 | 0.001 | — | — | — | — | 0.11 | 0.006 | — | — | |
| $Y_2O_3$ | 2.6 | 0.084 | 7.3 | 0.254 | 4.0 | 0.128 | 3.06 | 0.100 | 7.80 | 0.289 | |
| $La_2O_3$ | — | — | — | — | — | — | 0.09 | 0.002 | 0.19 | 0.005 | |
| $Ce_2O_3$ | 0.48 | 0.011 | 2.11 | 0.051 | 0.33 | 0.007 | 0.76 | 0.017 | 1.64 | 0.042 | |
| $Pr_2O_3$ | — | — | — | — | — | — | 0.07 | 0.002 | 0.55 | 0.014 | |
| $Nd_2O_3$ | 0.56 | 0.011 | 3.3 | 0.078 | 0.63 | 0.015 | 0.79 | 0.018 | 3.08 | 0.077 | |
| $Sm_2O_3$ | 0.22 | 0.004 | — | — | — | — | 0.36 | 0.008 | 1.74 | 0.042 | |
| †$Eu_2O_3$ | — | — | — | — | — | — | <0.01 | 0.000 | 0.13 | 0.003 | |
| $Gd_2O_3$ | 0.45 | 0.007 | 1.92 | 0.043 | 0.42 | 0.007 | 0.25 | 0.005 | 2.00 | 0.046 | B |
| $Tb_2O_3$ | — | — | — | — | — | — | 0.13 | 0.003 | 0.37 | 0.009 | |
| $Dy_2O_3$ | — | — | 1.61 | 0.035 | 0.67 | 0.015 | 0.79 | 0.016 | 2.64 | 0.059 | |
| $Ho_2O_3$ | — | — | — | — | — | — | <0.01 | 0.000 | 0.37 | 0.008 | |
| $Er_2O_3$ | — | — | — | — | — | — | 0.72 | 0.014 | 1.43 | 0.031 | |
| $Tm_2O_3$ | — | — | — | — | — | — | <0.01 | 0.000 | 0.01 | 0.000 | |
| $Yb_2O_3$ | 0.43 | 0.007 | — | — | — | — | 0.67 | 0.013 | 0.54 | 0.012 | |
| $Lu_2O_3$ | — | — | — | — | — | — | 0.21 | 0.004 | 0.33 | 0.007 | |
| $ZrO_2$ | 45.4 | 1.348 | 40.7 | 1.289 | 40.2 | 1.190 | 37.21 | 1.119 | 33.60 | 1.140 | |
| $HfO_2$ | — | — | 0.47 | 0.008 | 0.28 | 0.004 | 0.72 | 0.013 | 1.11 | 0.022 | |
| $UO_2$ | — | — | 0.21 | 0.004 | 0.22 | 0.004 | 0.30 | 0.004 | 0.40 | 0.006 | C |
| $ThO_2$ | 0.83 | 0.011 | 0.46 | 0.008 | 0.44 | 0.007 | 1.19 | 0.017 | 0.35 | 0.006 | |
| $PbO_2$ | — | — | 0.23 | 0.004 | 0.48 | 0.007 | 0.52 | 0.009 | 0.45 | 0.008 | |

| | | | | | | | | | | | |
|---|---|---|---|---|---|---|---|---|---|---|---|
| $TiO_2$ | 28.3 | 1.297 | 26.9 | 1.316 | 34.6 | 1.580 | 32.61 | 1.512 | 25.48 | 1.333 | D |
| †$SiO_2$ | 1.9 | 0.117 | 1.08 | 0.070 | 0.27 | 0.015 | 1.74 | 0.108 | — | — | D |
| †$Nb_2O_5$ | 0.46 | 0.011 | 0.40 | 0.012 | 0.63 | 0.018 | 1.85 | 0.052 | 2.75 | 0.086 | D |
| †$Ta_2O_5$ | — | — | — | — | — | — | 0.12 | 0.002 | 0.19 | 0.004 | D |
| $Al_2O_3$ | 1.6 | 0.114 | 1.14 | 0.086 | 1.34 | 0.095 | 1.07 | 0.078 | 0.48 | 0.040 | D |
| $Cr_2O_3$ | 0.59 | 0.029 | 0.44 | 0.023 | 0.52 | 0.026 | 0.56 | 0.027 | 0.07 | 0.004 | D |
| Total | 99.22* | 4.006* | 99.95 | 4.008 | 100.25 | 4.041 | 99.81 | 3.999 | 99.40 | 4.016 | |

† Significant Si unlikely in structure; Eu probably present in divalent state;
Nb and Ta may be in quadrivalent state.
a = weight precent of oxide; b = cation calculated for 7 oxygens.

| | | | | | | |
|---|---|---|---|---|---|---|
| $M_I^{2+}$ | A | 0.995 | 0.727 | 0.923 | 0.856 | 0.723 |
| | B | 0.045 | 0.273 | 0.077 | 0.144 | 0.277 |
| | | 1.000 | 1.000 | 1.000 | 1.000 | 1.000 |
| $M_{II}^{4+}$ | C | 0.921 | 0.812 | 0.905 | 0.942 | 0.663 |
| | B | 0.079 | 0.188 | 0.095 | 0.058 | 0.367 |
| | | 1.000 | 1.000 | 1.000 | 1.000 | 1.000 |
| $M_{III}^{4+}$ | D | 1.568 | 1.507 | 1.734 | 1.779 | 1.467 |
| | Zr | 0.438 | 0.501 | 0.307 | 0.220 | 0.540 |
| | | 2.006 | 2.008 | 2.041 | 1.999 | 2.016 |
| Total cations | | 4.006 | 4.008 | 4.041 | 3.999 | 4.016 |

[1] Roedder and Weiblen, *Geochim. Cosmochim. Acta* **37**, 1036. From a troctolite clast in a breccia fragment from a soil sample. *Analysis includes $K_2O$ = 0.01% and K, calculated for 7 oxygens, = 0.001.
[2] Busche et al, *EPSL* **14**, 316. From an ophitic basalt.
[3] ———, *ibid.* From a soil sample.
[4] Wark et al, *Lun. Sci. IV*, *Abstr.*, 766. From a mare basalt.
[5] ———, *ibid.* From a mare basalt.

**Table 5-1 Analyses of Zirconolite** (*Continued*)

| | [6] 15538,4 | | [7] 15555/39 | | [8] 14305,77 | | [9] 12013 | [10] Terrestrial Zirconolite | | |
|---|---|---|---|---|---|---|---|---|---|---|
| | a | b | a | b | a | b | a | a | b | |
| CaO | 3.55 | 0.259 | 3.2 | 0.171 | 6.15 | 0.437 | 2.9 | 12.61 | 0.869 | A |
| MgO | 0.03 | 0.003 | 0.1 | 0.006 | 1.15 | 0.115 | — | 0.14 | 0.012 | A |
| FeO | 7.44 | 0.423 | 11.4 | 0.477 | 4.23 | 0.234 | 13.8 | 2.30 | 0.124 | A |
| MnO | 0.08 | 0.005 | 0.3 | 0.012 | — | — | — | — | — | A |
| $Y_2O_3$ | 10.53 | 0.381 | 10.4 | 0.276 | 7.70 | 0.270 | 8.9 | 0.28 | 0.008 | B |
| $La_2O_3$ | 0.29 | 0.007 | 0.6 | 0.012 | 0.08 | 0.002 | 0.2 | 0.19 | 0.004 | B |
| $Ce_2O_3$ | 1.63 | 0.041 | 1.9 | 0.036 | 0.84 | 0.020 | 1.6 | 1.57 | 0.039 | B |
| $Pr_2O_3$ | 0.43 | 0.011 | 0.7 | 0.012 | 0.32 | 0.008 | — | 0.28 | 0.008 | B |
| $Nd_2O_3$ | 2.13 | 0.052 | 3.3 | 0.060 | 1.04 | 0.024 | 0.9 | 1.52 | 0.035 | B |
| $Sm_2O_3$ | 1.08 | 0.025 | 1.7 | 0.030 | 0.55 | 0.012 | — | 0.49 | 0.012 | B |
| †$Eu_2O_3$ | 0.09 | 0.002 | 0.4 | 0.006 | <0.01 | 0.000 | — | — | — | B |
| $Gd_2O_3$ | 1.45 | 0.033 | 2.1 | 0.036 | 0.33 | 0.008 | — | 0.18 | 0.004 | B |
| $Tb_2O_3$ | 0.42 | 0.009 | 0.3 | 0.006 | 0.24 | 0.004 | — | — | — | B |
| $Dy_2O_3$ | 2.09 | 0.046 | 0.9 | 0.015 | 1.59 | 0.036 | — | 0.22 | 0.004 | B |
| $Ho_2O_3$ | 0.38 | 0.008 | 0.2 | 0.003 | 0.03 | 0.001 | — | — | — | B |
| $Er_2O_3$ | 1.48 | 0.032 | — | — | 1.01 | 0.020 | — | 0.39 | 0.008 | B |
| $Tm_2O_3$ | 0.11 | 0.002 | — | — | 0.20 | 0.004 | — | — | — | B |
| $Yb_2O_3$ | 1.24 | 0.026 | — | — | 1.03 | 0.020 | — | — | — | B |
| $Lu_2O_3$ | 0.44 | 0.009 | — | — | 0.36 | 0.008 | — | 0.04 | 0.001 | B |
| $ZrO_2$ | 32.78 | 1.087 | 30.8 | 0.751 | 30.06 | 0.968 | 17.2 | 35.54 | 1.050 | C |
| $HfO_2$ | 0.89 | 0.017 | — | — | 0.86 | 0.016 | — | 0.55 | 0.012 | C |
| $UO_2$ | 0.14 | 0.022 | — | — | 1.16 | 0.016 | 3.6 | 0.70 | 0.001 | C |
| $ThO_2$ | 0.60 | 0.009 | — | — | 2.34 | 0.036 | 3.5 | 1.70 | 0.023 | C |
| $PbO_2$ | 0.22 | 0.004 | — | — | 2.19 | 0.040 | 4.2 | 0.15 | 0.004 | C |

| | | | | | | | | | | | |
|---|---|---|---|---|---|---|---|---|---|---|---|
| $TiO_2$ | | 27.48 | 1.406 | 27.1 | 1.018 | 29.62 | 1.472 | 22.1 | 31.57 | 1.525 | D |
| †$SiO_2$ | | 0.25 | 0.017 | — | — | 0.26 | 0.016 | 2.0 | — | — | D |
| †$Nb_2O_5$ | | 1.62 | 0.050 | — | — | 4.34 | 0.131 | 7.9 | 4.80 | 0.139 | D |
| †$Ta_2O_5$ | | 0.16 | 0.003 | — | — | 0.40 | 0.008 | — | 0.75 | 0.012 | D |
| $Al_2O_3$ | | 0.43 | 0.035 | 0.5 | 0.030 | 0.67 | 0.052 | — | 0.10 | 0.008 | D |
| $Cr_2O_3$ | | 0.24 | 0.013 | 0.5 | 0.021 | 0.50 | 0.028 | — | — | — | D |
| Total | | 99.70 | 4.017 | 96.4 | 4.165 | 99.25* | 4.006* | 91.0* | 96.85* | 4.037* | |
| $M_I^{2+}$ | A | | 0.690 | | 0.932 | | 0.786 | | | 1.005 | |
| | B | | 0.310 | | 0.068 | | 0.214 | | | −0.005 | |
| | | | 1.000 | | 1.000 | | 1.000 | | | 1.000 | |
| $M_{II}^{4+}$ | C | | 0.626 | | 0.380 | | 0.777 | | | 0.877 | |
| | B | | 0.374 | | 0.620 | | 0.223 | | | 0.123 | |
| | | | 1.000 | | 1.000 | | 1.000 | | | 1.000 | |
| $M_{III}^{4+}$ | D | | 1.524 | | 1.495 | | 1.707 | | | 1.824 | |
| | Zr | | 0.493 | | 0.670 | | 0.299 | | | 0.213 | |
| | | | 2.017 | | 2.165 | | 2.006 | | | 2.037 | |
| Total cations | | | 4.017 | | 4.165 | | 4.006 | | | 4.037 | |

[6] ———, *ibid.* From a mare basalt.
[7] Brown et al, *Suppl.* **3**, 151. From a basalt. Phase Y.
[8] Wark et al, *Lun. Sci. IV*, *Abstr.*, 766. From a recrystallized breccia. *Analysis recalculated from corrected total.
[9] Haines et al, *EPSL* **12**, 148. From a breccia. *Analysis includes other REE oxides = 2.2%. Total is too low to permit a significant calculation. Phase β.
[10] Wark et al, *Lun. Sci. Iv*, *Abstr.*, 766. From Arbarastkh Massif, Aldan, USSR. *Analysis, recalculated from corrected total, includes $Fe_2O_3$ = 2.78% and $Fe^{3+}$ calculated for 7 oxygens = 0.135.

**Table 5-2 X-Ray Data for Terrestrial Zirconolite from Arbarastkh Massif, Aldan, USSR**

*Cell Parameters*

$a_0$ = 12.343 Å
$b_0$ = 7.248 Å
$c_0$ = 11.470 Å

*Powder Data*

| *hkl* | Observed *d* (Å) | Calculated *d* (Å) | Observed *I* | Calculated *I** |
|---|---|---|---|---|
| 221 | 2.928 | 2.931 | VS | 1818 |
| $40\bar{2}$ | 2.899 | 2.899 | VS | 875 |
| 004 | 2.818 | 2.821 | W | 1069 |
| $22\bar{3}$ | 2.513 | 2.514 | M | 555 |
| 223 | 2.297 | 2.296 | W | 277 |
| $60\bar{4}$ } | 1.8025 | 1.8050 | M | 2 |
| 602 } | | 1.8047 | | 2 |
| 134 } | 1.7798 | 1.7839 | W | 139 |
| 315 } | | 1.7797 | | 101 |
| Not yet indexed | 1.7751 | | M | |
| 225 | 1.7507 | 1.7519 | M | 651 |
| $44\bar{2}$ | 1.5345 | 1.5366 | W | 443 |
| 800 } | | 1.5178 | | 88 |
| $24\bar{4}$ } | 1.5148† | 1.5138 | M | 108 |
| 623 } | | 1.5107 | | 201 |
| $22\bar{7}$ } | 1.4859 | 1.4883 | W | 309 |
| 406 } | | 1.4839 | | 133 |
| 442 | 1.4655 | 1.4657 | W | 329 |
| $80\bar{4}$ | 1.4505 | 1.4496 | W | 42 |
| $44\bar{6}$ | 1.2567 | 1.2569 | W | 190 |
| $84\bar{2}$ | 1.1728 | 1.1719 | W | 39 |
| 247 | 1.1522 | 1.1522 | W | 6 |
| $80\bar{8}$ | 1.1404 | 1.1403 | W | 28 |

$CuK_\alpha$ radiation; $\lambda$ = 1.5418.

* Calculated intensities from a single crystal of synthetic zirconolite ($CaZr_{1.3}Ti_{1.7}O_7$). It was noted that intensities varied considerably with change in composition in the synthetic zirconolites, and close agreement was not expected.

† Broad line; $d$ = 1.517 to 1.512 Å.

Measurements made on a sample that had been heated to 800°C for several days in vacuum.

*References*

Data (obtained through kindness of J. F. Lovering, School of Geology, University of Melbourne, Australia) will appear in Reid et al, "Characterisation of zirconolite (versus zirkelite)," in preparation.

*References*

1. Wark et al, *Lun. Sci. IV*, *Abstr*, 764–766.
2. Haines et al, *EPSL* **12,** 148–150.
3. Lovering et al, *Suppl.* **3,** 289.
4. Meyer and Boctor, *Lun. Sci. V*, *Abstr.*, 513–514.

## Perovskite $CaTiO_3$ (tentative)

It has been suggested that a water-clear isotropic mineral from breccia 14315,9, found together with troilite, is possibly perovskite.[1] No supportive data have been given.

*Reference*

1. Ramdohr, *EPSL* **15,** 114.

## Unidentified Zr-Rich Phases

### Zirconium-Rich Mineral with Th

Numerous micron-sized, Zr-rich crystals from breccia 10056-46 were examined with the microprobe. Thorium may be concentrated in this mineral.[1] The phase may be zirconolite, but the data are insufficient for identification.

*Reference*

1. Richardson et al, *Suppl.* **1,** 765–766.

### Zr- and Ti-Rich Mineral

It was suggested that a Zr- and Ti-rich mineral from coarse-grained basalt 10024,23 might be $ZrTiO_4$.[1] No supportive data were given.

*Reference*

1. Kushiro et al, *Sci.* **167,** 610.

### Zr Phase

In fines sample 10084, a Zr phase of grain size too fine (0.5 $\mu$) for analysis was regularly distributed in certain large pyroxene crystals or was concentrated in clusters of about 10 $\mu$ diameter near ilmenite. The grains, with U contents up to 400 ppm, may be zirconolite.[1]

*Reference*

1. Thiel et al, *EPSL* **16,** 32, 34.

## Armalcolite (Fe, Mg) $Ti_2O_5$

*Synonymy*

"Anosovite;" Ramdohr and El Goresy, *Sci.* **167,** 617.
Cr-Zr armalcolite; Steele and Smith, *Nature* **237,** 105.
Cr-Zr-Ca armalcolite; Haggerty, *Suppl.* **4,** 778.
Cr-Zr-Ca enriched armalcolite; Haggerty, *Lun. Sci. IV, Abstr.*, 329.
Cr-Zr-REE armalcolite; Dowty et al, *Lun. Sci. V, Abstr.*, 174.
Ferro-armalcolite; Anderson et al, *Suppl.* **1,** 55.
"Ferropseudobrookite"; Keil et al, *Sci.* **167,** 597.
Magnesian armalcolite; Anderson et al, *Suppl.* **1,** 55–56.
Magnesian-ferropseudobrookite; Cameron, *Suppl.* **1,** 221.
Orthoarmalcolite; Haggerty, *Suppl.* **4,** 778.
Para-armalcolite; Haggerty, *Suppl.* **4,** 778.
Phase X (Brown); Brown et al, *Lun. Sci. III, Rev. Abstr.*, 95.
Pseudo-armalcolite; Christophe-Michel-Lévy et al, *Lun. Sci. III, Rev. Abstr.*, 137.
Ti-Fe-Zr phase; Steele and Smith, *Nature* **237,** 106.
Zirconian armalcolite; Reid et al, *Geochim. Cosmochim. Acta* **37,** 1025.
Zr armalcolite; Haggerty, *Suppl.* **4,** 778.

Armalcolite, a new lunar mineral related to the terrestrial pseudobrookite series, was recognized simultaneously by a number of principal investigators. The name is derived from the initial letters of the three astronauts, *Arm*strong, *Al*drin, and Michael *Col*lins, who collected the Apollo 11 material and brought it back to Earth.

*Occurrence and Form*

In the Apollo 11 samples armalcolite was observed mostly in the fine-grained basalts coexisting with native iron.[1] Its usual occurrence, as minute grains that are cores in ilmenite (Figure 5–2), suggests that the armalcolite was primary and one of the first minerals to crystallize.[2] Early formed, discrete unreacted crystals, or grains mantled by Mg-ilmenite as a result of reaction with liquid at high temperatures, have been termed orthoarmal-

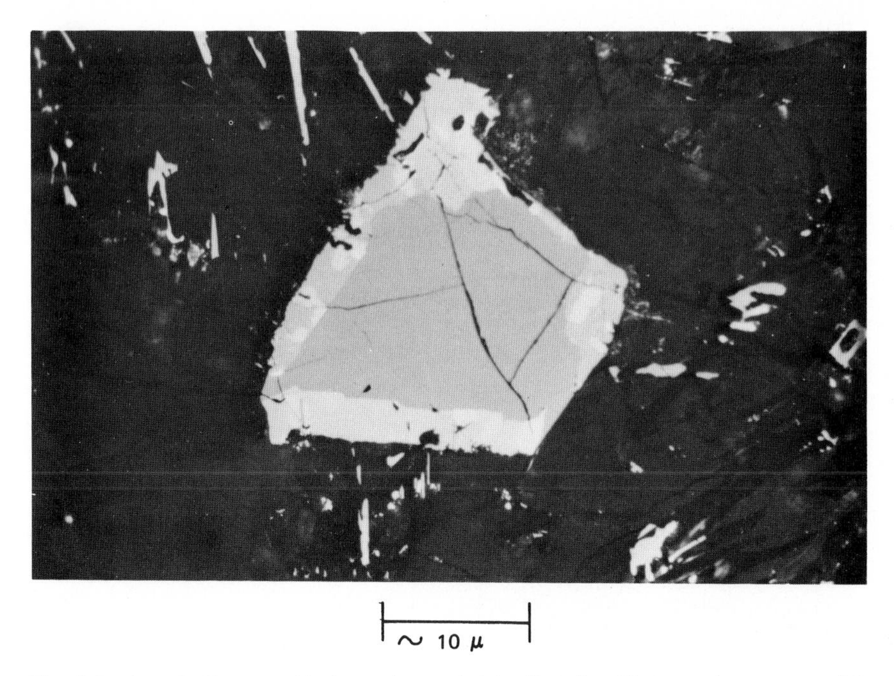


Fig. 5–2 Armalcolite core (dark gray) mantled by ilmenite. Photograph courtesy of S. Haggerty, Department of Geology, University of Massachusetts, Amherst.

colite.[3] The armalcolite reacts rapidly to ilmenite on cooling, and its preservation indicates that the rocks were quenched very quickly.[4] All stages of replacement of armalcolite cores by rims of ilmenite (Figure 5–3) can be seen, and the final product is a pseudomorph of ilmenite after armalcolite (Figures 5–4 to 5–6).[5] Another form, called para-armalcolite, also occurs as discrete crystals, or mantles early formed ilmenite. It has been suggested that para-armalcolite is a new polymorphic form of (Fe, Mg) $Ti_2O_5$, or that the two-phase assemblage—ilmenite + armalcolite—resulted from decomposition of a preexisting mineral intermediate between that of the armalcolite solid-solution series and the ilmenite-geikielite solid-solution series.[3] Alternatively, however, it is believed that the terms "ortho" and "para" refer simply to armalcolites that have occurred at two different times in the paragenetic sequence, and any minor optical and/or chemical variations are regarded as continuous and probably resulting from fractional crystallization. From present data there appear to be no consistent differences between the two "types." Crystallization probably was rapid, and compositional differences tended to be governed by local inhomogeneities in the parent liquid rather than by a well-defined

Fig. 5–3 Armalcolite (dark gray cores) partly replaced by ilmenite (lighter gray). Oil immersion. X 725. From *Suppl.* **1,** 234. Photograph courtesy of E. N. Cameron, Department of Geology and Geophysics, University of Wisconsin, Madison.

fractionation.[6] Both ortho- and para-armalcolites occur chiefly in mare basalts.[3]

Another variety of armalcolite, Zr-armalcolite, is the dominant opaque oxide in two particles from soil sample 15102,12 and also is present as the major Zr-bearing oxide in feldspathic basalt 61156. This variety of armalcolite often occurs together with still another type, Cr-Zr-Ca-armalcolite (Figure 5–7).[3] The latter, first discovered in Apollo 14 samples, was given the tentative name of pseudo-armalcolite,[7] but this was not approved by the IMA Commission on Nomenclature.[8] This mineral, noted in fines sample 14003-47, was included in a very frothy and scoriaceous material. In thin section the Cr-Zr-Ca-armalcolite was about 30 $\mu$ across and surrounded by ilmenite.[7] Both Zr-armalcolite and Cr-Zr-Ca-armalcolite are found in highly feldspathic nonmare basalts.[3] In crushed dunite samples 72415,11 and 72415,12, rare grains of Cr-Zr-Ca-armalcolite occur together with Fe-metal, troilite, and whitlockite.[9] This association is unusual, since armalcolite is an early mineral, whereas whitlockite generally is formed at a late stage.

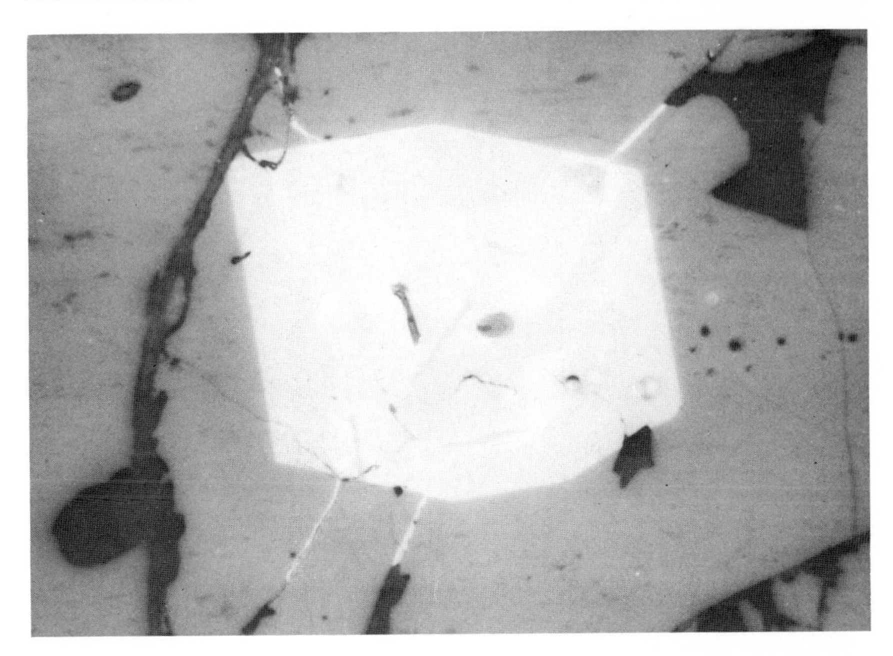

Fig. 5–4 Breakdown of armalcolite to ilmenite plus rutile, retaining original armalcolite morphology. Photography courtesy of R. Brett, U.S. Geological Survey, Reston, Va.

Armalcolite grains often occur in clusters; for the most part they are rectangular, with the longest dimension 100 to 300 $\mu$, and they are almost always homogeneous and free of exsolution, alteration, or zoning.[1] Tabular and skeletal crystals have been observed being replaced at the margins by magnesian ilmenite.[10] In coarse-grained basalt 10045-35, several crystals of magnesian ilmenite contain exsolution of bluish lamellae of armalcolite (Figure 5–8).[11]

The apparent stability of armalcolite with Mg-rich silicates suggests that it is the stable oxide phase in the Moon's mantle, if the latter consists of Mg-rich mafics. It is possible that the early crystallization of armalcolite served as an efficient means of concentrating Ti in the lunar mantle during an early lunar differentiation.[12]

*References*

1. Anderson et al, *Suppl.* **1,** 56.
2. Haggerty et al, *Suppl.* **1,** 521.
3. Haggerty, *Suppl.* **4,** 778–795.
4. Chao et al, *Suppl.* **1,** 292.

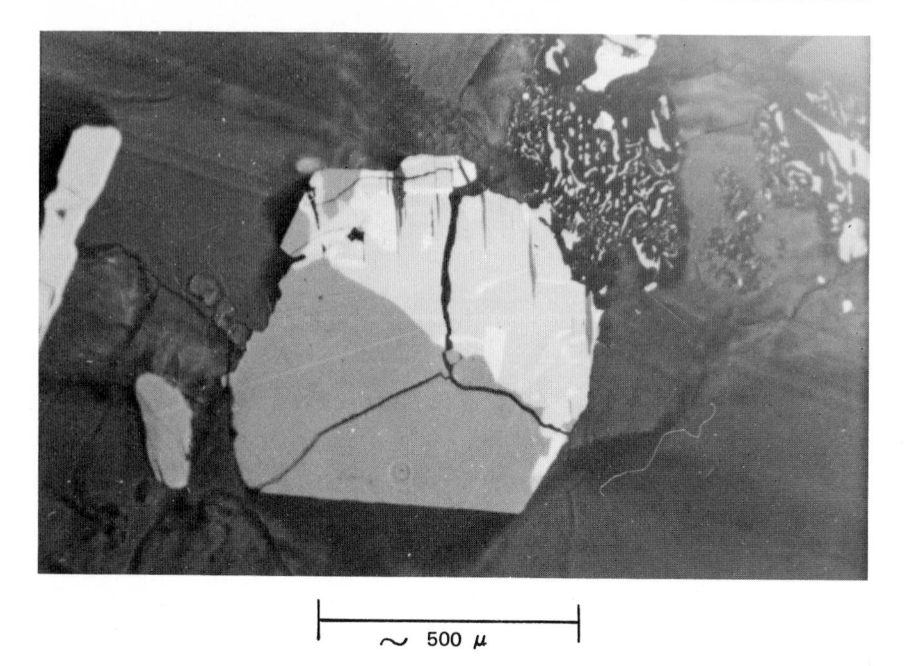


Fig. 5–5 In basalt 70035, a subhedral crystal of a possible pseudomorphic assemblage after a preexisting mineral. One half of the crystal is para-armalcolite (medium-gray) and the other half is ilmenite (light gray), with lamellae and blebs of rutile (white) and oriented rods of chromite (dark gray). From *Nature* **242,** 125. Photograph courtesy of S. Haggerty, Department of Geology, University of Massachusetts, Amherst.

5. Cameron, *Suppl.* **1,** 235.
6. Taylor and Williams, *Lun. Sci. V.*, *Abstr.*, 784.
7. Christophe-Michel-Lévy et al, *Lun. Sci. III*, *Rev. Abstr.*, 138.
8. Steele and Smith, *Nature* **237**, 106.
9. Albee et al, *Lun. Sci. V*, *Abstr.*, 3.
10. Agrell et al, *Suppl.* **1,** 112.
11. Brown et al, *Suppl.* **1,** 205.
12. Steel et al, *Lun. Sci. V*, *Abstr.*, 730.

### *Optics*

Armalcolite is opaque, blue-gray in reflected light, and distinctly anisotropic; pleochroism is pale gray to dark blue-gray.[1] Para-armalcolite is described as tan in reflected light.[2] In fines sample 14003-47, the Cr-Zr-Ca-armalcolite has a gray color less brownish than ilmenite and an estimated reflectance analogous to that of ilmenite.[3] From KREEP basalt 14310 the

Fig. 5–6 In microbreccia 14321-21, type 1 armalcolite is partially decomposed to rutile (white) and magnesian-ilmenite (medium gray). Larger subrounded crystals of chromite (black) are associated with the breakdown assemblage. From *Suppl.* **4,** 781. Photograph courtesy of S. Haggerty, Department of Geology, University of Massachusetts, Amherst.

Cr-Zr-Ca-armalcolite is dark brown,[4] and two grains of this same variety found in the Luna 20 soil are dark green.[5] The reflectivity and the microhardness (i.e., polishing hardness[6]) are less than those of ilmenite.[1]

## *References*

1. Anderson et al, *Suppl.* **1,** 56.
2. Haggerty, *Suppl.* **4,** 778.
3. Christophe-Michel-Lévy et al, *Lun. Sci. III*, *Rev. Abstr.*, 138.
4. Brown et al, *Suppl.* **3,** 149.
5. Reid et al, *Geochim. Cosmochim. Acta* **37,** 1025.
6. Cameron, *Suppl.* **1,** 235.

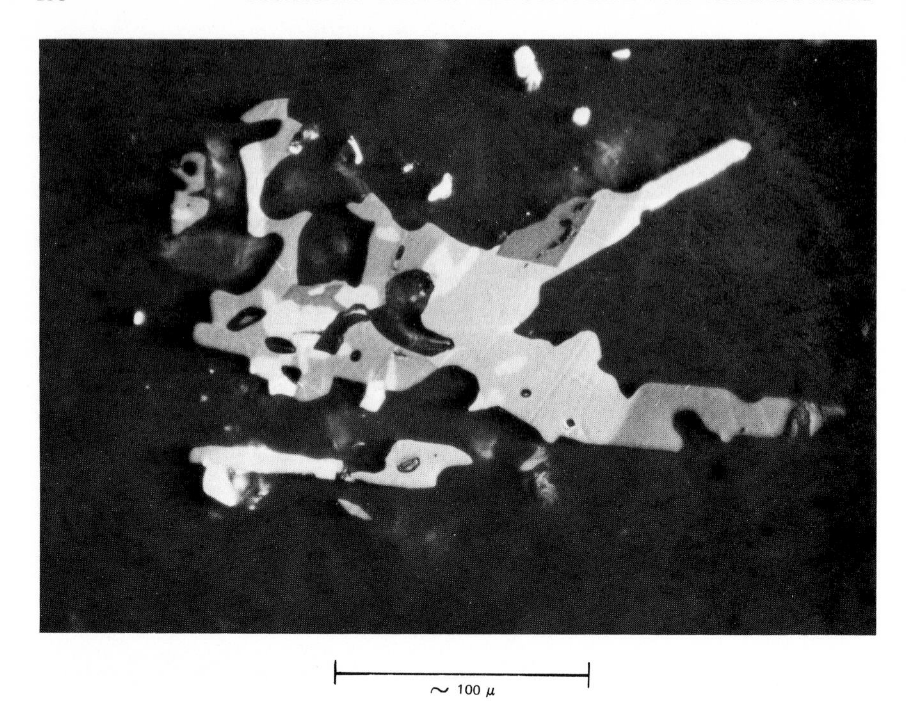


Fig. 5–7 Complex, irregular oxide assemblage of type 2 Cr-Zr-Ca armalcolite (medium gray), magnesian-ilmenite (light gray), chromite (dark gray), rutile (white), and bright flecks of metallic iron. From *Suppl.* **4,** 786. Photograph courtesy of S. Haggerty, Department of Geology, University of Massachusetts, Amherst.

*Chemical Composition*

Armalcolite may be regarded as the ferrous iron analog of terrestrial kennedyite ($Fe_2MgTi_3O_{10}$). Kennedyite is isostructural with pseudobrookite and contains 28.77% $Fe_2O_3$ and 60.33% $TiO_2$. Also the MgO content (6.45%) of kennedyite is within the range of armalcolite, but the latter is distinctive because of its higher $TiO_2$ content and lack of $Fe_2O_3$.[1] In breccia 10059, an early formed dark blue-grayish phase, surrounded by ilmenite, and with lamellae parallel to (0001) of ilmenite, was called possibly anosovite.[2] Anosovite ($Ti_3O_5$), however, is found terrestrially only in titaniferous blast furnace slags and is not recognized as a mineral. The occurrence, color, and chemical composition of the so-called lunar anosovite indicate that it is armalcolite.

Compositionally there are three types of armalcolite now recognized in the lunar samples. Type 1 includes both ortho- and para-armalcolite discussed previously. This group is made up of intermediate members in the

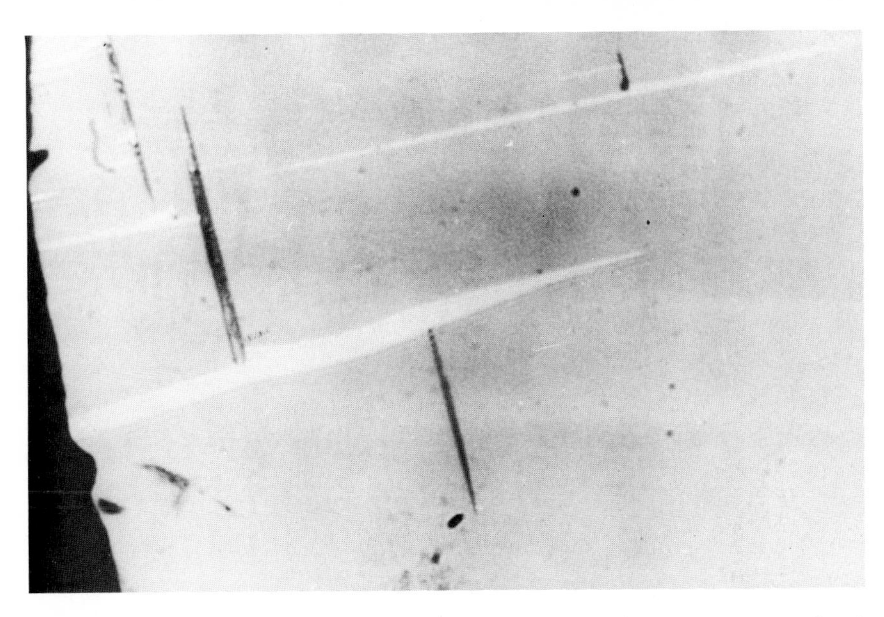

Fig. 5–8 Lamellae 5 $\mu$ thick of armalcolite in magnesium ilmenite. X-1350. From *Suppl.* **1,** 209. Photograph courtesy of M. Brown, Department of Geological Sciences, Durham University, England.

solid-solution series $FeTi_2O_5$-$MgTi_2O_5$.[3] In most of the type 1 armalcolite, the range of FeO is from approximatley 11 to 18%, the range of MgO is from approximately 5 to 8%; $ZrO_2$ is lacking or present only in trace amounts, and CaO has a range of from 0.00 to about 0.4%. With increase in FeO (as much as 23.5% FeO has been reported: Table 5–3; [1]) there is a decrease in MgO, and this variety has been called ferroarmalcolite. A magnesian armalcolite (MgO = 13.1%) has been found in an Apollo 15 breccia (Table 5-3; [6]).[4] Zoning is rare in armalcolite, but in microgabbro 10072-49 an elongated inclusion of armalcolite in ilmenite, partly in contact with clinopyroxene, is zoned, with the core containing more Mg and Cr and less Fe than the margin (Table 5–3; [3], [4]).[5]

Type 2 armalcolite is the Cr-Zr-Ca-armalcolite already mentioned. It has high $ZrO_2$ ($\sim$ 4 to 6.5%) and $Cr_2O_3$ (up to 11.6%) contents, with CaO present in a range of approximately 3 to 4% (Table 5-3; [9]–[13]). The Zr, Cr, and Ca replace the Ti, Fe, and Mg as follows:[3]

$$Ca^{2+} \text{ for } (Fe, Mg)^{2+}$$

$$Zr^{4+} \text{ for } Ti^{4+}$$

$$2Cr^{3+} \text{ for } (Fe, Mg)^{2+} + Ti^{4+}$$

**Table 5-3 Analyses of Armalcolite and Phase $Z_1$**

| | [1] | [2] | [3] | [4] | [5] | [6] | [7] |
|---|---|---|---|---|---|---|---|
| | | | 10072–49 | | | | |
| | 10045 | Luna 20 22003,2 | Rim | Core | 10065, 15–1a | 15445, 10 | 61156,5 |
| $SiO_2$ | 0.16 | 0.47 | 0.08 | 0.09 | 0.20 | — | 0.23 |
| $Al_2O_3$ | 3.05 | 0.70 | 1.53 | 1.62 | 2.14 | — | 0.94 |
| $Cr_2O_3$ | 1.27 | 1.68 | 1.40 | 1.94 | 2.43 | 0.5 | 1.49 |
| $TiO_2$ | 66.1 | 69.6 | 73.0 | 74.0 | 75.31 | 76.9 | 71.84 |
| FeO | 23.5 | 20.2 | 19.5 | 16.2 | 12.75 | 9.6 | 14.08 |
| MgO | 4.70 | 6.0 | 5.07 | 6.84 | 7.89 | 13.1 | 8.80 |
| MnO | 0.34 | — | 0.13 | 0.11 | 0.11 | — | 0.08 |
| CaO | — | 0.41 | 0.03 | 0.02 | 0.14 | <0.1 | 0.33 |
| $ZrO_2$ | — | 0.19 | — | — | — | <0.1 | 2.76 |
| Total | 99.12 | 99.33* | 100.74 | 100.82 | 100.97 | 100.1* | 100.75* |

*References*

[1] Brown et al, *Suppl.* **1,** 202. From a fine-grained basalt. Ferroarmalcolite.

[2] Brett et al, *Geochim. Cosmochim. Acta* **37,** 769. From a coarse-grained soil sample. *Analysis includes $Ce_2O_3$ = 0.08%. Ferroarmalcolite.

[3] Kushiro and Nakamura, *Suppl.* **1,** 619. From a fine-grained basalt. Zoned armalcolite crystal: Ferroarmalcolite rim.

[4] ———, *ibid.* From a fine-grained basalt. Zoned armalcolite crystal: Armalcolite core.

[5] Agrell et al, *Suppl.* **1,** 111. From a breccia. Armalcolite.

[6] Anderson, *J. Geol.* **81,** 221. From a breccia. *Analysis includes $Y_2O_3$ = <0.01%. Magnesian armalcolite.

[7] Haggerty, *Suppl.* **4,** 789. From a plagioclase-rich clast. *Analysis includes $Nb_2O_5$ = 0.20% and $Y_2O_3$ = <0.01%. Zr-armalcolite.

This seems to suggest cross-coupling substitution in several sites, but the actual substitutional mechanism is not yet understood.

Type 3, Zr-armalcolite, is compositionally intermediate between types 1 and 2. In Zr-armalcolite the CaO content is low (< 1%), and $Cr_2O_3$ also is low (~ 1.5%), but $ZrO_2$ ranges from approximately 2 to 4% (Table 5-3; [7], [8]). Most armalcolites (whether they be type 1, 2, or 3) contain minor amounts of $Nb_2O_5$, $Y_2O_3$ and/or REE, ranging from a few tenths to approximately 1.5%.[3]

| | [8] | [9] | [10] | [11] | [12] | [13] | [14] |
|---|---|---|---|---|---|---|---|
| | | | | | | Luna 20 | |
| | 15102,12 | 60335 | 68415,137 | 15102 | 15445 | 22001,16 | 15102,12 |
| $SiO_2$ | 0.23 | 0.2 | — | 0.23 | — | 0.27 | 0.24 |
| $Al_2O_3$ | 0.97 | 1.2 | 1.89 | 1.49 | 1.6 | 1.48 | 0.15 |
| $Cr_2O_3$ | 1.49 | 3.8 | 3.41 | 10.31 | 11.3 | 7.67 | 0.29 |
| $TiO_2$ | 68.16 | 71.1 | 70.18 | 66.52 | 64.2 | 65.42 | 42.26 |
| FeO | 17.33 | 11.9 | 13.9 | 9.33 | 8.5 | 10.66 | 22.02 |
| MgO | 6.78 | 1.7 | 0.97 | 2.31 | 2.1 | 1.98 | 4.33 |
| MnO | 0.02 | 0.3 | 0.23 | 0.13 | — | 0.10 | 0.30 |
| CaO | 0.35 | 3.1 | 4.31 | 3.40 | 4.0 | 3.40 | 0.32 |
| $ZrO_2$ | 3.92 | 4.8 | not analyzed | 6.01 | 6.4 | 6.55 | 30.17 |
| Total | 100.36* | 99.6* | 94.89* | 100.10* | 98.1 | 97.53 | 100.33* |

*References*

[8] ———, *ibid.* From a soil sample. *Analysis includes $Nb_2O_5$ = 0.58% and $Y_2O_3$ = 0.53. Zr. armalcolite.

[9] Brown et al, *Suppl.* **4,** 508. From a plagioclase-rich troctolite. *Analysis includes $RE_2O_3$ = 1.5% (chiefly as $Ce_2O_3$). Cr-Zr-Ca armalcolite.

[10] Helz and Appleman, *Suppl.* **4,** 654. From an anorthositic gabbro. *Low summation due to lack of analysis for $ZrO_2$. If $ZrO_2$ content is assumed to be between 5 and 6%, this is Cr-Zr-Ca armalcolite.

[11] Haggerty, *Apollo* 15 *Lunar Samples*, 91. From a soil sample. *Analysis includes $Nb_2O_5$ = 0.37%. Cr-Zr-Ca armalcolite.

[12] Ridley et al, *J. Geol.* **81,** 623. From a breccia Cr-Zr-Ca-armalcolite.

[13] Reid et al, *Geochim. Cosmochim. Acta* **37,** 1015. From a fine-grained soil sample. Cr-Zr-Ca armalcolite.

[14] Haggerty, *Apollo* 15 *Lunar Samples*, 91. From a soil sample. *Analysis includes $Nb_2O_5$ = 0.25%. Phase $Z_1$.

Type 1 is part of a simple oxide assemblage (i.e., armalcolite–ilmenite). Types 2 or 3, on the other hand, occur either as discrete grains or in intimate contact with a complex mineral assemblage including ilmenite, baddeleyite, zirconolite, rutile, chromite, troilite, and metallic iron. In feldspathic basalt 61156,5, the complex nature of the oxide intergrowths is strongly suggestive of partial or complete decomposition of either the Cr-Zr-Ca-armalcolite or a preexisting mineral or mineral assemblage. In soil sample 15102, the presence of an additional unidentified phase (referred to as phase

**Table 5-4 X-Ray Data for Synthetic and Lunar Armalcolite and Terrestrial Kennedyite Obtained Using Fe Radiation and Mn Filter**

| | [1] | | | [2] | | | [3] | | |
|---|---|---|---|---|---|---|---|---|---|
| *hkl* | $d_{obs}$ | $d_{calc}$ | *I* | $d_{obs}$ | $d_{calc}$ | *I* | *hkl* | *d* | *I* |
| 020 | 5.019 | 5.024 | 40 | | | | | | |
| 200 | 4.879 | 4.876 | 80 | | | | 200 | 4.88 | 80 |
| 101, 220 | 3.493 | 3.499 | 100 | 3.468 | 3.483 | 100 | 101, 220 | 3.485 | 100 |
| | | | | | | | 121 | 2.855 | 20 |
| 230 | 2.762 | 2.761 | 80 | 2.763 | 2.755 | 25 | 230 | 2.743 | 80 |
| 301 | 2.452 | 2.452 | 10 | 2.454 | 2.452 | 25 | 301 | 2.450 | 60 |
| 400 | 2.438 | 2.438 | 5 | | | | | | |
| 131 | 2.415 | 2.416 | 10 | 2.414 | 2.428 | 10 | 131 | 2.403 | 40 |
| 240 | 2.233 | 2.233 | 15 | 2.235 | 2.228 | 15 | 240 | 2.217 | 40 |
| 420 | 2.194 | 2.193 | 4 | 2.199 | 2.191 | 15 | 321 | 2.195 | 40 |
| 430 | 1.972 | 1.971 | 17 | 1.958 | 1.968 | 80 | 331 | 1.970 | 60 |
| | | | | | | | 002 | 1.865 | 80 |
| 250 | 1.858 | 1.858 | 8 | | | | 250 | 1.843 | 40 |
| 341 | 1.755 | 1.755 | 8 | 1.751 | 1.752 | 10 | 022 | 1.746 | 40 |
| 060 | 1.675 | 1.675 | 10 | 1.669 | 1.669 | 10 | 060 | 1.661 | 40 |
| 521 | 1.634 | 1.635 | 28 | 1.632 | 1.634 | 10 | | | |
| 600 | 1.625 | 1.625 | 13 | | | | 600 | 1.631 | 40 |
| $a_0$ | 9.752 ±0.003 Å | | | 9.743 ±0.03 Å | | | | 9.77 Å | |
| $b_0$ | 10.048 ±0.003 Å | | | 10.024 ±0.02 Å | | | | 0.95 Å | |
| $c_0$ | 3.736 ±0.004 Å | | | 3.738 ±0.03 Å | | | | 3.77 Å | |

*References*


[1] Anderson et al, *Suppl.* **1,** 61. Synthetic armalcolite.

[2] ———, *ibid.* Lunar armalcolite.

[3] Von Knorring and Cox, *Mineral. Mag.* **32,** 680. Terrestrial kennedyite.

$Z_1$—Table 5-3; [14]) leads to further speculation: phase $Z_1$ may be the last unreacted remanent in the decomposition assemblage, or there may be some special significance in the observed trend from type 1 armalcolite through intermediate Zr-armalcolite to Cr-Zr-Ca-armalcolite through phase $Z_1$ to zirconolite. If the Cr-Zr-Ca-armalcolite is related compositionally to zirconolite, the practice of using the term armalcolite to describe intermediate members may need revision.[3] Further research, with structural study, is needed.

### *X-Ray Data*

A comparison of the *d* spacings and cell constants of armalcolite and kennedyite shows their structural relationship (Table 5-4). The theoretical density of armalcolite is 4.94. The value was determined for synthetic armalcolite ($Fe_{0.5}Mg_{0.5}Ti_2O_5$) and the calculation was based on the unit cell volume.[1]

### *References*

1. Anderson et al, *Suppl.* **1,** 61.
2. Ramdohr and El Goresy, *Sci.* **167,** 617.
3. Haggerty, *Suppl.* **4,** 778–795.
4. Anderson, *J. Geol.* **81,** 219, 221.
5. Kushiro and Nakamura, *Suppl.* **1,** 619–621.

# CHAPTER SIX | CARBONATES AND PHOSPHATES

## Carbonates: Aragonite (Tentative) and Calcite (Doubtful)

### ARAGONITE $CaCO_3$ (tentative)

*Occurrence and Form*

A lathlike fragment[1] was found in the container that held crystals of plagioclase and other minerals. It is believed the fragment fell out of a vug in medium-grained basalt 10058.[2] The fragment may be a contaminant, or it may have come from a carbonaceous chondrite.[1]

*Optics*

The fragment is pink.[1]

*X-Ray Data*

X-Ray investigation revealed a randomly oriented aggregate of aragonite crystallites.[1]

*References*

1. Gay et al, *Suppl.* **1,** 484.
2. Agrell et al, *Suppl.* **1,** 97.

### CALCITE $CaCO_3$ (doubtful)

Calcite has been reported to occur in Apollo 11 soil.[1] No supportive data were given. It is possible that the above-mentioned aragonite was intended.

*References*

1. Agrell, *Nature* **225,** 325.

## Unknown Carbonate

From soil sample 66081, one fragment is possibly a carbonate with CaO = 57%. It is noted that $CaCO_3$ has 56% CaO.[1]

### *Reference*

1. Taylor and Carter, *Suppl.* **4,** 302.

## Phosphates: Apatite, Unidentified Apatite-like Phases, Whitlockite, Unidentified Whitlockite-like Phases, Monazite, and Tentative Farringtonite

Small grains of phosphate minerals are common in the lunar rocks and typically are associated with interstitial high-K phase and pyroxferroite.[1] Apatite and whitlockite have been identified positively as trace constituents. Several other calcium phosphate phases that may or may not be apatite or whitlockite also have been observed. One grain of monazite has been identified from an Apollo 11 coarse-grained basalt.[2] It is possible that a magnesium orthophosphate from spinel troctolite 65785 is farringtonite.[3] The presence of the phosphate graftonite is doubtful.

In Luny Rock I (10085-LR No. 1), of Apollo 11 the whitlockite is more abundant than the apatite, and both occur in elongate grains approximately 2 to 5 $\times$ 5 to 20$\mu$.[1] In some rocks, such as KREEP-rich rock 65015, apatite is lacking, although whitlockite is very abundant, and it is possible that the apatite was converted to whitlockite during metamorphic recrystallization.[4]

Although not abundant in the dark part of the rock, phosphates are present in all areas of 12013, which is a mixed light and dark breccia presumed to have been permeated by a once-fluid "granitic" component. In this breccia the phosphates occur in an intergrowth of a high-Ca phase (chlorofluorapatite) and a lower-Ca phase (whitlockite). The contact between the two minerals is sharp and fairly straight. None of the grains is large enough to determine whether one phosphate forms a core and the other a rim, but many grains show whitlockite partially enclosing apatite. This situation could be interpreted as whitlockite rimming an apatite core.[5] In basaltic rock 14276,[13] the apatite occurs in the mesostasis typically as needles 0.05 to 0.07 mm long, and electron beam scans indicate that distinct grains of the apatite may be included within the whitlockite.[6] Both apatite and whitlockite are late phases occurring with late-crystallizing ilmenite and fayalite.

*References*


1. Albee and Chodos, *Suppl.* **1,** 139–140.
2. Lovering et al, *EPSL* **21,** 164–168.
3. Dowty et al, *Lun. Sci. V, Abstr.*, 174.
4. Albee et al. *Lun. Sci. IV, Abstr.*, 24–26.
5. Drake et al, *EPSL* **9,** 106, 121.
6. Gancarz et al, *EPSL* **16,** 321.


## APATITE $(Ca, X)_5 ([P, Si] O_4)_3 (F, Cl)$

*Synonymy*

Chlorapatite; Keil et al, *Suppl.* **1,** 565.
Chlorofluorapatite; Albee and Chodos, *Suppl.* **1,** 135.
Chlorine-bearing fluorapatite; Keil et al, *Suppl.* **2,** 332.
Fluor-chlorapatite; Fuchs, *Suppl.* **1,** 477.
Fluorapatite; Fuchs, *Suppl.* **1,** 475–477.
Cerian-fluorapatite; Bunch et al, *Meteorit.* **7,** 246.
Ce-apatite; Ramdohr and El Goresy, *Naturwiss.* **57,** 102.
Mineral B (Ramdohr); Ramdohr and El Goresy, *Sci.* **167,** 617.
Phase D (in part); Lovering and Ware, *Suppl.* **2,** 156.
Ca-Fe-phosphate (in part); Jedwab, *Apollo 15 Lunar Samples*, 108.

*Occurrence and Form*

Apatite occurs as elongate grains,[1] as hexagonal prisms,[2] and as small $150 \times 10\ \mu$ needles.[3] In the granular interstitial areas of fine-grained basalt 10017[4] and medium-grained basalt 10058, small crystals of apatite are only a few microns in size and have been noted to occur together with brown needles of (posibly) rutile.[5] Apatite has been observed also as irregular patches of the fine vermicular blebs in the silicate minerals pyroxene, pyroxferroite (Figure 6–1), cristobalite, tridymite, and plagioclase.[6] In coarse-grained basalt 10047, together with cristobalite and tridymite, there are apatite needles up to 1.5 mm long.[7] Apatite is commonly associated with pyroxferroite, and in coarse-grained basalt 10044, fluorapatite inclusions in pyroxferroite constitute about 5% of the volume of the mineral. Here the apatite forms euhedral crystals up to $100 \times 25\ \mu$.[1] A cerian-fluorapatite is among the exotic late-stage minerals of microgabbro 12039.[8]

Of all the minerals found as crystals lining the vugs in highly metamorphosed or recrystallized Apollo 14 breccias, the apatite crystals are the most perfectly formed and are concentrated preferentially in the vugs relative to the matrix of the breccias (Figure 6–2). First order $\{10\bar{1}0\}$ and second order $\{11\bar{2}0\}$ prisms; first-, second-, and third-order hexagonal dipyramids

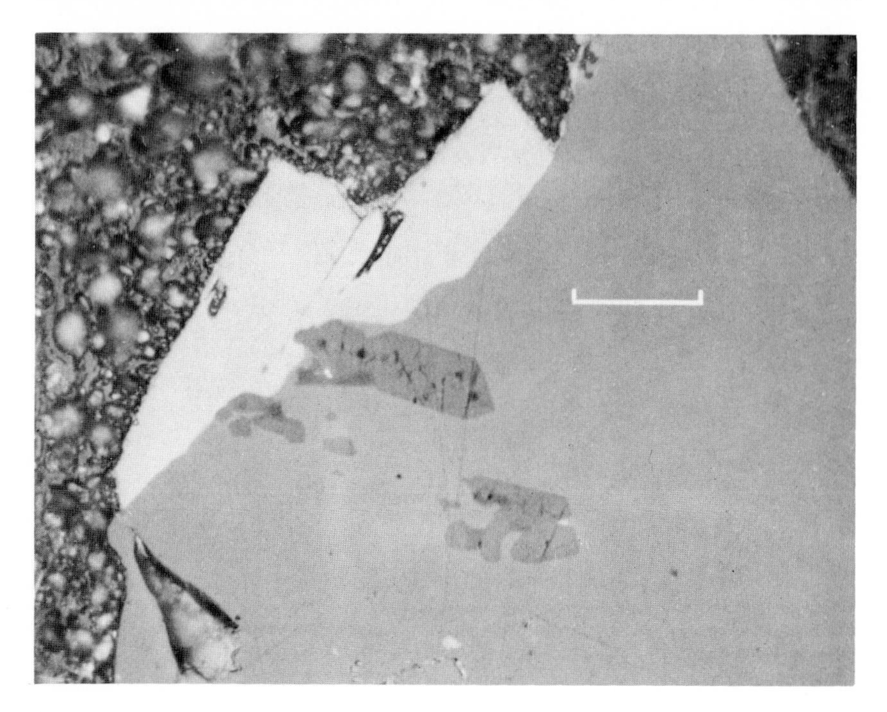

Fig. 6–1 Euhedral and subhedral fluorapatite crystals (dark gray, center of photograph) included in a pyroxferroite grain (light gray area). White grains are ilmenite. Mottled dark gray area is epoxy. Scale bar is 25 $\mu$. From *Suppl.* **1,** 476. Photograph courtesy of L. Fuchs, Argonne National Laboratory, Illinois.

$\{hh^{h}/_{2l}\}$; and basal pinacoids $\{0001\}$ are observed regularly on the crystals. These crystals appear to be late-forming and are found only on a substrate of plagioclase and pyroxene (Figure 6–3).[9]

*Optics*

The apatite is colorless, but in coarse-grained basalt 10044 a few purplish hexagonal crystals, found as inclusions in cristobalite, may be apatite; no supportive chemical data have been given, however.[10] The mean index of refraction is 1.644.[11]

*Chemical Composition*

Most lunar apatites are fluorapatites, although, in general, some Cl also is present, and in many cases there are sufficient amounts to justify the varietal name chlorofluorapatite (Table 6-1; [6], [9], [11]).[1] From fine-

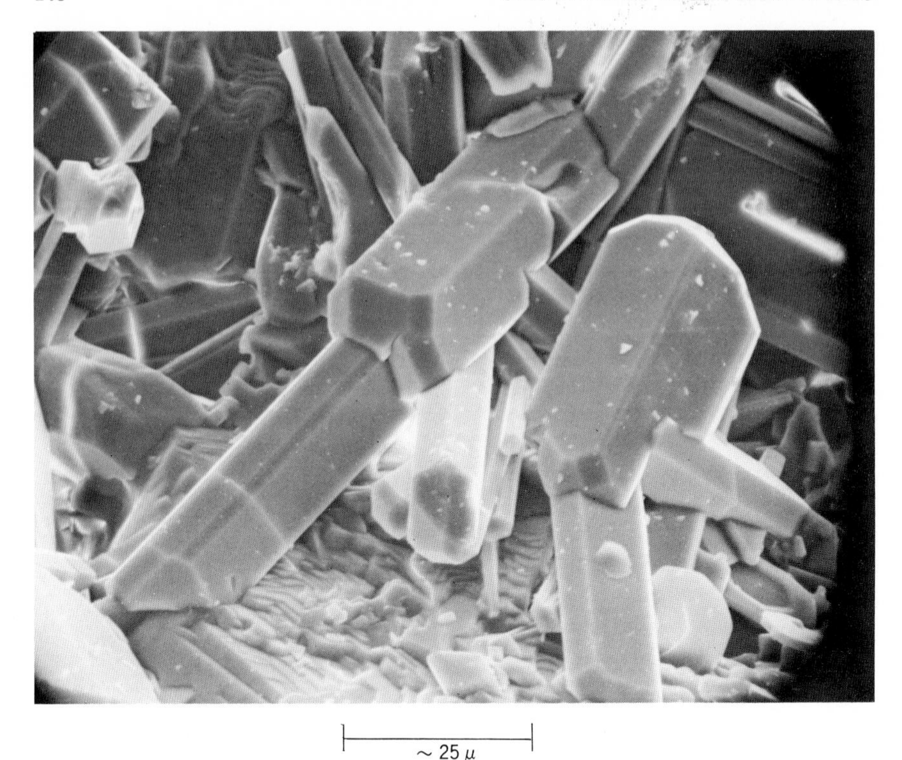


Fig. 6–2 SEM photograph of a nest of apatite crystals. NASA photograph S-72-17292, courtesy of U. Clanton, Johnson Space Center, Houston.

grained basalt 10045-29, one grain of apatite with a Cl content of 0.34% and no detected F has been referred to as chlorapatite.[12] Terrestrial chlorine-rich apatites are characteristic of high-temperature occurrences, and meteoritic apatites are exclusively chlorapatite. The latter, however, have a lower Si content than Apollo 11 apatites, and this may be related to the difference in REE content. Terrestrial fluorapatite from alkaline igneous rocks, pegmatites, and cavities in iron formations has high REE content, but in mafic igneous rocks the fluorapatites contain no REE.[1] The Si and Fe content of Apollo 11 apatite is higher than in typical terrestrial apatite.

Most lunar apatites contain both F and Cl in amounts high enough to indicate that hydroxyl is either absent or minor.[4] In the analyses of apatite from fines samples 10085-LR No. 1, 10085-1-11 and coarse-grained basalt 10044,30 (Table 6-1; [1], [4], [9]) the atomic proportions are normalized to 25 total cation charges (i.e., oxygen plus equivalent F and Cl = 12.5). This corresponds rather closely to $(Ca, X)_5$ $([P, Si]O_4)_3$ (F, Cl), and

Fig. 6–3 SEM photograph of doubly terminated apatite crystal in a breccia vug. Analysis showed apatite had 2% Cl. NASA photograph S-73-30448, courtesy of U. Clanton, Johnson Space Center, Houston.

hydroxyl is not required to fill the F—Cl position.[1] The 3.3 to 3.4% F in the apatites from 10044 (Table 6-1; [2], [3]) is close to the maximum of 3.5% F for pure fluorapatite. Hence hydroxyl content must be negligible if not nonexistent. Hydroxyl content cannot be determined with the microprobe, and the accuracy of the F content determination is not great enough to establish that insufficient F is present for pure fluorapatite, which would permit the assumption of hydroxyl presence. The high Y and REE contents are of interest (Table 6-1; [2]), for their general distribution in this lunar apatite is similar to some terrestrial varieties enriched in the lighter REE. The silica content indicates the substitution of $SiO_4$ groups for $PO_4$ groups in the apatite structure and is a consequence of the substitution of $REE^{3+}$ and $Y^{3+}$ for $Ca^{2+}$. Valence requirements are satisfied by the substitution of $Si^{4+}$ for $P^{5+}$ in the anion groups. Ideally the Si atoms should equal the

**Table 6-1 Apatite Analyses**

| | [1] | [2] | [3] | [4] | [5] | [6] | [7] |
|---|---|---|---|---|---|---|---|
| | 10085–1–11 | 10044 | | 10044–30 | Unit IV | 12013,10 | 12039,3 |
| $SiO_2$ | 1.9 | 2.3 | 2.2 | 1.2 | — | 1.38 | — |
| $Al_2O_3$ | <s* | — | — | <s | — | 0.04 | 0.49 |
| FeO | 1.7 | 1.5 | 1.5 | 1.9 | — | 0.35 | — |
| MgO | <s | — | — | <s | 0.24 | 0.11 | 0.05 |
| CaO | 51.7 | 52.1 | 52.5 | 52.2 | 53.96 | 52.14 | 52.80 |
| $Na_2O$ | <s | — | — | <s | 0.02 | 0.10 | 0.20 |
| $K_2O$ | <0.1 | — | — | <s | — | — | 0.05 |
| $P_2O_5$ | 38.4 | 38.7 | 39.0 | 39.3 | 39.30 | 39.40 | 39.60 |
| Cl | 0.1 | 0.03 | 0.03 | <0.1 | — | 1.38 | 0.30 |
| F | 3.3 | 3.3 | 3.3 | 3.4 | 2.95 | 2.59 | 2.50 |
| $Y_2O_3$ | 1.9 | 1.2 | — | 1.3 | 2.38 | 0.81 | — |
| $Ce_2O_3$ | — | 0.77 | — | — | — | 0.66 | 2.60 |
| $CeO_2$ | 0.5 | — | — | 0.4 | — | — | — |
| $La_2O_3$ | — | 0.21 | — | <s | — | 0.28 | 0.90 |
| $Nd_2O_3$ | ~0.5 | 0.56 | — | ~0.4 | — | — | 1.40 |
| Subtotal | 99.4 | 101.67* | 98.53 | 100.1 | 98.85 | 99.24 | 101.39* |
| F, Cl ≡ 0 | | 1.4 | 1.4 | | | | 1.13 |
| Total | | 100.27 | 97.13 | | | | 100.26 |

* <s = less than sensitivity of wave scan.

*References*

[1] Albee and Chodos, *Suppl.* **1,** 140. From a coarse-grained basalt fragment in coarse fines.

[2] Fuchs, *Suppl.* **1,** 477. From a coarse-grained basalt. Average of analyses on four grains. *Analysis includes $Pr_2O_3$ = 0.14%, $Sm_2O_3$ = 0.20%, and $Gd_2O_3$ = 0.66%.

[3] Fuchs, *Suppl.* **1,** 477. From a coarse-grained basalt. Average of analyses on three grains.

[4] Albee and Chodos, *Suppl.* **1,** 140. From a coarse-grained basalt.

[5] Sellers et al, *Suppl.* **2,** 671. From fines of Apollo 12 double-core tube.

[6] Lunatic Asylum, *EPSL* **9,** 142. From a variegated breccia. Chlorfluorapatite.

[7] Keil et al, *Suppl.* **2,** 332. From a fragment in loose fines. Averages of analyses on two grains. *Analysis includes $Pr_2O_3$ = 0.50%. Cerian-fluorapatite.

| | [8] | [9] | [10] | [11] | [12] | [13] |
|---|---|---|---|---|---|---|
| | | 10085–LR | | | | |
| | 14310 | No. 1 | 14001,7,3 | 12040 | Terrestrial | Apatites |
| $SiO_2$ | 1.22 | 0.93 | 0.40 | 0.21 | 0.20 | — |
| $Al_2O_3$ | — | 0.14 | <0.01 | — | — | 0.24 |
| FeO | 0.70 | 0.59 | 0.01 | — | 0.05 | 0.14 |
| MgO | 0.10 | 0.23 | <0.01 | 0.15 | 0.04 | — |
| CaO | 51.3 | 52.45 | 53.95 | 54.18 | 53.1 | 55.16 |
| $Na_2O$ | <0.01 | 0.14 | <0.01 | — | <0.01 | — |
| $K_2O$ | — | 0.15 | — | — | — | — |
| $P_2O_5$ | 39.8 | 40.0 | 40.56 | 42.17 | 41.4 | 41.30 |
| Cl | 0.29 | 1.14 | 0.04 | 1.94 | 1.60 | 0.09 |
| F | 3.35 | 3.13 | 3.96 | 2.70 | 0.29 | 3.67 |
| $Y_2O_3$ | 0.57 | 0.13 | 0.20 | 0.35 | 0.27 | — |
| $Ce_2O_3$ | 0.51 | — | <0.01 | 0.15 | 0.24 | — |
| $CeO_2$ | — | 0.05 | — | — | — | — |
| $La_2O_3$ | 0.31 | 0.09 | <0.01 | — | 0.01 | — |
| $Nd_2O_3$ | 0.52 | 0.20 | <0.01 | — | 0.14 | — |
| Subtotal | 99.10* | 99.49* | 99.12 | 101.85 | 97.53* | 102.14* |
| F, Cl ≡ 0 | 1.47 | | | 1.58 | 0.48 | 1.56 |
| Total | 97.63 | | | 100.27 | 97.05 | 100.58 |

*References*


[8] Griffin et al, *EPSL*, **15,** 55. From a KREEP basalt. Average of analyses on five grains. *Analysis includes MnO = <0.01%, $Pr_2O_3$ = 0.09%, $Sm_2O_3$ = 0.14%, and $H_2O$ (calculated only) = 0.08%.

[9] Albee and Chodos, *Suppl.* **1,** 140. From a gray mottled basalt fragment in lunar soil. Average of analyses on two grains. *Analysis includes $TiO_2$ = 0.10% and BaO = 0.02%. Chlorfluorapatite.

[10] Gancarz et al, *EPSL* **12,** 9. From a basalt fragment in coarse fines.

[11] Brown et al, *Suppl.* **2,** 595. From a picritic basalt. Chlorfluorapatite.

[12] Griffin et al, *EPSL* **15,** 55. From Ödegarden, Norway. *Analysis includes $Pr_2O_3$ = 0.04%, $Sm_2O_3$ = 0.04%, $Gd_2O_3$ = 0.08%, and $Dy_2O_3$ = 0.03%.

[13] Palache et al, *Dana VII* **2,** 883. From Faraday Township, Ontario. *Analysis includes $Fe_2O_3$ = 0.63%, insol. = 0.28%, $CO_2$ = 0.50%, $H_2O$ = 0.01% and MnO = 0.12%.

sum of the Y and REE atoms, but 50% more Si atoms are present than required.[11] In another analysis of a fluorapatite from the same rock sample (Table 6-1; [3]) the lack of REE may be the reason for the low summation. The cause for the low analytical total in fluorapatite from KREEP basalt 14310 (Table 6-1; [8]) is unknown. The structural formula (normalized to $P + Si = 2$) shows a deficiency in the *A* position, suggesting an error in the Ca position or the presence of an unanalyzed element.[13]

Many apatite grains contain U from 144 $\pm 22$ to 270 $\pm 30$ ppm.[14] Uranium in other apatites runs as follows:

from Apollo 14 breccias $U = 20$ to 700 ppm

from Apollo 14 mare-type basalts $U = 100$ to 600 ppm

from Apollo 14 nonmare-type basalts $U = 20$ to 200 ppm

from "granitic" fraction of breccia 12013 $U = \leq 1000$ ppm.[15]

### *X-Ray Data*

The X-ray powder pattern of the apatite from 10044 is weak, but it contains 12 of the strongest lines for fluorapatite, enough to distinguish it from chlorapatite and hydroxyl apatite.[11]

## Unidentified Apatite-Like Phases

### *Mineral B*

A phase rich in Ce and Ca, called mineral B,[16] is no doubt the Ce-apatite reported to occur with baddeleyite and dysanalyte. This apatite has indices of refraction and reflectivity close to those for pyroxene.[17] The apatite from fines sample 12070 (Table 6-1; [7]) could be considered a Ce-apatite[17] or a cerian-fluorapatite such as has been reported from microgabbro 12039.[8]

### *Phase D*

A phosphate phase, called phase D, characteristically occurs as euhedral crystals with roughly hexagonal cross sections, high relief, and low birefringence. The phase contains major Ca and P, with two distinct compositional variants, high REE and low REE. The latter, poor in Y and REE, contains Cl and may be apatite.[18]

*Ca-Fe-phosphate*

In basalt 15065 a high concentration of phosphates was found deposited on a free-growing ilmenite crystal. The phosphates consist of:

1. Hexagonal platelets, interfering with the growth of ilmenite lamellae
2. Rounded, partially faceted crystals forming linear rows
3. A continuous layer stratified between ilmenite and a late Ca-silicate.

Probe analyses indicate a Ca-Fe-phosphate containing about 5% Fe and traces of Y and Ce. The phase may be apatite or whitlockite. Other finds of Ca-rich, Fe-poor phosphates, but with different morphologies, suggest a rather diverse phosphate mineralogy for 15065.[19]

*References*


1. Albee and Chodos, *Suppl.* **1,** 139–141.
2. Bailey et al, *Sci.* **167,** 594.
3. Brown et al, *Suppl.* **1,** 204.
4. French et al, *Suppl.* **1,** 441.
5. Agrell et al, *Suppl.* **1,** 97.
6. James and Jackson, *J. Geophys. Res.* **75,** 5799.
7. Dence et al, *Suppl.* **1,** 324.
8. Bunch et al, *Meteorit.* **7,** 246, 249.
9. McKay et al, *Suppl.* **3,** 739–741.
10. Bailey et al, *Suppl.* **1,** 187.
11. Fuchs, *Suppl.* **1,** 475–477.
12. Keil et al, *Suppl.* **1,** 565, 587.
13. Griffin et al, *EPSL* **15,** 55–56.
14. Haines et al, *EPSL* **12**, 147.
15. Lovering et al, *Suppl.* **3,** 284, 287.
16. Ramdohr and El Goresy, *Sci.* **167**, 617.
17. Ramdohr and El Goresy, *Naturewiss.* **57,** 102.
18. Lovering and Wark, *Suppl.* **2,** 156–157.
19. Jedwab, *Apollo 15 Lunar Samples*, 108.


WHITLOCKITE $(Ca, x)_3 ([P, Si] O_4)_2$

*Synonymy*

Ca-Fe-phosphate (in part); Jedwab, *Apollo* **15** *Lunar Samples*, 108.
Ca, K phosphate phase; El Goresy et al, *Suppl.* **2,** 219.
Cerian-iron whitlockite; Griffin et al, *EPSL* **15,** 56.

"Ferro" whitlockite; Brown et al, *Suppl.* **3,** 145.
"Magnesio" whitlockite; Brown et al, *Suppl.* **3,** 145.
Phase D (in part); Lovering and Wark, *Suppl.* **2,** 156.
Phosphate mineral; El Goresy et al, *Suppl.* **2,** 232.
Rare-earth calcium phosphate mineral; Fuchs, *EPSL* **12,** 172–173.
Rare-earth-rich whitlockite; Papanastassiou et al, *EPSL* **8,** 4.
Sr-whitlockite; Brown et al, *Suppl.* **3,** 141.
Yttrian-cerian whitlockite; Busche et al; *Spec. Publ.* #**3,** 59.
Yttrian-iron whitlockite; Mason, *Mineral. Rec.* **2,** 278.
Yttrian-magnesian whitlockite; Mason, *Mineral. Rec.* **2,** 278.
Yttrian whitlockite; Griffin et al, *EPSL* **15,** 56.
Yttrium-bearing calcium phosphate; Reid et al, *Suppl.* **1,** 753.
Yttrium-bearing whitlockite; Kushiro et al, *Suppl.* **2,** 487.

### *Occurrence and Form*

In the lunar rocks, whitlockite is much more abundant than apatite and occurs typically in the mesostasis as needles 0.05 to 0.07 mm long. Distinct grains of apatite may be included in the whitlockite.[1] Crystals up to 0.2 mm long have been found in fines samples 14001,7,2.[2] Whitlockite is rare in Apollo 14 breccias, but it is one of the minerals found in vugs. It occurs as stubby, discoidal crystals with hexagonal outlines (Figure 6–4). A tabular rhombohedral habit is suggested, but crystals found to date have somewhat rounded faces that make indexing difficult.[3] In 65015, a Ba-KREEPUTH-rich rock, whitlockite is very abundant, forming slender needles up to 40 $\mu$ long, and the crystals are not restricted to the high-K areas in the rock. The lack of apatite in this sample is attributed to a loss of Cl and F during metamorphic recrystallization, with conversion of apatite to whitlockite.[4]

### *Chemical Composition*

The whitlockite chemical composition, normalized to a total cation charge of 16 (i.e., oxygen = 8.0), corresponds rather closely to $[Ca, X][(P, Si)O_4]_2$. In some analyses the substitution of $Na^{1+}$ for $Ca^{2+}$ and $Si^{4+}$ for $P^{5+}$ does not appear sufficient to balance the substitution of $(REE)^{3+}$ for $Ca^{2+}$. This may reflect analytical uncertainties in the Ca position.[5]

The whitlockites fall into two groups, REE-rich and REE-poor. It has been noted that in noritic fragments found in Apollo 14 soil, the REE-rich whitlockite had Nd concentrations between 1.5 and 2.5%, whereas the Nd content of REE-poor whitlockite was less than 0.4%.[6] The REE content of whitlockite may be as much as 10% (Table 6-2; [2], [3]) and is highly concentrated relative to the coexisting apatite.[7] In type A norite-anorthosite fragments from soil samples of Apollo 12, abundant whitlockite, making

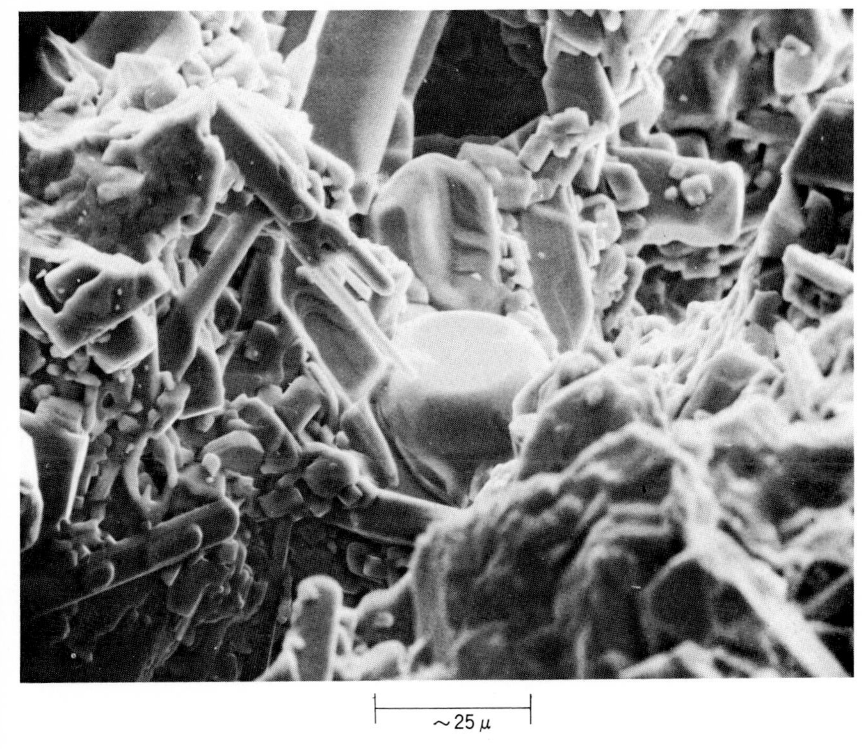


Fig. 6–4 SEM photogtaph of stubby discoidal whitlockite crystal (center of picture). From *Suppl.* **3,** 744. NASA photograph S-71-58457, courtesy of U. Clanton, Johnson Space Center, Houston.

up to 3% of the rock's volume,[8] contains 10% Y and REE oxides.[9] High-yttrian whitlockite may be restricted to lunar rocks.[7] Apparently REE have been detected only in trace amounts in terrestrial and meteoritic whitlockites.[5] An analysis of whitlockite from basalt 10069,26 is chemically more in accord with a terrestrial whitlockite from Ödegarden (Table 6-2; [10] and [11], respectively.) Both the Ödegarden and lunar whitlockite could be fitted, also, to a suggested formula of $Ca_9MgH(PO_4)_7$, but there is an apparent excess of both (Ca,REE) and (Mg,Fe) in the lunar whitlockite. The terrestrial whitlockite is a magnesium wihtlockite, but the lunar whitlockite may be regarded as an yttrian-iron whitlockite, or perhaps more appropriately, a "cerian-iron whitlockite."[7] So-called "magnesio" whitlockite (Table 6-2; [8]), occurring in breccia and fines samples of Apollo 14, possibly may be derived from the Mg-rich noritic crustal rocks.[10] "Ferro" whitlockite has been found in both KREEP

**Table 6-2 Analyses of Lunar Whitlockite and Terrestrial Whitlockite and Graftonite**

| | [1] 12032 | [2] 12013–10 | [3] 14310/20B | [4] 10085–LR No. 1 | [5] 14163 | [6] 12040–39 |
|---|---|---|---|---|---|---|
| F | — | — | — | 0.07 | — | — |
| Cl | — | — | — | 0.01 | — | — |
| $SiO_2$ | 3.9 | 0.20 | 0.39 | 0.31 | — | 0.46 |
| $TiO_2$ | — | — | — | 0.11 | 0.1 | — |
| $Al_2O_3$ | — | — | — | 0.10 | — | — |
| FeO | 4.8 | 1.38 | 3.35 | 0.90 | 0.95 | 2.30 |
| MgO | 0.63 | 2.45 | 1.88 | 3.38 | 3.2 | 3.32 |
| MnO | — | — | — | n.a. | 0.1 | — |
| $K_2O$ | — | — | — | 0.04 | 0.01 | — |
| $Na_2O$ | 0.02 | — | — | 0.60 | 0.45 | — |
| CaO | 33.5 | 38.61 | 38.68 | 40.9 | 41.0 | 42.26 |
| $P_2O_5$ | 36.5 | 44.45 | 43.22 | 44.1 | 42.6 | 44.27 |
| $Y_2O_3$ | — | 3.31 | 3.01 | 2.98 | 1.85 | 2.17 |
| $Ce_2O_3$ | — | 3.24* | 2.86 | 2.45 | 3.25 | 1.38 |
| $La_2O_3$ | — | 1.09 | 1.02 | 0.92 | 0.75 | — |
| $Nd_2O_3$ | — | 2.15 | 2.05 | 2.16 | 3.0 | 1.37 |
| $Pr_2O_3$ | — | 0.76 | 0.74 | — | — | 0.44 |
| $Sm_2O_3$ | — | 1.00 | 1.02 | — | — | 0.30 |
| $Gd_2O_3$ | — | 1.11 | 0.84 | — | 0.45 | 0.60 |
| $Tb_2O_3$ | — | 0.08 | 0.20 | — | — | 0.20 |
| $Dy_2O_3$ | — | 0.38 | — | — | — | 0.51 |
| $Ho_2O_3$ | — | 0.20 | — | — | — | 0.30 |
| $Eu_2O_3$ | — | 0.08 | — | — | — | — |
| Total | 99.05* | 100.49 | 99.26 | 99.08* | 97.75* | 100.08* |

*References*

[1] Mason, *Mineral. Rec.* **2,** 278. From a fines sample. *Analysis includes 19.7% $R_2O_3$ where $R = Y + \text{REE}$. Yttrian-iron whitlockite.

[2] Peckett, *Moon* **3,** 405, From a variegated breccia. *Ce reported as $CeO_2$.

[3] Brown et al, *Suppl.* **3,** 153. From a KREEP basalt. "Ferro"-whitlockite.

[4] Albee and Chodos, *Suppl.* **1,** 140. From a KREEP basalt fragment in soil sample. *Analysis includes BaO = 0.05%, $Cr_2O_3$ = 0.06%, and Gd, Dy, and Er ~ 0.04%.

[5] Fuchs, *EPSL* **12,** 172. From a fragment in soil sample. *Analysis includes $Cr_2O_3$ = 0.04%.

[6] Brown et al, *Suppl.* **3,** 153. From a picritic basalt. *Analysis includes $Er_2O_3$ = 0.20%.

| | [7]<br>12036/9 | [8]<br>14310/20A | [9]<br>15475/125 | [10]<br>10069,26 | [11]<br>Terrestrial<br>Whitlockite | [12]<br>Terrestrial<br>Graftonite |
|---|---|---|---|---|---|---|
| F | 0.05 | — | — | <s* | 0.29 | — |
| Cl | 0.02 | — | — | <s | <0.01 | — |
| $SiO_2$ | 0.54 | 0.21 | 0.60 | <s | <0.01 | — |
| $TiO_2$ | 0.04 | — | — | <s | — | — |
| $Al_2O_3$ | 0.25 | — | — | <s | — | 2.60 |
| FeO | 2.13 | 2.70 | 6.02 | 0.4 | 0.87 | 32.58 |
| MgO | 3.8 | 3.63 | 0.28 | 3.4 | 3.20 | 0.53 |
| MnO | — | — | 0.03 | <s | <0.01 | 15.65 |
| $K_2O$ | — | — | — | <0.1 | — | — |
| $Na_2O$ | 0.33 | — | — | 2.4 | 0.22 | — |
| CaO | 42.3 | 42.74 | 43.72 | 46.5 | 47.1 | 7.95 |
| $P_2O_5$ | 42.5 | 45.46 | 45.82 | 45.3 | 46.7 | 40.81 |
| $Y_2O_3$ | 2.28 | 1.64 | 0.79 | <0.1 | 0.02 | — |
| $Ce_2O_3$ | 1.69 | 1.56 | 0.20 | <s | 0.02 | — |
| $La_2O_3$ | 0.50 | 0.47 | 0.87 | <s | <0.01 | — |
| $Nd_2O_3$ | — | 1.25 | 0.30 | <s | 0.01 | — |
| $Pr_2O_3$ | 0.30 | 0.39 | 0.20 | — | <0.01 | — |
| $Sm_2O_3$ | — | 0.62 | 0.35 | — | <0.01 | — |
| $Gd_2O_3$ | 0.50 | — | — | — | <0.01 | — |
| $Tb_2O_3$ | — | — | — | — | — | — |
| $Dy_2O_3$ | 0.37 | — | — | — | <0.01 | — |
| $Ho_2O_3$ | 0.17 | — | — | — | — | — |
| $Eu_2O_3$ | — | 0.07 | — | — | — | — |
| Total | 98.05* | 100.74 | 100.19* | 98.0 | 98.61* | 100.12* |

* <s = below sensitivity of wavelength scan.

*References*


[7] Keil et al, *Suppl.* **2,** 332. From a coarse-grained basalt. Average of analyses on nine grains. *Analysis includes $Er_2O_3$ = 0.18% and $Lu_2O_3$ = 0.10%.

[8] Brown et al, *Suppl.* **3,** 153. From a KREEP basalt. "Magnesio" whitlockite.

[9] Brown et al, *Suppl.* **3,** 153. From a mare-type basalt. *Analysis includes SrO = 1.01%. Sr-whitlockite.

[10] Albee and Chodos, *Suppl.* **1,** 140. From a fine-grained vesicular basalt.

[11] Griffin et al, *EPSL* **15,** 55. From Ödegården, Norway. *Analysis includes $V_2O_5$ = 0.3%, giving a subtotal of 98.73%, and O ≡ F, Cl = 0.12%. In addition, calculated $H_2O$ = 0.84%.

[12] Palache et al, *Dana VII* **2,** 687. From Brissago, Switzerland. *Analysis includes $H_2O$ = Tr.

and mare basalts,[10] and whitlockite from fines sample 12032 (Table 6-2; [1]), with Y + REE = 19.7% and FeO = 4.8%, has been called an yttrian-iron whitlockite.[11] In a mare basalt 15475/125, whitlockite with an FeO content exceeding 6% (Table 6-2; [9]) has been called Sr-whitlockite, rather than ferro-whitlockite, because of its SrO content of 1.01%. This is the highest recorded concentration of strontium in any lunar mineral.[10]

It has been suggested that the Th and U found in lunar samples probably are concentrated in whitlockite and apatite. Although other minerals contain greater amounts of these elements, fission track work indicates the presence of U and Th in phosphate grains from fines sample 14001,7,3, KREEP basalt 14310, and basalt 14053,17.[2] Analyses of whitlockite show the following uranium contents:

from breccia 14305,77 U = 40 to 166 ppm

from mare basalt 14072 U = $\sim$ 50 ppm

from "granitic" fraction of 12013 U = $\leq$ 46 ppm.[12]

*X-Ray Data*

All known terrestrial and meteoritic whitlockites have X-ray powder patterns identical to synthetic $\beta$-$Ca_3(PO_4)_2$.[7] Data on whitlockite from breccia 12013,10 suggests $\beta$-$Ca_3(PO_4)_2$ (e.g., trigonal, $a$ = 10.4 Å) with a pseudo-halving of the $c$ axis repeat and a probable space group $R\bar{3}C$.[13] After 30 hours of exposure (Cu radiation and a 57 mm camera used), whitlockite from a fragment in soil sample 14163 gave an X-ray pattern of only the 10 strongest lines of whitlockite. The line intensities and $d$ spacings of the pattern were in essential agreement with meteoritic whitlockite and synthetic $\beta$-$Ca_3(PO_4)_2$. This whitlockite was intergrown with and less abundant than an amorphous (or metamict) rare-earth calcium phosphate with the same chemical composition (Table 6-2; [5]). It is not certain that the amorphous material is structurally akin to whitlockite.[14] It is possible that the lunar whitlockite originally formed in another crystallographic state [e.g., as $\alpha$-$Ca_3(PO_4)_2$] more favorable than either apatite or ordinary whitlockite to the incorporation of REE. In experimental systems the $\alpha \rightarrow \beta$ transition is very sluggish even at 1180°C, and it probably is still more sluggish where excess CaO (or REE) are present and the transition temperature is lower. In experimental systems this excess CaO may be removed during the inversion by reaction with water to give apatite, but excess water probably was not available in the dry lunar rocks. Moreover, the whitlockite from KREEP basalt 14310 shows an excess of (Ca,REE) and (Mg,Fe) in the formula, relative to the (P + Si) = 7, as would be predicted by this mechanism. Therefore, lunar whitlockite possibly could have crystallized (simultaneously with apatite) as the high-temperature $\alpha$ polymorph with an excess of REE, and it now exists as a disordered, poorly crystalline $\beta$ polymorph. This would explain the weak X-ray pattern, the presence of the amorphous phase of the same chemical composition, and a noted instability of the lunar whitlockite under the electron beam.[7]

There was a question of whether the high-REE phosphate phase from Apollo 11 samples was whitlockite or graftonite [$(Fe,Mn,Ca)_3(PO_4)_2$], since both minerals have a chemical composition closely corresponding to the $A_3(XO_4)_2$ group.[5] However, the chemical composition of terrestrial graftonite (Table 6-2; [12]) differs greatly from the lunar metarial, and subsequent X-ray investigation has verified the phosphate phase as whitlockite.

## Unidentified Whitlockite-Like Phases

### *Y-Bearing Ca Phosphate*

A Y-bearing Ca phosphate, occurring in the last liquid to crystallize and revealed by partial analysis, most probably is whitlockite.[15]

### *Ca, K Phosphate Phase*

A Ca,K phosphate phase, always associated with a fayalite–glass assemblage, was observed in fine-grained porphyritic basalts 12018 and 12063.[16] The mineral occurs in long crystals (50–80 $\mu$). Its composition is similar to that of whitlockite, with no detectable F or Cl, but it has a $K_2O$ content of about 3%. The REE contents have not been determined, and there are not enough data at present to constitute a new mineral species. The term *phosphate mineral* has been used synonymously for this Ca,K phosphate phase.[17]

The compositional variant of *phase D* (described under apatite) that is relatively high in Y and REE, with minor Mg and U ($\leq$ 50–$\leq$ 100 ppm), probably is whitlockite.[18]

### *References*


1. Gancarz et al *EPSL* **16,** 321.
2. Gancarz et al *EPSL* **12,** 9, 12, 14.
3. McKay et al, *Suppl.* **3,** 741–742.

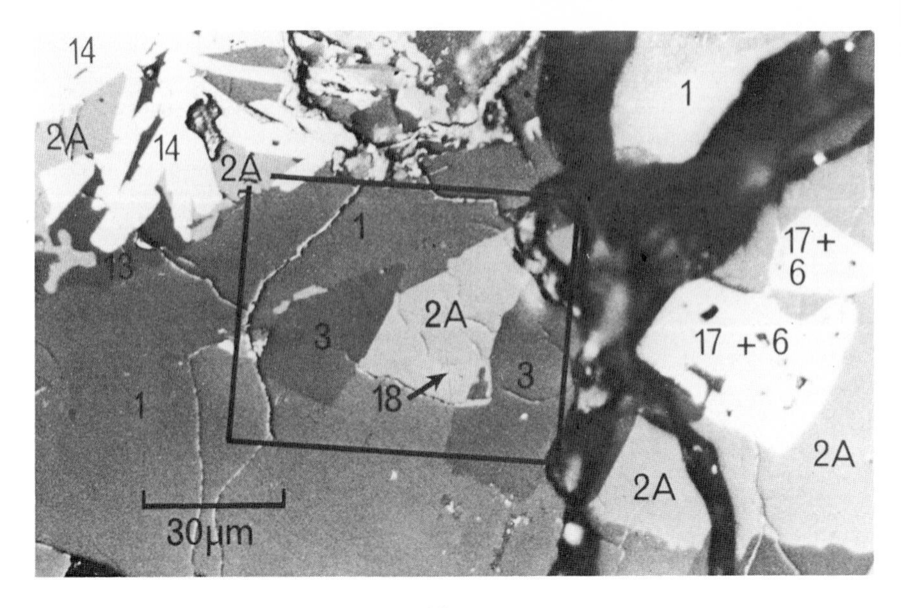


*(a)*

Fig. 6–5 (*a*) Monazite (18), in mesostasis of basalt 10047,68, in association with ferrohedenbergite (2A), potash feldspar (3), plagioclase (1), troilite (6), ulvöspinel (17), $SiO_2$-phase (13), and tranquillityite (14). (*b*) enlargement of ferrohedengergite grain (2A of Fig. 6–5*a*) with monazite inclusion. From *EPSL* **21,** 165. Photographs courtesy of Elsevier Press and J. Lovering, School of Geology, University of Melbourne, Parkville, Victoria, Australia.

4. Albee et al, *Lun. Sci. IV*, *Abstr.*, 24–26.
5. Albee and Chodos, *Suppl.* **1,** 140–141.
6. Taylor et al, *Suppl.* **3,** 997.
7. Griffin et al, *EPSL* **15,** 55–57.
8. Wood et al, *Spec. Rep.* **333,** 76.
9. Marvin et al, *Suppl.* **2,** 684.
10. Brown et al, *Suppl.* **3,** 145, 151–153.
11. Mason, *Mineral Rec.* **2,** 278.
12. Lovering et al, *Suppl.* **3,** 284, 287.
13. Gay et al, *EPSL* **9,** 124.
14. Fuchs, *EPSL* **12,** 172–173.
15. Reid et al, *Suppl.* **1,** 753.
16. El Goresy et al, *Suppl.* **2,** 232.
17. Ramdohr et al, *Lun. Sci. Conf. '71*, *Abstr.*, 97.
18. Lovering and Wark, *Suppl.* **2,** 156–157.

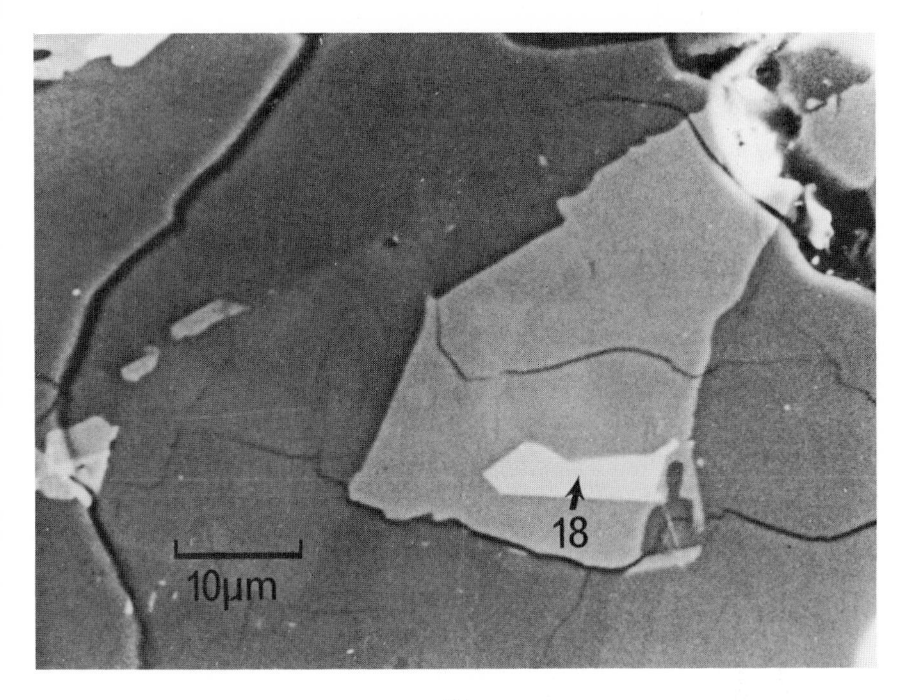


*(b)*

Fig. 6–5 (*continued*)

## MONAZITE (Ce, La, Y, Th) $(PO_4)$

### *Occurrence and Form*

One grain of monazite was found as an inclusion in a ferrohedenbergite grain in the mesostasis of coarse-grained basalt 10047,68 (Figure 6–5). The mesostasis contained some of the largest grains of late-stage apatite, whitlockite, tranquillityite, and zirconolite that have been observed in the lunar rocks.

### *Chemical Composition*

The monazite is thorium-poor and shows a distinct europium anomaly that is absent from terrestrial monazite. Crystallization of the monazite from late-stage liquids that formed during crystallization of the lunar igneous rocks could lead to these liquids becoming increasingly depleted in the light REE relative to the heavy REE. The high total of the probe analysis probably is due to some contribution from the surrounding ferrohedenbergite matrix.[1]

| | | |
|---|---|---|
| $SiO_2$ | = | 1.30 ±0.08% |
| $Al_2O_3$ | = | 0.11 ±0.04% |
| FeO | = | 1.17 ±0.04% |
| MnO | = | 0.19 ±0.02% |
| MgO | = | <0.02% |
| CaO | = | 0.60 ±0.02% |
| $P_2O_5$ | = | 26.36 ±0.23% |
| PbO | = | 0.05 ±0.09% |
| $UO_2$ | = | 0.03 ±0.03% |
| $ThO_2$ | = | 0.76 ±0.10% |
| $Y_2O_3$ | = | 1.11 ±0.23% |
| $Ce_2O_3$ | = | 30.08 ±0.30% |
| $La_2O_3$ | = | 16.51 ±0.15% |
| $Nd_2O_3$ | = | 14.91 ±0.20% |
| $Pr_2O_3$ | = | 4.30 ±0.15% |
| $Sm_2O_3$ | = | 4.09 ±0.13% |
| $Eu_2O_3$ | = | 0.36 ±0.04% |
| $Gd_2O_3$ | = | 1.55 ±0.12% |
| $Tb_2O_3$ | = | 0.13 ±0.03% |
| $Dy_2O_3$ | = | 0.33 ±0.07% |
| $Er_2O_3$ | = | 0.18 ±0.09% |
| Total | = | 104.12 |

*Reference*

1. Lovering et al, *EPSL* **21,** 164–168.

FARRINGTONITE $(Mg, Fe)_3(PO_4)_2$ (tentative)

In spinel troctolite 65785, a magnesium-orthophosphate has been found together with other accessories (i.e., ilmenite, metallic Ni-Fe, troilite, whitlockite, zirconian rutile, chromite, Cr-Zr-REE-armalcolite, and K-feldspar). It has been suggested that this phosphate phase is farringtonite,[1] but no data have been given.

*Reference*

1. Dowty et al, *Lun. Sci. V, Abstr.*, 174.

# CHAPTER SEVEN | THE $SiO_2$ MINERALS

## Cristobalite, Tridymite, Quartz, Unidentified Polymorphs, and Silica Glass

Of the three lunar silica minerals, cristobalite occurs most frequently, tridymite is next in quantity, and quartz (almost ubiquitous terrestrially) is present in such small amounts that early reports of it were suspect. All these minerals are only minor accessories. Even where it is most abundant, cristobalite makes up only about 5% of the rocks' volume. This amount was considered significant, however, in that, early reports on Apollo 11 material termed such rocks cristobalite basalts. Their general texture is coarse grained, ophitic, and vuggy, and the cristobalite is also coarse grained.[1]

In some rocks cristobalite occurs together with tridymite (in one instance appearing to replace part of a tridymite crystal.[2] In such cases the two polymorphs can be distinguished by cathodoluminescence as well as by their optical properties.[3] Quartz and cristobalite are found together in finer-grained rocks. Both high and low cristobalite and high and low tridymite have been reported, although in most cases the low phase pseudomorphous after the high phase is probably being observed. Only low (or $\alpha$) quartz has been identified positively by X-ray diffraction.[4]

Counterparts of the terrestrial large and widespread granites have not been found on the Moon. Portions of the lunar rocks that have been termed granite, or (more properly) rhyolite and granophyre, are "granitic" only in their composition.[5] Breccia 12013, the much-studied "black and white" rock, contains such portions. In the "white" portion of 12013 there are skeletal intergrowths of elongated quartz crystals, minor plagioclase of intermediate composition, and elongate rods and granules of light brown pyroxene. The quartz crystals are in optical continuity over irregular patches within the potash feldspar, and the texture is somewhat similar to the micrographic intergrowths commonly seen in terrestrial rocks. In intermediate portions of 12013 (between the "white" and "black" members), large amounts of potash feldspar and quartz appear to displace the finer pyroxene, plagioclase, and opaque minerals of the black matrix. In typical "black" areas, spherical vesicles have been filled by the potash feldspar-quartz intergrowths. Such textural relations suggest that the progenitor of the white end member was a fluid phase of low viscosity. This fluid appears to have been essentially of high-K, low-Ca "granitic" composition, and it is similar to the interstitial residuum present in many of the crystalline rocks of Apollo 11 and 12.[6]

The silica minerals generally are late-crystallization phases found in the mesostasis of the rocks. In a few instances some earlier crystallization of tridymite has been suggested on the basis of its textural relationship with plagioclase and clinopyroxene.[7]

In addition to the crystalline silica phases just named, silica glass has been found.[8]

### *References*

1. Schmitt et al, *Suppl.* **1,** 6–7, 10, 31, 50.
2. Roedder and Weiblen, *Suppl.* **1,** 813.
3. Weill et al, *Suppl.* **2,** 425.
4. Bailey et al, *Suppl.* **1,** 171.
5. Anderson et al, *Suppl.* **3,** 820.
6. Lunatic Asylum, *EPSL* **9,** 138.
7. Dollase et al, *Suppl.* **2,** 141.
8. Frondel et al, *Suppl.* **1,** 464.

## CRISTOBALITE $SiO_2$

### *Occurrence and Form*

Cristobalite ranges in size from up to 2 mm (in the coarse-grained rocks, where it is more abundant), down to sparsely occurring patches of only 10 to 100 $\mu$.[1] Cristobalite is a late-crystallizing component in Apollo 11 samples, occurring as aggregates interstitial to major minerals. In the microgabbros (especially those low in or lacking olivine), there are euhedral to subhedral crystals of high cristobalite up to 0.2 mm long.[2] (Actually, the single crystals are pseudomorphs of low cristobalite after high cristobalite (Figure 7–1), and the cracked appearance (Figure 7–2) of the euhedral crystals is due to the inversion from the high to the low phase.[3]) The crystals apparently are octahedrons (Figure 7–3), tabular on a (111) face, and contact twins presumably on this plane have been observed. The high

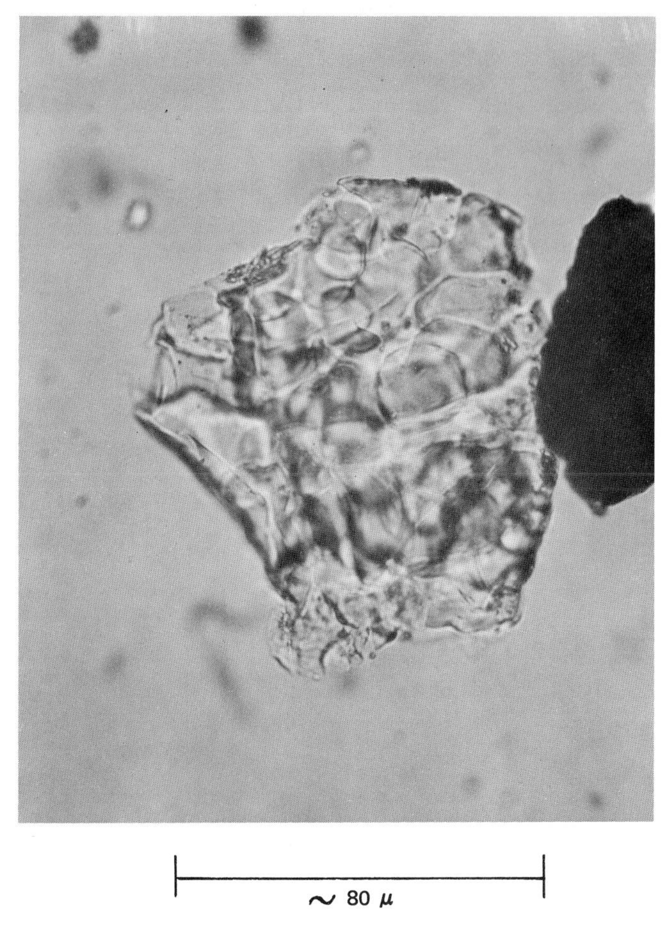


Fig. 7–1 Microgranular grain of high inverted to low cristobalite. Photograph courtesy of C. Frondel, Harvard University, Cambridge.

cristobalite appears to be a primary constituent of the microgabbros, although it also has been reported as seen in vugs[2] (e.g., associated with the mesostasis in Apollo 14 samples).[4] Cristobalite is found also in coarse patches in the residuum occasionally intergrown with a plagioclase that is more sodic than most of the plagioclase in the lunar rocks.[5] In some Apollo 12 samples cristobalite has been observed forming clusters on ilmenite and pyroxene grains.[6]

Cristobalite, distributed fairly uniformly throughout coarse-grained basalt 10044, occupies the interstices between plagioclase and pyroxene grains and generally does not include other minerals.[7] In fine-grained basalt

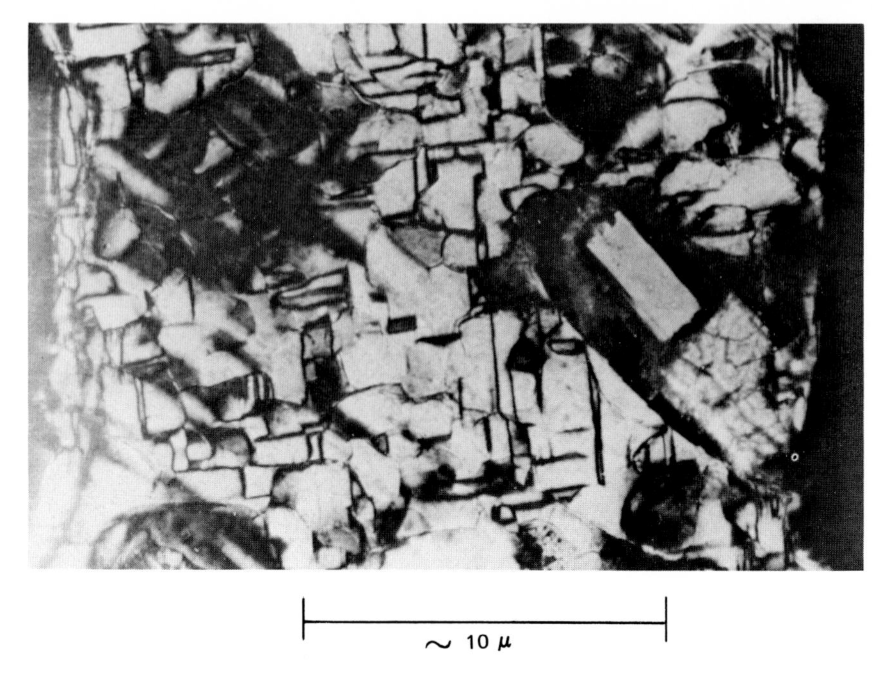


Fig. 7–2 Inversion twinning in low cristobalite. Crossed nicols. From *Suppl.* **1**, 208. Photograph courtesy of M. Brown, Department of Geological Sciences, Durham University, England.

10017 and medium-grained basalt 10058, abundant subhedral crystals form a mosaic of twinned anisotropic grains of low cristobalite.[8] In 10017, the 0.5-mm cristobalite grains, occurring between (and rarely included in) plagioclase grains, contain stacking faults, twin boundaries, and boundaries separating enantiomorphic parallel axis twins. The twins, as well as the mosaic texture, originated during the high to low cristobalite transformation at 268°C.[9] Cristobalite with typical granular texture, found in coarse-grained basalt 10047, has numerous inclusions of minor minerals.[10]

In fine-grained basalts 10025 and 10045, cristobalite occurs along with quartz and is intergrown with small flecks of potassium feldspar or glassy mesostasis. In porphyritic basalt 12021 and ophitic basalt 12038 the silica mineral assemblage is an intergrowth of cristobalite-tridymite.[11] In 12021 the cristobalite grains are as large as 75 μ across. They are subhedral and clear, and they display characteristic mosaic twinning and curved fractures. In porphyritic basalt 12022, the cristobalite is in equant anhedral grains.[12] In 12038,22 and porphyritic basalt 12053, the cristobalite forms 0.5-mm subhedral crystals and interstitial patches and consistently shows subgrain

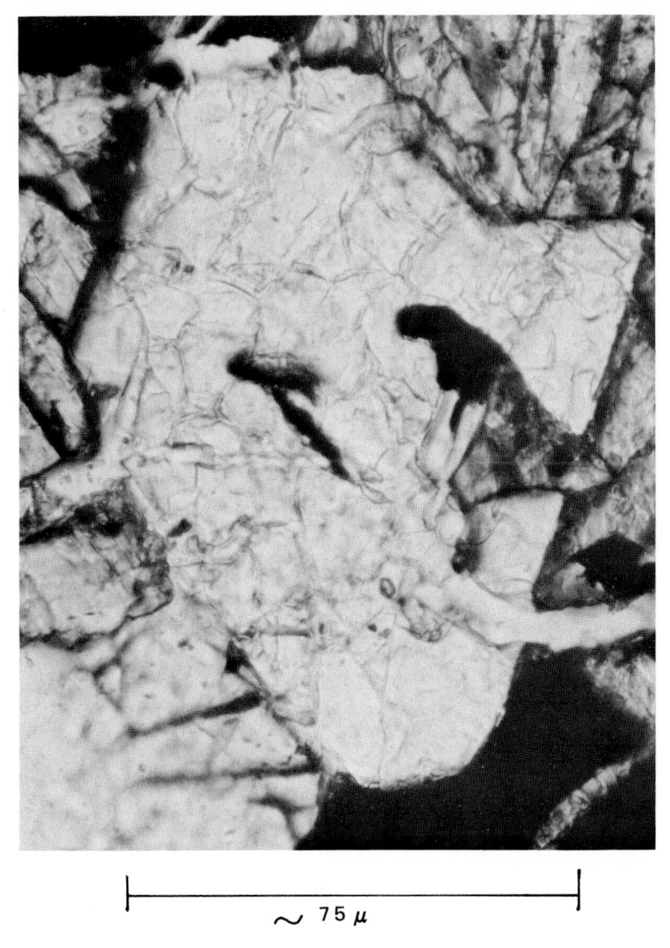


Fig. 7–3 Cristobalite crystal. Photograph courtesy of C. Frondel, Harvard University, Cambridge, Mass.

structure of approximately 50-μ size.[13] In ophitic basalt 12064, cristobalite makes up 1.1% of the rock's volume and occurs as anhedral to subhedral crystals showing pronounced inversion twinning.[14]

Mosaic cristobalite in crystalline clast 14072 forms a spongy network with native iron within iron-rich olivines and pyroxenes (Figure 7–4).[15] Cristobalite makes up as much as 1.5% of the volume of many Apollo 15 rocks that are olivine-cristobalite basalts.[16] In mare basalt 15475, an interstitial cristobalite crystal contained more than 12 ($< 10$ μ-sized) crystals of tranquillityite.[17] Cristobalite, together with rhyolitic glass containing

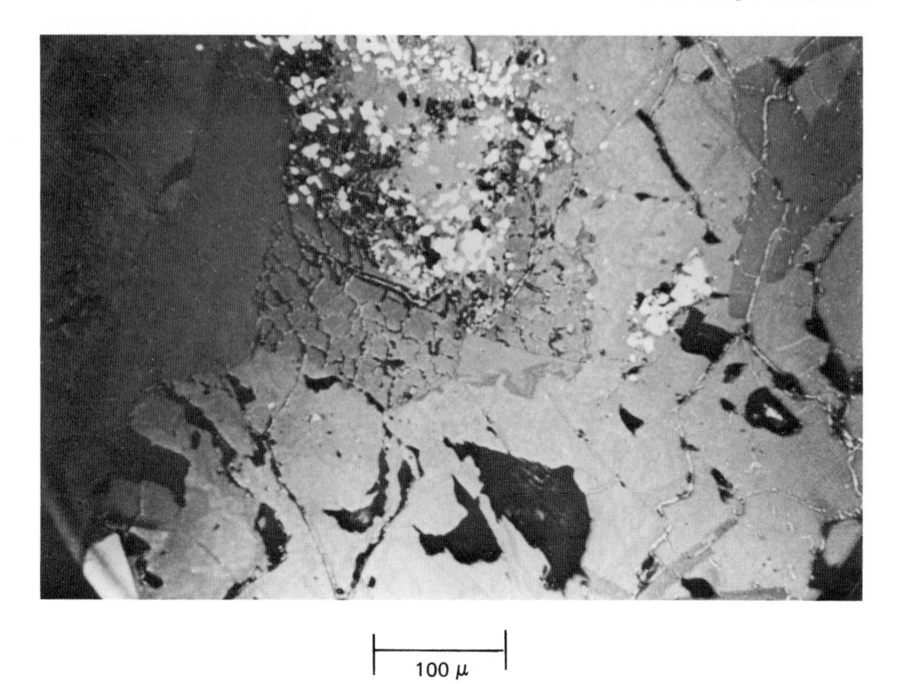


Fig. 7–4 Late-stage assemblage in basalt 14072. Dark gray, mosaic field is cristobalite; white grains are iron surrounded by a thin layer of tridymite (?); plain dark gray field is plagioclase. From *Suppl.* **3,** 138. Photograph courtesy of D. Walker, Harvard University, Cambridge, Mass.

phosphate phases, is found in the interstices of fine plagioclase laths in anorthositic gabbro 68415.[18]

*References*


1. Brown, *J. Geophys. Res.* **75,** 6484.
2. Frondel et al, *Suppl.* **1,** 465.
3. Smith et al, *Suppl.* **1,** 909.
4. *Apollo* 14 *preprint*, 11.
5. Agrell et al, *Sci.* **167,** 584.
6. Papanastassiou and Wasserburg, *EPSL* **8,** 271.
7. Bailey et al, *Suppl.* **1,** 185.
8. Brown et al, *Suppl.* **1,** 204.
9. Nord et al, *Lun. Sci. V, Abstr.*, 556.
10. Dence et al, *Suppl.* **1,** 317, 324.
11. Sippel; *Suppl.* **2,** 258–259.

12. Weill et al, *Suppl.* **2,** 425.
13. Christie et al, *Suppl.* **2,** 72.
14. Klein et al, *Suppl.* **2,** 280.
15. Longhi et al, *Suppl.* **3,** 137.
16. Powell et al, *Lun. Sci. IV, Abstr.*, 597–598.
17. Brown et al, *Suppl.* **3,** 149.
18. Walker et al, in preprint for *Lun. Sci. IV, Abstr.*

### *Optics*

Cristobalite is colorless[1] and optically clear.[2] Extinction is generally patchy (revealing the mosaic structure), but the individual grains or patches of the microgranular aggregates extinguish uniformly between crossed nicols and show a parallel alignment of the inversion twinning lamellae.[1]

The refractive indices of cristobalite range from 1.486 to 1.490.[3] The cristobalite from coarse-grained basalt 10044 has a mean index of refraction of 1.487 $\pm 0.002$,[4] and from Apollo 11 fines its mean index of refraction is 1.486, with a birefringence of 0.002.[1]

A luminescence petrographic study indicates that the spectrum exhibited by cristobalite is identical to that of tridymite, but the emission is much duller. The luminescence is bluish. Cristobalite is stable in the luminescence microscope, but under intense bombardment in the microprobe, red-emitting luminescence centers develop.[5]

### *References*

1. Frondel et al, *Suppl.* **1,** 465.
2. Dence et al, *Suppl.* **1,** 324.
3. Roedder and Weiblen, *Suppl.* **1,** 813.
4. Bailey et al, *Suppl.* **1,** 185.
5. Sippel, *Suppl.* **2,** 258–259.

### *Chemical Composition*

The lunar cristobalite is characteristically impure and contains significant amounts of $Ti^{4+}$, $Al^{3+}$, $Fe^{2+}$, Ca, and Na ions "stuffed" into its relatively open structure (Table 7-1; [1]–[4]).[1] Analyses of density separates of cristobalite from some Apollo 15 basalts yield traces of Rb and Sr as follows:[2]

| | Rb ($10^{-8}$ mg) | Sr ($10^{-8}$ mg) |
|---|---|---|
| Gabbro 15065 | 9.13 | 111.6 |
| Gabbro 15076 | 5.081 | 82.3 |
| Gabbro 15076 | 5.900 | 96.1 |
| Porphyritic basalt 15682 | 3.839 | 41.97 |

**Table 7-1 Analyses of Lunar Silica Minerals: Cristobalite, Tridymite, and Quartz**

| | [1] 15085 | [2] 12022,22 | [3] 10058 | [4] 10024–33 | [5] 15085 | [6] Luna 20 B–A2 | [7] 12021 | [8] Luna 16 G 38/2 |
|---|---|---|---|---|---|---|---|---|
| $SiO_2$ | 99.13 | 98.44 | 98.0 | 96.9 | 99.05 | 98.86 | 98.25 | 97.99 |
| $Al_2O_3$ | 0.18 | 0.33 | 0.92 | 1.12 | 0.34 | 0.48 | 0.35 | 0.81 |
| $TiO_2$ | 0.38 | 0.41 | 0.27 | 0.35 | 0.28 | 0.16 | 0.92 | — |
| FeO | 0.09 | 0.69 | 0.05 | 0.17 | <0.02 | 0.30 | 0.26 | 0.54 |
| CaO | <0.02 | 0.13 | 0.16 | 0.29 | 0.02 | 0.30 | 0.32 | 0.23 |
| MgO | <0.03 | 0.06 | — | — | <0.03 | — | 0.02 | 0.02 |
| $Na_2O$ | 0.05 | 0.01 | 0.15 | 0.25 | 0.05 | — | 0.07 | — |
| $K_2O$ | 0.17 | — | — | 0.01 | 0.26 | — | 0.01 | — |
| Total | 100.00* | 100.07 | 99.55 | 99.09 | 100.00* | 100.10 | 100.20 | 99.59 |

*References*

[1] Mason, *Am. Mineral.* **57,** 1532. From a coarse-grained basalt. *$SiO_2$ determined by difference. Analysis includes $MnO_2$ = 0.02% and $P_2O_3$ < 0.03%. Ba and Sr sought for but not detected. Cristobalite.

[2] Weill et al, *Suppl.* **2,** 417. From a porphyritic basalt. Cristobalite.

[3] Brown et al, *Suppl.* **1,** 203. From a medium-grained basalt. As a rim on tridymite. Cristobalite.

[4] Kushiro and Nakamura, *Suppl.* **1,** 619. From a medium-grained basalt. In an intergrowth with plagioclase. Cristobalite.

[5] Mason, *Am. Mineral.* **57,** 1532. From a coarse-grained basalt. *$SiO_2$ determined by difference. Analysis includes $P_2O_5$ <0.03%. Tridymite.

[6] Kridelbaugh and Weill, *Geochim. Cosmochim. Acta* **37,** 918. From a basalt fragment. Tridymite.

[7] Weill et al, *Suppl.* **2,** 417. From a porphyritic basalt. Tridymite.

[8] Grieve et al, *EPSL* **13,** 237. From a fine-grained basalt fragment. Quartz.

*References*

1. Ware and Lovering, *Sci.* **167,** 518.
2. Papanastassiou et al, *EPSL* **17,** 332.

*X-Ray Data*

Cell parameters for cristobalite have been determined using the X-ray diffraction powder method (Table 7-2; [1]–[5]). Very small grains of cristobalite, from the groundmass of porphyritic basalt 12052, yielded a diffraction pattern showing tetragonal symmetry. This indicates that the phase is in the low structural state.[1]

The very fine pattern of striations in intersecting planes that appears in electron micrographs of cristobalite is due to coherent microtwins produced during the change from high to low cristobalite. While being observed under the electron beam, the cristobalite developed a speckled contrast and became amorphous. Extensive streaks in several different directions on the diffraction pattern also faded along with the diffraction spots as the amorphous state developed. Rather than being associated with the

**Table 7-2 Cell Parameters for Cristobalite and Tridymite**

| Sample No. | $a$ (Å) | $b$ (Å) | $c$ (Å) | Space Group |
|---|---|---|---|---|
| [1] Apollo 11 | 4.97 | — | 6.95 | — |
| [2] 10047–44 | 4.970 | — | 6.931 | — |
| [3] 12051,36 | 4.9784 ±0.0004 | — | 6.9312 ±0.0009 | — |
| [4] 12038,72 | 4.9793 ±0.0004 | — | 6.9372 ±0.0009 | — |
| [5] 12052,57 | 4.9811 ±0.0009 | — | 6.936 ±0.0002 | — |
| [6] 12021,29 | 8.660 ±0.0002 | 4.999 ±0.0002 | 8.205 ±0.003 | — |
| [7] 12021 | 8.65 ±0.02 | 5.01 ±0.01 | 8.21 ±0.02 | $C222_1$ |

*References*

[1] Frondel et al, *Suppl.* **1,** 465. From a fines sample. Cristobalite.
[2] Dence et al, *Suppl.* **1,** 324. From a coarse-grained basalt. Cristobalite.
[3] Appleman et al, *Suppl.* **2,** 130. From an ophitic basalt. Cristobalite.
[4] ———, *ibid.* From an ophitic basalt. Cristobalite.
[5] ———, *ibid.* From a porphyritic basalt. Cristobalite.
[6] ———, *ibid.* From a porphyritic basalt. Tridymite.
[7] ———, *ibid*, 129. From a porphyritic basalt. Single crystal, studied by precession X-ray diffraction method, has an apparent orthorhombic subcell. Tridymite.

nucleation of the amorphous phase or with the high-temperature form, the streaks are believed to be associated with thermal diffuse scattering.[1]

Because the cubic ($F_{d3m}$) symmetry of the high cristobalite structure changes to tetragonal symmetry ($P4_12_1$, $P4_32_1$) at approximately 268°C, with only slight atomic displacements, it is possible to index diffraction patterns on either the cubic or tetragonal structures. The tetragonal indices given below are based on the primitive cell in which the *c* axis is one of the cube axes and the *a* and *b* axes are rotated 45° from the remaining cube axes. In a transmission electron microscope (TEM) study of lunar cristobalite, electron micrographs revealed a complex substructure with three distinct features:

1. The diffraction patterns are characterized by marked streaking of many spots and groups of nonrational spots replacing single spots in regions of large deviation from the Bragg condition. These features are due to the presence of closely spaced stacking faults on $\{111\}c \equiv \{101\}t$ and are probably growth phenomena related to mistakes in the sequence of sheets of six-membered rings of $SiO_4$ tetrahedra on $\{111\}_c$ planes.*
2. Twin boundaries parallel to $\{101\}_c = \{112\}_t$ originated during the $\beta \rightarrow \alpha$ transformation in which the cubic structure may transform to one of three tetragonal orientations, related by reflection across the $\{101\}_c$ planes. The boundaries are quite widely spaced ($\sim$ 10 $\mu$) and are probably responsible for the optical heterogeneity of the crystals.
3. Antiphase boundaries separate domains of approximately 0.5-$\mu$ size and there is no change of contrast across the domain boundaries, indicating no change in orientation of the diffracting planes. These structures are analogous to the parallel axis twins in quartz. Since low cristobalite belongs to an enantiomorphic space group and high cristobalite does not, these structures are probably parallel axis twins (or enantiomorphic domains) that originated on cooling through the transformation.

Similar to the damage observed in quartz and other similar framework structures, the cristobalite suffered damage in the electron beam, resulting ultimately in loss of crystallinity (as mentioned earlier). It was noted that the damage in cristobalite from ophitic basalt 12038 was considerably less than that of cristobalite from porphyritic basalt 12053. This may be related to impurity content.[2]

*References*

1. Champness et al, *Suppl.* **2,** 373–374.
2. Christie et al, *Suppl.* **2,** 72–73.

* The *c* and the *t* refer to cubic and tetragonal cristobalite structures, respectively.

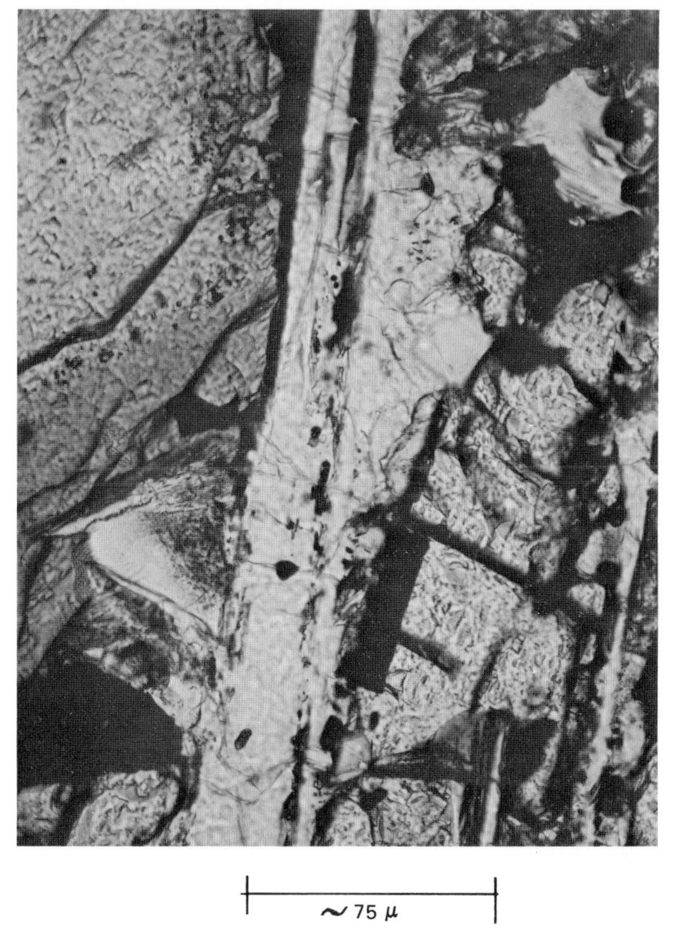


Fig. 7–5 Tridymite (irregular grain to upper right of plagioclase lath in center of picture) with pyroxene (left side of picture) and olivine (right side of picture). Photograph courtesy of C. Frondel, Harvard University, Cambridge, Mass.

## TRIDYMITE $SiO_2$

### *Occurrence and Form*

Tridymite occurs as thin plates, tabular on (0001), or as lathlike crystals, generally two or more in parallel intergrowth with pyroxene and plagioclase (Figure 7–5). In one case part of a lathlike crystal appeared to be replaced by cristobalite.[1] The thin plates show few inversion-induced cracks but have large numbers of melt inclusions, frequently as a parallel array of rods.[2] Anhedral grains and crystal fragments are found in the lunar fines.[3] Sheaflike grains, occurring as cores in cristobalite, have been classed as tridymite. In the mesostasis of medium-grained basalt 10050, tridymite prisms were recognized by their habit.[4]

In coarse-grained basalt 10047, acicular grains of tridymite are associated with apatite and potassium feldspar. Sector twinning is not visible, and presumably the tridymite has not inverted to the low-temperature form.[5] In Apollo 11 microgabbros, however, inversion pseudomorphs of low tridymite after high tridymite occur, although less abundantly than does high cristobalite. The mineral apparently is a primary constituent that tends to occur in or near vuggy areas.[3]

Planar features parallel to the hexagonal prism have been observed in tridymite adjacent to diaplectic plagioclase glass in basalt fragment 12057, 14. It is possible that these features are shock produced.[6] In porphyritic basalt 12021 tridymite forms grains up to 200 $\mu$ long that usually are clouded with small inclusions of pyroxene and ilmenite.[7] In 12021, in contrast to the cristobalite that occurs interstitially, the tridymite forms large skeletal to subhedral grains intergrown with plagioclase and Ca-rich clinopyroxene. Typical subhedral grains are 200 to 500 $\times$ 50 $\mu$, whereas skeletal grains are up to 1 mm long. Textural relationships between tridymite, plagioclase, and granular clinopyroxene suggest that all three were crystallizing simultaneously and possibly at an earlier time than rosettes of acicular clinopyroxene and plagioclase.[8] In porphyritic basalt 12022, tridymite, together with cristobalite, occurs as equant anhedral grains.[7] In the dark lithic portions of breccia 12013, large (up to 1 mm) crystal fragments of tridymite are almost always surrounded by a narrow rim of the "granitic" material.[9] In the coarse-grained Apollo 12 rocks, some tridymite laths are as long as 2 mm.[10] In microgabbro 12064, tridymite makes up to 4% of the rock's volume and occurs in thin, lathlike, subhedral to euhedral crystals up to 1.5 mm long. The crystals are surrounded by granular clinopyroxene and relatively equant plagioclase grains. This texture seems to indicate that tridymite was very early in the silicate crystallization sequences.[11]

In coarse-grained dolerite 14053, a unique texture has been tentatively attributed to the breakdown of fayalite as a result of subsolidus reduction processes. The late fayalite ($Fa_{86\text{-}96}$) broke down to a spongy mass of pure Fe metal plus tridymite and silica glass.[12]

In pyroxene basalt fragment #116 from Apollo 15 rake samples, tridymite needles are 1 mm long.[13] In gabbro 15116 the only silica phase is tridymite. It occurs as fairly long laths (> 1 mm) that sometimes are in parallel groups.[14]

*References*

1. Bailey et al, *Suppl.* **1,** 185.
2. Roedder and Weiblen, *Suppl.* **1,** 813.
3. Frondel et al, *Suppl.* **1,** 465.
4. Brown et al, *Suppl.* **1,** 204.
5. Dence et al, *Suppl.* **1,** 324.
6. von Engelhardt et al, *Suppl.* **2,** 835.
7. Weill et al, *Suppl.* **2,** 425.
8. Dollase et al, *Suppl.* **2,** 141–142.
9. Drake et al, *EPSL* **9,** 122.
10. Sippel, *Suppl.* **2,** 247.
11. Klein et al, *Suppl.* **2,** 283.
12. El Goresy et al, *Suppl.* **3,** 343.
13. Dowty et al, *Lun. Sci. IV*, *Abstr.*, 181.
14. ———, *Suppl.* **4,** 439.

*Optics*

The low indices of refraction of tridymite aid in differentiating it from cristobalite. The indices of refraction of low tridymite from Apollo 11 fines are:

$$n_\alpha = 1.470 \pm 0.001$$

$$n_\beta \sim 1.472$$

$$n_\lambda = 1.476 \pm 0.002 \text{ (with a few grains up to 1.481).}$$

The 2V(+) is medium-large[1] to large,[2] and the acute bisectrix is perpendicular to a large face on the tabular crystals.[1] Extinction is parallel, with length-fast elongation.[3] Some crystals show undulatory or patchy extinction, sometimes distributed in elongate areas, with occasional indistinct lamellar twinning in three sets at about 60°.[1] The large tridymite crystals in breccia 12013 have areas of variable extinction giving rise to a mosaic texture.[4]

The cathodoluminescence of tridymite is bright blue and brilliant (greater than that of cristobalite and about 10 times as intense as the plagioclase luminescence).[5]

*References*

1. Frondel et al, *Suppl.* **1,** 465.
2. Bailey et al, *Suppl.* **1,** 185.
3. Roedder and Weiblen, *Suppl.* **1,** 813.

4. Drake et al, *EPSL* **9,** 122.
5. Sippel; *Suppl.* **2,** 258–259.

### *Chemical Composition*

Although it has been stated that lunar tridymite differs from the cristobalite by containing $K_2O$ but no $Na_2O$,[1] many analyses show that the two silica phases do not differ significantly in their composition (Table 7-1; [6], [7]).[1]

From medium-grained basalt 10045-29 and in lithic fragments and matrices of breccias 10019-22 and 10067-8, a silica phase (presumably not a glass) contained appreciable amounts of Al, Ti, Fe, Mg, Na, and K. On the basis of this composition, the phase was considered to be tridymite, but the identification was not confirmed by X-ray. The $TiO_2$ content of this material is high in comparison to terrestrial and meteoritic tridymite, but the amounts of the other trace elements are not unusual.[3]

### *References*

1. Brown et al, *Suppl.* **1,** 204.
2. Roedder and Weiblen, *Suppl.* **1,** 813.
3. Keil et al, *Suppl.* **1,** 583.

### *X-Ray Data*

Identification of tridymite by X-ray diffraction has been made both by the powder method[1] and on single crystals.[2] The X-ray powder pattern shows minor differences from the known polytypes of low tridymite; the five darkest lines are:[1]

| *d* (Å) | *I* |
|---|---|
| 4.34 | 7 |
| 4.06 | 10 |
| 3.82 | 6 |
| 3.17 | 4 |
| 2.50 | 5 |

Cell parameters are given in Table 7-2, [6].

A single crystal of tridymite from porphyritic basalt 12021 had an apparently orthorhombic subcell of space group *C*222 (Table 7-2; [7]). These parameters agree well with those for orthorhombic material obtained by heating monoclinic, meteoritic tridymite, which transformed reversibly to the orthorhombic phase on heating to 180°C.[2]

Grains taken from thin sections 12021,135 and 12021,51 yielded tridymite with a prominent (pseudo) hexagonal subcell with:

$$a = 5.0\ \text{Å}$$
$$c = 8.2\ \text{Å}$$

and, in addition, a weaker monoclinic cell with:

$$a = 18.52\ \text{Å}$$
$$b = 4.98\ \text{Å}$$
$$c = 23.79\ \text{Å}$$

space group *Cc* or *C*2/*c*.

With very long exposure (Cu radiation), there developed on a precession photograph of the 12021 tridymite a few unexplained faint diffuse reflections and streaks; appearance and locations were somewhat reminescent of the superstructure reflections seen with terrestrial tridymites. The similarity of the X-ray diffraction pattern of this tridymite and that of synthetic tridymite (crystallized at $\sim$ 1400°C) is consistent with the idea that this lunar tridymite crystallized within its stability field, rather than metastably like most vesicle-filling terrestrial tridymite.[3]

A tridymite crystal from gabbro 15076,55 yielded (from Weissenberg pictures) a subcell with:

| | Hexagonal Setting | Orthohexagonal Setting |
|---|---|---|
| $a$ = | 4.98 ±0.02 Å | 4.98 ±0.02 Å |
| $b$ = | — | 8.64 ±0.02 Å |
| $c$ = | 8.27 ±0.02 Å | 8.27 Å |

These parameters are in agreement with those for terrestrial high tridymite.[4] Single crystals of 15076 tridymite show the superstructure reflections of low tridymite and a submicroscopic twinning of this superstructure. All the crystals studied thus far have a diffraction pattern with diffuse streaks typical of a one-dimensional disorder. Only (001) reflections are free from diffuse scattering. Tridymite single crystals are well suited for diffraction study of radiation damage, since the diffuse streaks are well defined and may be separated easily from the diffuse halves.[5]

It is pointed out that a periodic phase change from low to high tridymite takes place every lunar day. There is no indication of an additional disorder correlated with this phase change.[5]

*References*

1. Frondel et al, *Suppl.* **1,** 465.
2. Appleman et al, *Suppl.* **2,** 129–130.
3. Dollase et al, *Suppl.* **2,** 141–142.
4. Jagodzinski and Korekawa, *Suppl.* **4,** 947.
5. Jagodzinski, *Lun. Sci. IV, Abstr.*, 411.

## QUARTZ $SiO_2$

### *Occurrence and Form*

Generally absent in the coarse-grained rocks, quartz is widely though sparsely distributed in the soils of Apollo 11 and 12 samples.[1] From an Apollo 11 fines sample, a few small grains of a material that is possibly quartz were found in a heavy liquid separation of density 2.6.[2] Rare quartz grains were found in Apollo 14 soils, together with tridymite, red spinel, and potash feldspar.[3]

In breccia 12013, quartz occurs in several forms: fragment # O6A1, consisting largely of potash feldspar, contains minor amounts of quartz, andesine plagioclase ($An_{50-55}$), ilmenite, and other minerals. In both the light and dark portions of 12013, the crystals of plagioclase, potash feldspar, ilmenite, and quartz are all encrusted with a sugary coating of quartz.[4] The discrete quartz grains are a primary crystallization product of the late-stage residual liquid. Small round quartz crystals, up to 200 $\mu$, are enclosed by a rim of additional silica material. Projecting from this rim, into a potash feldspar matrix, are more common acicular silica crystals. The textural relations suggest that the larger quartz crystals initially were resorbed by a silica undersaturated "granitic" liquid and subsequently acted as sites for silica crystallization.[5] The presence of quartz, rather than tridymite, in the light veins of 12013, suggests crystallization at depth, although it also has been argued that the quartz may be the result of devitrification during impact metamorphism.[6]

The late-stage quartz may be either inverted tridymite or a primary low-temperature equilibrium phase.[7]

*References*

1. Sippel, *Suppl.* **2,** 249, 257.
2. Frondel et al, *Suppl.* **1,** 469.
3. McKay et al, *Suppl.* **3,** 986.
4. Gay et al, *EPSL* **9,** 124.
5. Drake et al, *EPSL* **9,** 122.
6. Charles et al, *Suppl.* **2,** 658.
7. Brown, *J. Geophys. Res.* **75,** 6489.

*Optics*

Lunar quartz is colorless,[1] uniaxial positive, and (from breccia 12013) has a birefringence of 0.006.[2] In breccia 14306, minor quartz (possibly including some cristobalite and tridymite) has been identified on the basis of its relief, reflectivity, and uniaxial (+) optic figure.[3]

Except for its pink and bluish cathodoluminescence, the quartz easily could be overlooked. It exhibits a very marked luminescence color change when examined through a rotating nicol prism. As the upper nicol of the microscope is rotated, the color changes distinctly from red to blue.[4] From fragment G 38/2 of Luna 16, areas up to 20 $\mu$ in diameter of nearly pure silica show a reddish-blue luminescence, suggesting that the phase is quartz.[5] The rim of additional silica material surrounding quartz grains in 12013 was identified by contrast in cathodoluminescence.[2]

In thin section of microbreccia 10060-20, three large (0.1–0.3 mm) single-crystal grains of a colorless low refringence and low birefringence have been observed. Each grain consists of a large number of small domains showing considerably variation in birefringence; thus each grain appears mottled between crossed nicols and does not exhibit a single extinction position. The grains show no cleavage and are biaxial (+) with 2V variable (approximately 25–35°). At high magnification they show as many as six closely spaced sets of planar features that strongly resemble those observed in shocked quartz or shocked plagioclase. The grains are either shock-strained quartz, shocked plagioclase in which the optical indicatrix has been changed markedly, or (though less likely) shocked clinopyroxene. Shocked quartz can have a biaxial positive 2V as large as 28°.[6] Silica grains from another thin section of the same breccia (10060-30) show features like those described previously and similar to those commonly observed in terrestrial quartz (i.e., four or five discrete sets of planar features appearing on rotation of the universal stage).[7] Still another investigation of silica grains from 10060 reports the same features.[8]

*References*


1. Bailey et al, *Suppl.* **1,** 171.
2. Drake et al, *EPSL* **9,** 122.
3. Anderson et al, *Suppl.* **3,** 821.
4. Sippel, *Suppl.* **2,** 249, 259.
5. Grieve et al, *EPSL* **13,** 241.
6. Sclar, *Suppl.* **1,** 857.
7. Short, *Suppl.* **1,** 867.
8. Dence et al, *Suppl.* **1,** 327.

### *Chemical Composition*

Analysis of quartz from Luna 16 fragment G38/2 (Table 7-1; [8]) show that the material contains impurities, just like tridymite and cristobalite.[1] The quartz cores from breccia 12013 contain greater than 99% $SiO_2$, but the $SiO_2$ content of the acicular crystals is only 95%. These crystals may be an unknown polymorph of $SiO_2$. They may be quartz, however, since it is possible that difficulty in resolving the fine needles with the probe is responsible for the apparent low $SiO_2$ content.[2]

#### *References*

1. Grieve et al, *EPSL* **13,** 237.
2. Drake et al *EPSL* **9,** 122.

### *X-Ray Data*

A single crystal from Apollo 11 fines was identified as $\alpha$ quartz by X-ray diffraction as well as optics.[1] Quartz grains, intermixed with cristobalite aggregates in fine-grained basalt 10072, were definitely identified by X-ray. After 3 days of mechanical concentration, the X-ray powder pattern was that of quartz with only minor cristobalite admixture.[2]

#### *References*

1. Bailey et al, *Suppl.* **1,** 171.
2. Fuchs, *Suppl.* **1,** 478.

## Unidentified Polymorphs of $SiO_2$

The acicular crystals associated with alkali feldspar in breccia 12013 may be an unknown polymorph of $SiO_2$, if they are not quartz (see previous discussion).[1]

From Apollo 16 fines, a hexagonal dipyramidal crystal, smaller than 1 $\mu$ and containing only Si, was observed in a clublike glass body that was heavily loaded with blue rutile of various sizes. The specific $SiO_2$ polymorph has not been determined.[2]

#### *References*

1. Drake et al, *EPSL* **9,** 122.
2. Jedwab, *Suppl.* **4,** 864.

Fig. 7–6 Intergrowth of colorless silica glass with brown glass, of higher index of refraction, containing a bubble. Photograph courtesy of C. Frondel, Harvard University, Cambridge, Mass.

## Silica Glass

Silica glass has been observed in Apollo 11 fines (Figure 7–6),[1] and a small amount of silica glass was found in fines sample 12003. The glass was recovered from the float fraction of heavy liquid of density 2.40. The glass is colorless to pale brown and has a slightly variable index of refraction of approximately 1.462.[2] Synthetic silica glass has a density of 2.203 and (in white light) an index of refraction of about 1.462. The indices of refraction increase with increasing content of the common impurities Al, Fe, Mg, Ca, and alkalies.[3] It would appear, then, that the lunar silica glass is close to pure $SiO_2$.

*References*

1. Frondel et al, *Suppl.* **1,** 464.
2. Frondel et al, *Suppl.* **2,** 724.
3. *Dana VII* **3,** *Part* 1, 320.

# CHAPTER EIGHT | SILICATES: THE FELDSPARS

## Plagioclase Feldspars

Although lunar plagioclase does not differ markedly from chemically equivalent terrestrial plagioclase,[1] there is a predominance of extreme calcic plagioclase in the lunar rocks. In fact, An-rich plagioclase is the most important lunar mineral[2] not only because of its quantity but also because of the information it contributes to an understanding of lunar rock genesis. Most of the plagioclases are in the range of calcic bytownite to sodic anorthite, but microprobe analyses have revealed compositions from essentially pure anorthite to labradorite,[3] even some andesine,[4] and, possibly, albite.[5]

The optical properties of the lunar plagioclases (i.e., their Euler angle measurements) indicate a slow cooling, yet when their structural properties (determined by TEM and X-ray precession photographs) are compared with those of the terrestrial plagioclases, rapid quenching is suggested.[6] Much of the lunar plagioclase has an anomalous composition; that is, it departs somewhat from the composition of the normal terrestrial plagioclase series by having a surplus of Si and a deficiency of Al. The dominant factor giving rise to this anomaly appears to be the composition of the liquid from which the plagioclase crystallized, and the departure from ideal composition is greatest for plagioclases that have crystallized from melts poorest in normative feldspar (Figure 8–1).[7] If the An content is inferred from the Al/Si ratio rather than from the Ca/Na ratio, lunar plagioclase corresponds more closely to normal volcanic plagioclase (see below under Chemical Composition).[6]

The position of plagioclase in the paragenetic sequence is a function of its activity in the melt, and its appearance is recorded in the crystallization trends of the pyroxenes by their compositional discontinuities.[8] A study

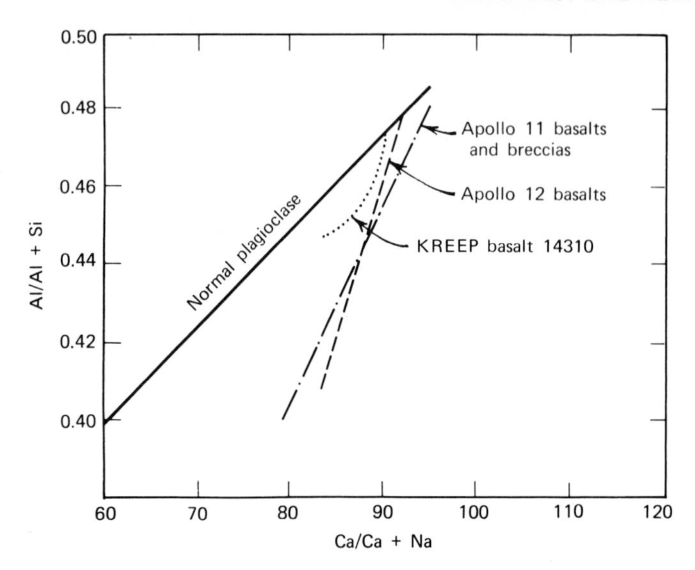


Fig. 8–1 Deviation of (both heated and unheated) lunar plagioclase for normal terrestrial plagioclase composition. Diagram is a composite of data from Storey et al, *Lun. Sci. V, Abstr.*, 754.

of the plagioclase indicates that the mare basalts were erupted as liquids and crystallized on or very close to the lunar surface.[9]

Deformation and shock effects from both large and small impacts are observed in the plagioclase. They are also evidenced by the occurrence of much maskelynite having a composition essentially that of the crystalline plagioclase in the same rock.[10]

Potash feldspar is a minor and late phase in the mesostasis and often contains appreciable barium. Potash feldspar has been termed sanadine by some investigators,[11] but this term should be used only if the data show the phase to be sanadine rather than orthoclase (the general terms—potash feldspar or, simply, K-feldspar—are used more commonly). Potash feldspar and silica appear as dominant constituents in a proposed rhyolitic component of some lunar rocks.[12] The so-called "granite problem" of the Moon has stemmed from the characteristics of the mesostasis of some basalts, in which potassic rhyolitic glass patches are associated with KREEP-type minerals. These are barian K-feldspar (containing also ppm of Rb), whitlockite, tranquillityite, and zirconolite. Rhyolite and tranquillityite-bearing granophyre fragments are common in Apollo 14 breccias. It is suggested that the "granite" may have formed (more than 4 billion years ago) by fractionation from KREEP basalt generated by low-volume partial melting of a feldspathic crust; alternatively, it could have formed in smaller volume by fractional crystallization of mantle-derived basalt. In either case the presence of a granitic component is inferred from the Rb and Ba contents in the K-feldspar in the fines, and from the Y, Zr, and Nb of the Zr-rich minerals.[13] Both KREEP basalt and prophyritic basalt contain late-stage Ba-K phases, approximating the composition of alkali feldspars, which have invited the speculation that such phases may be trapped remnants, within the basalt, of a larger granitic fraction occurring on some unsampled portion of the lunar surface.[14]

Some K-rich phases close to potash feldspar in composition are isotropic and irregularly distributed in patches interstitial to the pyroxene. They may be an interstitial residual glass rather than shock-isotropized feldspar.[10]

*References*

1. Stewart et al, *Suppl.* **1,** 929.
2. Jagodzinski and Korekawa, *Lun. Sci. IV, Abstr.*, 409.
3. Gay et al, *Suppl.* **2,** 382.
4. Drake et al, *EPSL* **9,** 118.
5. Bown and Gay, *EPSL* **11,** 23–26.
6. Wenk et al, *Suppl.* **3,** 573–578.
7. Storey and O'Hara, *Lun. Sci. V, Abstr.*, 752.
8. Bence and Papike, *Suppl.* **3,** 467.
9. Crawford, *Suppl.* **4,** 715.
10. Albee and Chodos, *Suppl.* **1,** 138.
11. Kim et al, *Suppl.* **2,** 749.
12. Wood et al, *Spec. Rep.* **333,** 107.
13. Brown et al, *Suppl.* **3,** 154–156.
14. Trzcienski and Kulick, *Suppl.* **3,** 600.

## Albite ($NaAlSi_3O_8$)–Anorthite ($CaAl_2Si_2O_8$) Series and Maskelynite (Plagioclase Glass), Essentially $CaAl_2Si_2O_8$)

*Synonymy*

Andesine; Mason et al, *Lun. Sci. Conf. '71, Abstr.*, 257.
Bytownite; Kushiro and Nakamura, *Suppl.* **1,** 618.
Calcic bytownite; Stewart et al, *Suppl.* **1,** 927.
Calcium labradorite; Gay et al, *EPSL* **9,** 125.
Ca-rich feldspar; Gancarz et al, *EPSL* **12,** 12.
Diaplectic glass; Quaide and Bunch, *Suppl.* **1,** 721.
Diaplectic plagioclase glass; Quaide and Bunch, *Suppl.* **1,** 721.
Feldspar glass thetomorphs; Short, *Suppl.* **1,** 865.
Isotropic "plagioclase glass"; Albee et al, *Sci.* **167,** 464.

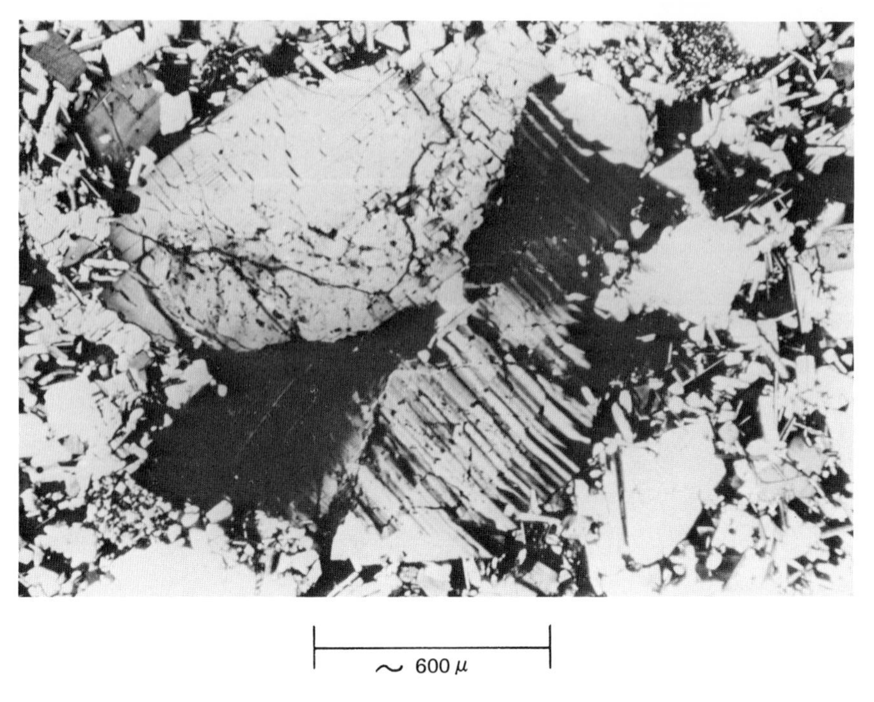


Fig. 8–2 Plagioclase metacryst in feldspathic basalt 60335. From *Suppl.* **4,** 1019. Photograph courtesy of D. Walker, Harvard University, Cambridge, Mass.

Labradorite: Jagodzinski and Korekawa; *Lun. Sci. IV, Abst.*, 409.
Maskelynite; Dence et al, *Suppl.* **1,** 327.
Sodic anorthite; Keil et al; *Suppl.* **1,** 577.
Sodic bytownite; Stewart et al, *Suppl.* **1,** 927–928.
Thetomorphic glass; Chao et al, *Suppl.* **1,** 295.

*Occurrence and Form*

Compared with terrestrial material, the plagioclase in the lunar rocks is remarkably clear and free of inclusions and alteration products.[1] It occurs in a wide range of grain sizes (< 0.05–> 3 mm) and shapes (e.g., tabular, lathlike, platy, or equant: Figure 8–2).[2] Some crystals grow into hollow square tubes (extreme hopper structure: Figure 8–3) enclosing devitrified liquid and other phases;[3] for example, in the Luna 20 KREEP basalt fragments, plagioclase crystals commonly have hollow or glass-filled cores.[4] Euhedral crystals of anorthite in the fines and in the vugs of microgabbro (Figure 8–4) are flattened on (010) with a rhomboidal outline determined

Fig. 8–3 Blocky plagioclase grain (center of picture) with $An_{84}$ inclusion—charged core and clear $AN_{96}$ rim. From a noritic anorthositic fragment in recrystallized breccia 60335. From *Suppl.* **4,** 1019. Photograph courtesy of D. Walker, Harvard University, Cambridge, Mass.

by (001) and (101); the acute internal angle is 52°18′. On (010) the extinction angle $X$ to the (010) ($\bar{1}$01) edge is about 5°.[2] In coarse-grained basalt 10062-35, the plagioclase forms an interlocking or radiating network of lathlike crystals up to 0.2 mm long.[5] In a gabbroic anorthosite chip 14276,13, euhedral plagioclase laths (~ 0.5 mm long) form an interlocking framework, the interstices of which contain anhedral pyroxene grains. Locally the plagioclase laths are in radial aggregates or in finer-grained patches, which could represent cognate inclusions.[6] In fine-grained basalt 10020, delicate crystals of plagioclase, together with ilmenite, project into vuggy cavities. In coarse-grained basalt 10047, these crystals are joined together by other minerals of late crystallization. Also in 10020, as well as in fine-grained basalt 10017, plagioclase crystals up to 1 mm long are either subhedral or poikilitic blades enclosing small grains of clinopyroxene and

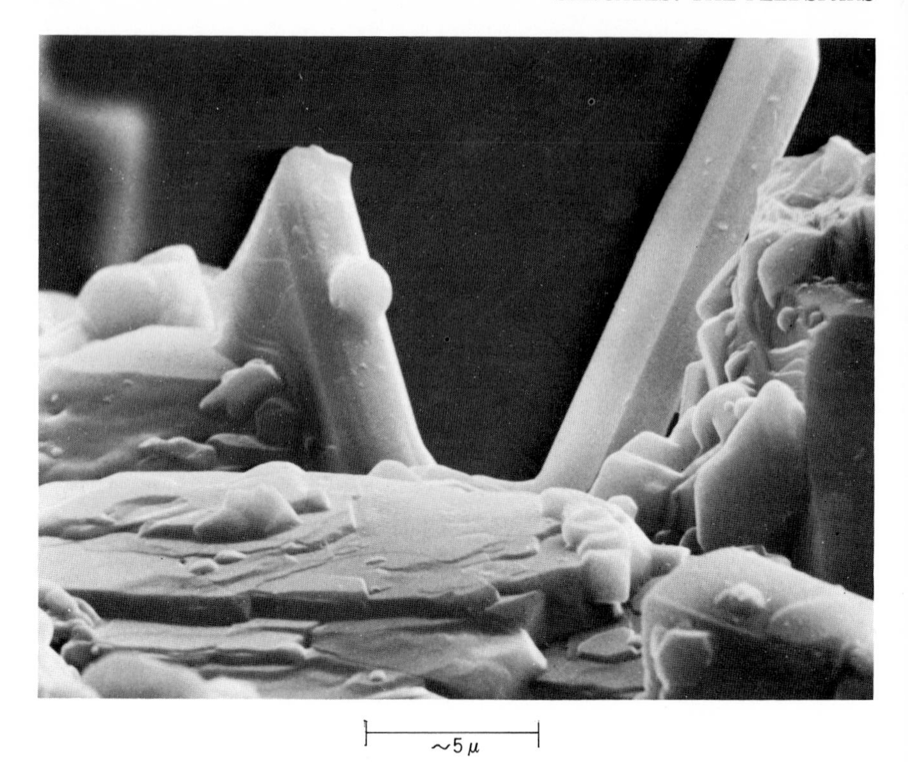


Fig. 8–4 Tabular plagioclase crystal (left of center) and prismatic pyroxene crystal (right). NASA photograph S-73-30446, courtesy of U. Clanton, Johnson Space Center, Houston.

rare ilmenite. In 10047 the plagioclase shows a large range of grain size; 0.1-mm euhedral crystals are enclosed in clinopyroxene, and larger platy crystals with pointed terminations occur in stellate, radiating, and interpenetrating clusters.[1] From gabbro 15076,55, two calcic bytownite crystals in vugs had well-developed faces and no twinning, indicating that they grew in the presence of a vapor phase. On their surface there are peculiar mound shapes, pillars, needles, and whiskers (Figure 8–5).[7] In Apollo 14 fines, the plagioclase, ranging from sodic bytownite to extreme calcic anorthite, is more abundant among the monomineralic grains than in other lunar samples.[8]

Plagioclase crystals are twinned mostly on the albite and Carlsbad laws,[9] but twinning on the pericline and Baveno laws also has been observed.[10] In many crystals, twinning with vicinal composition planes subtending angles of about 1.5° with (010) is common. Pericline twinning occurs frequently, with an angle $\sigma = -6°$, in plagioclase of coarse-grained basalt 10058. In the same rock, a plagioclase with a Banat twin was observed enclosed in pyroxferroite.[11] In KREEP basalt 14310, clear, inclusion-free crystals of plagioclase are as long as 2 mm but range in size mostly between 0.3 and 0.5 mm. Except for one very large crystal, no zoning has been found. Small areas of fine-grained plagioclase needles are compositionally the same as the large ones showing polysynthetic twinning. Twin laws are albite, Carlsbad, albite-Carlsbad (common), pericline (less common) and Baveno-*r* (only one crystal: Figure 8–6). Cruciform intergrowth is also common.[12] From porphyritic basalt 68416,77, the plagioclase forms crystal laths, mostly 0.6 × 0.2 mm to 0.2 × 0.04 mm, twinned on albite, pericline, and Carlsbad laws. There are a few large crystals (megacrysts) 2.0 × 0.3 mm.[13] In polymict breccia 60016,95, a 3-mm extreme calcic anorthite, with Fe and Mg contents corresponding to those of nonmare rocks, has the following polysynthetic twinning and inclusions:

1. Needles 1 to 2 μ long, parallel to the albite twin plane and either directly on the twin walls or within the narrow twin layers,
2. Vermicules of silica, probably located in the pericline twin plane,
3. Small birefringent crystals, probably quartz, accompanied by equant pyroxene grains and concentrated in what may be fracture zones.[14]

Both textural relationships and various types of zoning yield much information on the growth of the plagioclase. Normal, oscillatory, and sector zoning have been observed. The plagioclase in porphyritic basalt 12063 initially formed as hollow crystals when the rock was 60 to 75% crystalline. With continued growth that progressed both outward and inward from the initial "shell," sector zoning as well as normal zoning developed (Figure 8–7).[15] In ophitic basalt 12021,148, coarse-grained, clearly zoned feldspar occurs in 1.2 to 2 mm laths, mostly hollow, elongated along the *a* crystallographic direction. They have some sector zoning with (010), (001) and possibly (0*kl*) sectors developed. The nucleation zone has a euhedral outline and is the most sodic portion of the crystal. Zoning away from the nucleation zone is oscillatory, and the plagioclase growth proceeded both inward, into the hollow core, and outward, becoming more calcic (up to $An_{95-96}$) in both directions. In the outer portion of the grains there is a reverse trend toward a more sodic composition, and in places still another calcic zone has developed (Figure 8–8). Some cores are filled with plagioclase, others with pyroxene, ilmenite, and tridymite. This is different from the growth pattern in basalt samples 10024, 15555, and 70035. In these samples the initial plagioclase crystal, a lath platy parallel to (010), continued to grow by addition of material on the outside

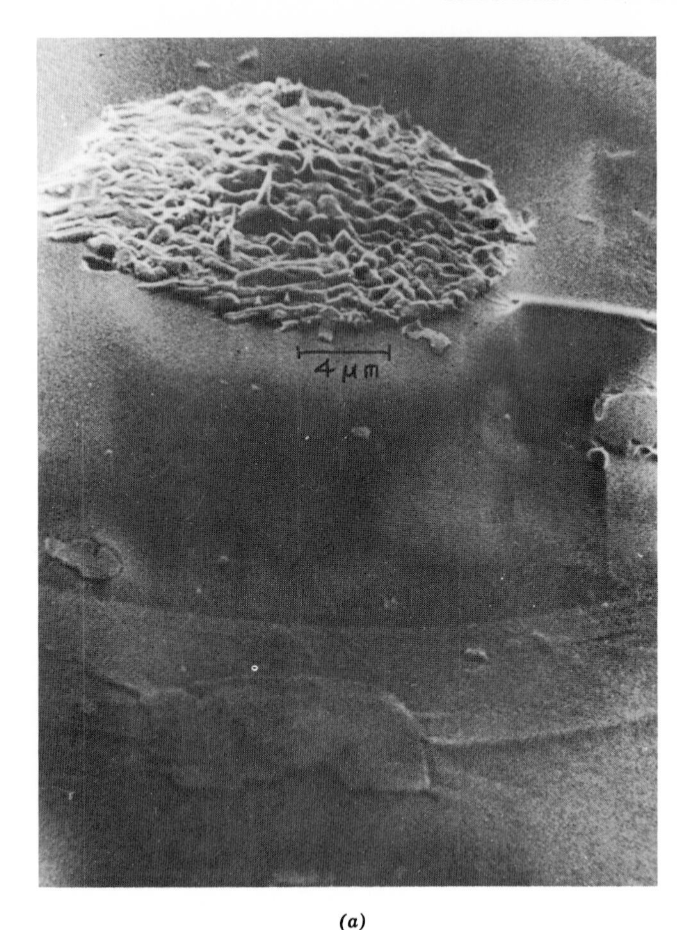


*(a)*

Fig. 8–5 (*a*) Mound on plagioclase crystal. From *Lun. Sci. V, Abstr.*, 426. (*b*) Cone-shaped pillar on mound on plagioclase crystal. From *Lun. Sci. V, Abstr.*, 426. Photographs courtesy of M. Korekawa, Institute für Kristallographie der Universität Frankfurt, W. Germany.

of the lath.[16] The plagioclase in KREEP basalt 14310 has two distinct habits: (1) 25% is in large subhedral crystals, and (2) the remainder is small, polysynthetically twinned laths. Textural relations suggest that the large phenocrysts are an earlier generation of plagioclase than the laths, and the two generations of growth may be related to liquid immiscibility.[15] Fine-scale exsolution in the plagioclase from sample 14310,145 suggests a slower cooling history than for Apollo 11 and 12 basalts.[17] In spinel-microtroctolite 62295, olivine-plagioclase ($An_{91-96}$) growth relationships

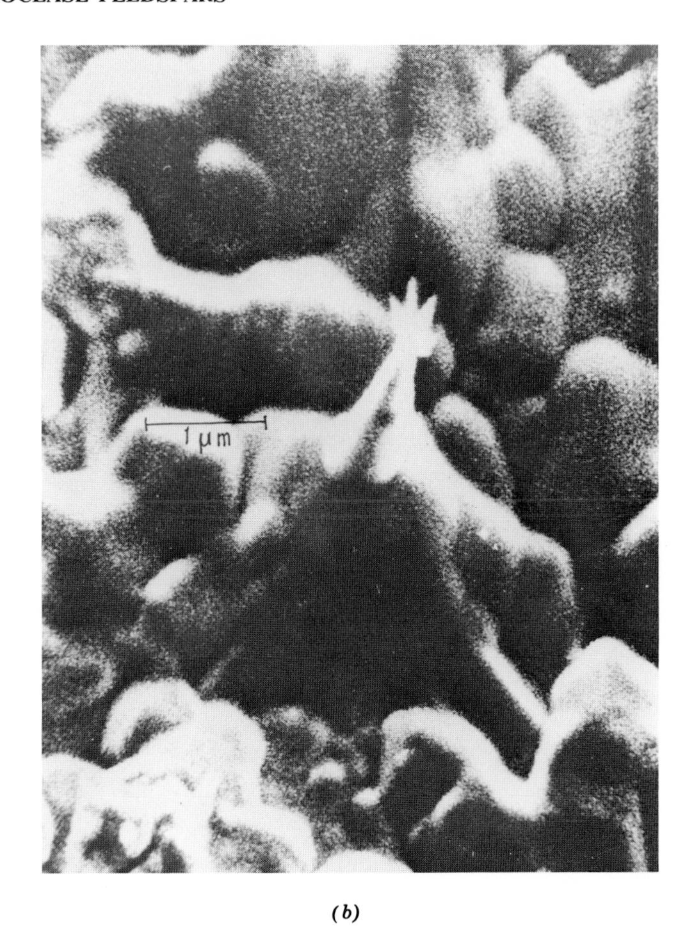


*(b)*

Fig. 8–5 (*continued*)

suggest extremely rapid crystallization. Sprays of hollow plagioclase laths (0.2 to 0.6 mm), set in glass, radiate out from olivine nuclei. Hollow plagioclase laths are intergrown with hollow olivine crystals, and bundles of fasciculate olivine and plagioclase are intergrown on a scale of 20 to 100 $\mu$. In poikiloblastic norite breccia 60315, the rims of the plagioclase show both textural and compositional signs of equilibration with the matrix. In porphyritic basalt 68415, plagioclase makes up 73% of the rock's volume.[18] Most of the plagioclase ($An_{97-71}$) occurs in laths with well-developed faces, heavily twinned and strongly zoned near the margins.[19] Blocky euhedral crystals with albite twinning grade into stubby laths up to

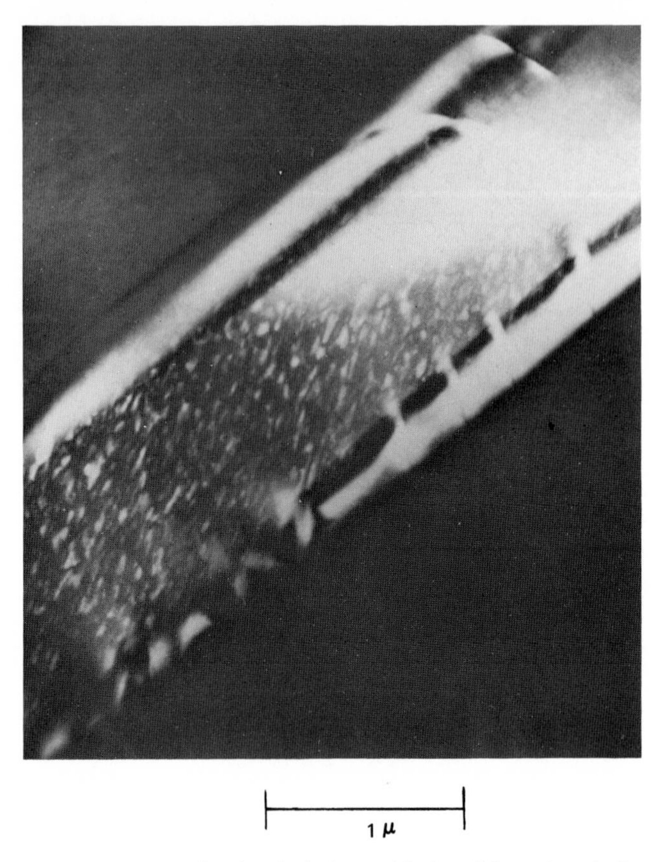


Fig. 8–6 Baveno-*r* twin lamellae in plagioclase with *b*-antiphase-domain boundaries in contrast. From *Suppl.* **3,** 577. Photograph courtesy of .H-R. Wenk, Department of Geology, Geophysics, and Space Science Laboratory, University of California, Berkeley.

1 mm long, associated with olivine and Mg-pigeonite. They also grade down in size to thin laths 50 to 100 μ long, associated with ferroaugite and a glassy residuum.[18] Many larger grains are zoned with an optical discontinuity near the rim, generally reflecting a reversal in composition.[6] The most sodic material is in thin rims on the medium to large laths, and in discrete small laths in the groundmass (Figure 8–9). There are some irregular grains, 0.5 mm to several millimeters in size, with approximately the same compositional range. Usually subequant, with no crystal faces, these grains display curved fractures, wavy extinction, and poorly developed twinning. There are inclusions in the plagioclase of polyphase P-rich globules made up of schreibersite and a metallic Fe phase.[19]

Fig. 8–7 Zoned "hollow" plagioclase crystal in basalt 12063. Zoning progresses outward toward the rim and inward toward the center from the area (lighter band) of highest anorthite content. From *Suppl.* **3**, 592. Photograph courtesy of W. Trzcienski, University of Montreal, Canada.

Moderately shocked plagioclase has fine lamellar mechanical twins, and, locally, deformation or shock lamellae (producing a mosaic structure: Figure 8–10). Strongly shocked plagioclase is partly or wholly converted to thetomorphic glass containing only sparse microfractures. Partially birefringent grain areas commonly contain shock lamellae. Very strong shock is demonstrated by plagioclase glass with flow structure (Figure 8–11) and

Fig. 8–8 Reverse and sector-zoned hollow plagioclase with cores filled by groundmass clinopyroxene and ilmenite. From picritic basalt 12002. From *Suppl.* **4,** 997. Photograph courtesy of D. Walker, Harvard University, Cambridge, Mass.

chemical homogenization. Although shock and deformation features are more dominant in the lunar breccias and highly feldspathic rocks than in the mare basalts, there is evidence of shock in some Apollo 11 plagioclase (Figure 8–12).[20] The most dominant substructural features noted by transmission electron microscopy in the plagioclase of coarse-grained basalt 10029-1 are microtwins, commonly in groups or bundles that may correspond to the fine twins observed optically. Dislocations frequently are seen in the twin boundaries and within the twin lamellae themselves,

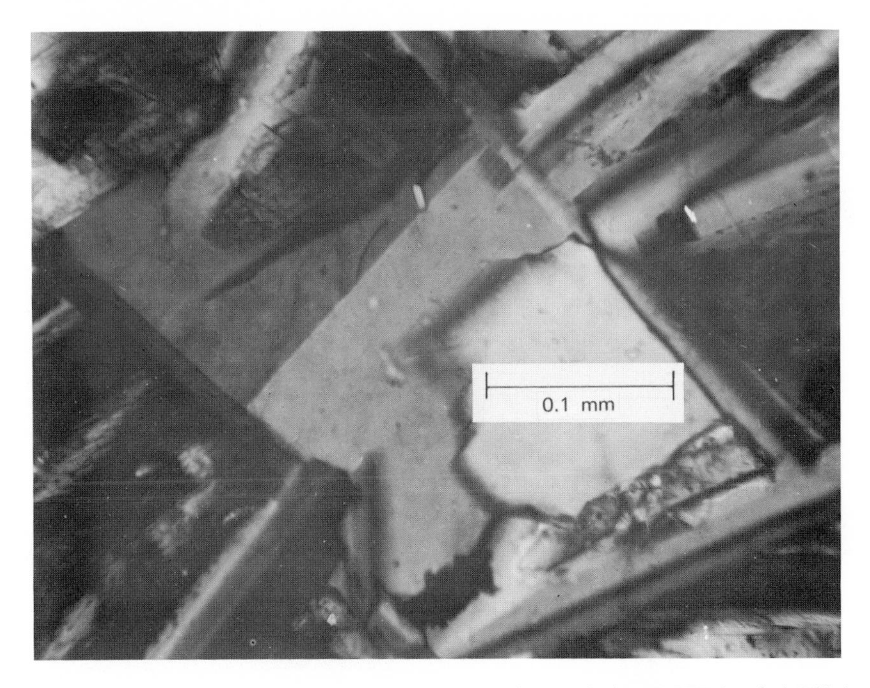


Fig. 8–9 Rare irregular zoning in an equant plagioclase grain in KREEP basalt 14310,4. Main part of grain is $AN_{93}$, light corner is $An_{79}$. From *Suppl.* **3,** 392. Photograph courtesy of A. G. Plant, Geological Survey of Canada, Ottawa.

suggesting that many dislocations may be due to deformation. The presence of the microdeformation twins is consistent with the existence of slip dislocations in the same grains, and these features seem to indicate that lunar rock 10029-1 was subjected to mild deformation at some stage in its history.[21] In Apollo 14 breccias, recognition of twins as deformation twinning is through association with (and progressively more intense development near) faults, mosaic zones, and zones of maskelynite.[22] In anorthosite 15415, the anorthite is more heavily deformed than plagioclase from Apollo 11 and 12 basalts, and offset twinning has been observed (Figure 8–13).[17] In recrystallized polymict breccia fragment 22006, large plagioclase clasts, 250 $\mu$ in diameter, exhibit undulatory extinction and ragged twinning suggestive of mild shock; in addition, compositional zoning is weak. Variation from core to rim is from sodic anorthite to calcic bytownite. The clasts occur in a groundmass of plagioclase of the same compositional range, although the individual grains generally are uniform in composition. From a troctolitic recrystallized polymict breccia fragment 22007, the compositional range of the plagioclase is restricted to sodic

(a)

Fig. 8–10 (*a*) Mechanical twins parallel to (010) in plagioclase. From *Suppl.* **1,** 297. Photograph courtesy of E. C. T. Chao, U.S. Geological Survey, Reston, Va. (*b*) Well-developed mosaic structure in shocked plagioclase crystal. From *Suppl.* **3,** 497. Photograph courtesy of H. Avé Lallement, Department of Geology, Rice University, Houston.

anorthite, and the euhedral form of the small groundmass plagioclase crystals indicates that they are products of recrystallization and not fragments originally in the progenitor.[23] From poikilitic basalt 62235, shocked and partially or totally annealed plagioclase grains and aggregates are interpreted as xenocrysts. The earliest phase, together with Fe-Ni globules and chromian pleonaste, consists of cores of almost pure anorthite (0.1–>0.3 mm) with reaction rims of later phases that are bronzite on Fe-Ni globules, plagioclase on spinel, and bytownite on anorthite.[24] In anorthositic cataclasite 60215/13, clasts of very calcic anorthite of uniform composition make up about 97% of the rock. The grains are from less than 0.001 to more than 6 mm. Twinning and undulatory extinction are common, and there is evidence of shock effects in many grains. Diaplectic glass, in some instance recrystallized, has produced a cherty appearance in the grains.[25] (Although the term "diaplectic glass" is used synonymously with

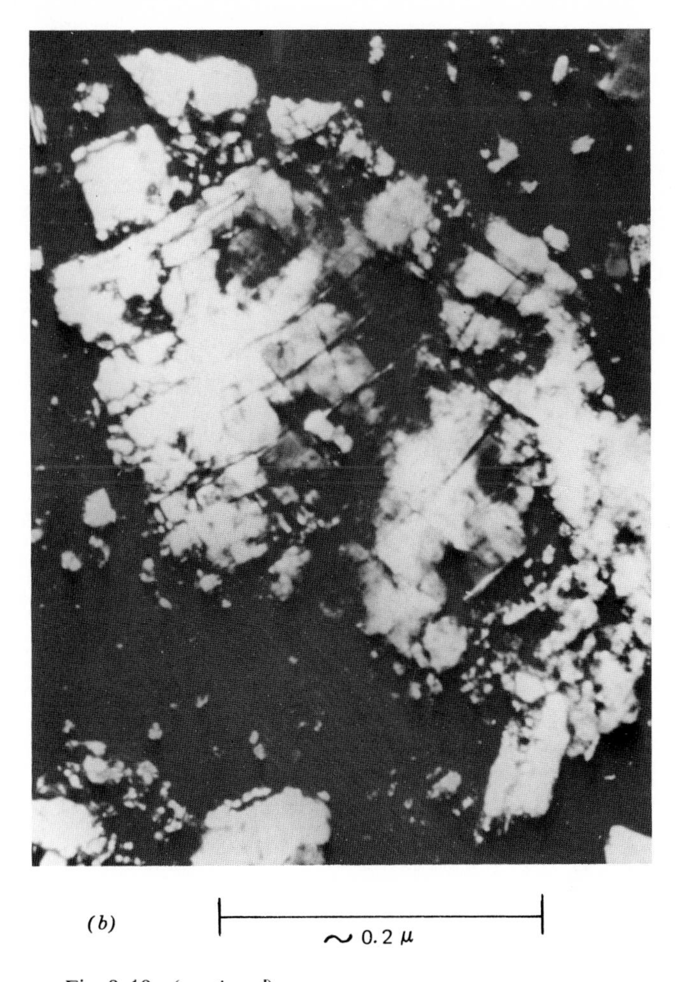


Fig. 8–10 *(continued)*

maskelynite here, it has been referred also to the amorphous state that preserves the morphology of the previous crystalline state—e.g., cleavage, twin, and grain boundaries.[26]) Size has no relationship to the condition of the grains, be they shocked or unshocked. Some have been granulated down to submicron sizes in a fine-grained clastic matrix, and much of the feldspar has been recrystallized.[27] The major component (82%) of anorthositic breccia 60015,126 is plagioclase, commonly shattered, fractured, and pulverized with bending and dislocation of original twin lamellae. The cryptocrystalline structure of the glassy matrix may represent a melting product of plagioclase now partially devitrified. Two-thirds of anorthositic

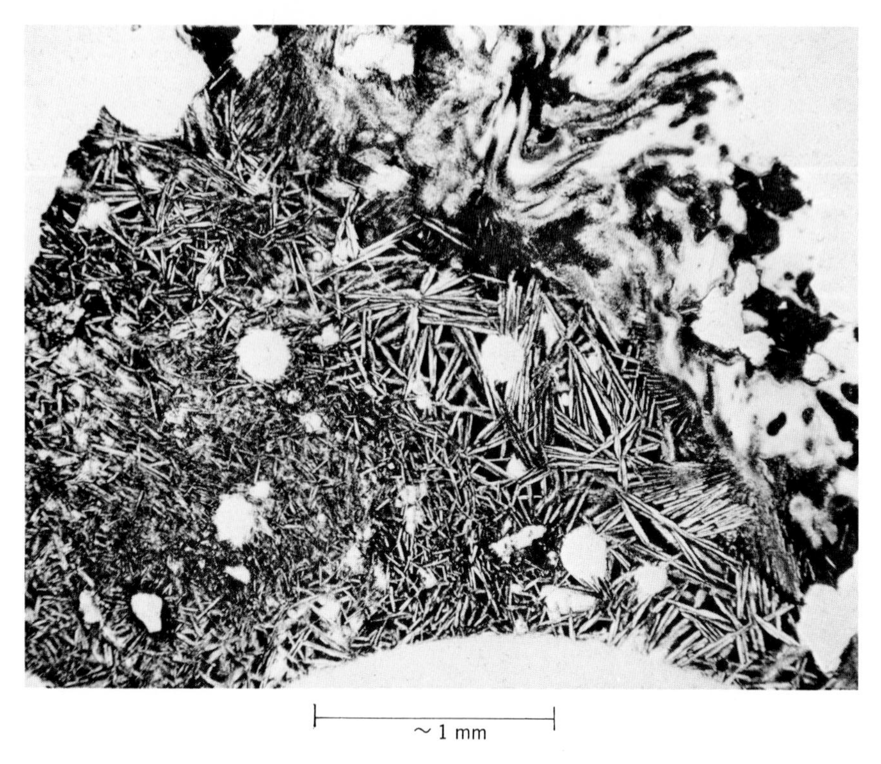


Fig. 8–11 Fibrous plagioclase crystals resulting from devitrification of glass (note flow structure in rimming glass) in metamorphosed breccia 15418. NASA photograph S-71-52202, courtesy of R. Laughan, Johnson Space Center, Houston.

breccia 61016,217 is composed of a single piece of maskelynite, and the other third is 38% maskelynite. (Here "maskelynite" refers to extreme-shock-isotropized plagioclase, and this is the general usage of the term. A suggestion was made, however, that usage be limited to a diaplectic glass of labradorite composition.[26] Moreover, the term "thetomorphic glass" is also used synonymously with maskelynite,[20] although it could be applied to "any glassy phase transformed by shock from the crystalline host mineral in the solid state."[26]) In the central portions of the maskelynite grains there is some relict feldspar, suggesting a weakening of shock intensity from the outer rim toward the interior. Some of the maskelynite in the breccia matrix has recrystallized.[28] In breccia 10060, maskelynite has been observed also in polycrystalline fragments, in one case with silica grains that probably are quartz. Two grains of the silica mineral (confirmed by probe) are separated by a grain of completely isotropic maskelynite close to $An_{95}$.[1]

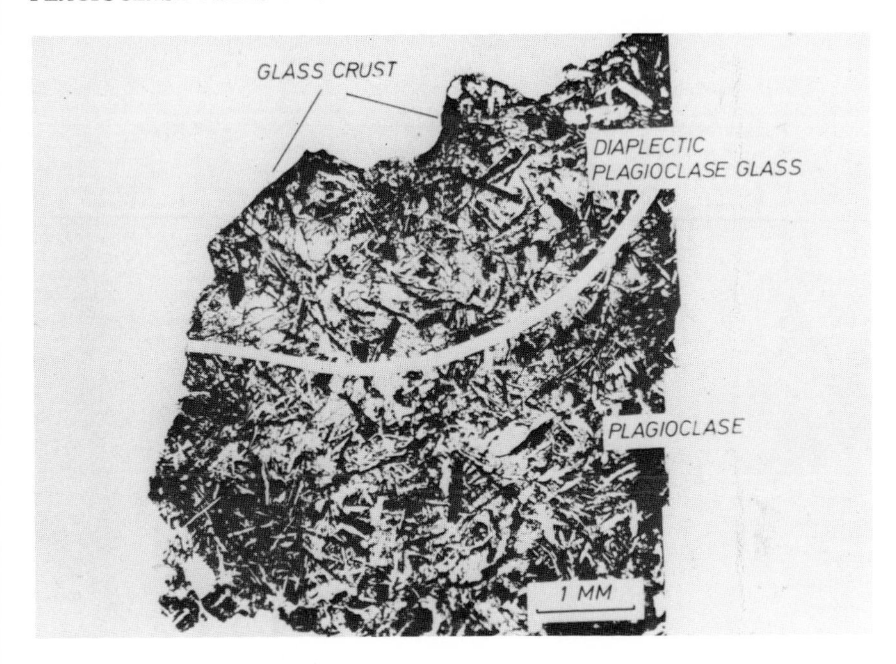


Fig. 8–12 Zones of progressive shock metamorphism in basaltic rock coated with vesicular glass. From a fragment M61 in fines sample 10085-25. From *Suppl.* **1,** 381. Photograph courtesy of W. von Engelhardt, University of Tübingen, W. Germany.

Almost pure albite ($An_3$) was found in 22002,2,7*d*, a handpicked single crystal from a soil sample. It has been verified by X-ray (see below), but its origin is considered uncertain, and contamination of the sample is possible. However, the poor quality of the crystal (i.e., diffuse and smeared-out reflections) suggest a shocked feldspar.[4]

*References*

1. Dence et al, *Suppl.* **1,** 317–318, 320, 327.
2. Frondel et al, *Suppl.* **1,** 461–462.
3. Anderson et al, *Sci.* **167,** 588.
4. Cameron et al, *Geochim. Cosmochim. Acta* **37,** 779, 789.
5. Carter and MacGregor, *Suppl.* **1,** 248.
6. Gancarz et al, *EPSL* **16,** 320–321.
7. Korekawa and Jagodzinski, *Lunar Sci. V, Abstr.*, 426.
8. Quaide and Wrigley, *Suppl.* **3,** 772.
9. Agrell et al, *Sci.* **167,** 585.
10. Bailey et al, *Suppl.* **1,** 184.

Fig. 8–13 Offset twinning of heavily deformed plagioclase in anorthosite 15415. NASA photograph S-71-52629, courtesy of R. Laughan, Johnson Space Center, Houston.

11. Agrell et al, *Suppl.* **1,** 108–109.
12. Wenk et al, *Suppl.* **3,** 569–570.
13. Juan et al, *Lun. Sci. IV, Abstr.*, 421.
14. Johan and Christophe, *Lun. Sci. V, Abstr.*, 385.
15. Trzcienski and Kulick, *Suppl.* **3,** 595–596, 600.
16. Crawford, *Suppl.* **4,** 705–707, 712.
17. Lally et al, *Suppl.* **3,** 401–402.
18. Walker et al, *Suppl.* **4,** 1014, 1016, 1020.
19. Helz and Appleman, *Lun. Sci. IV, Abstr.*, 352–353.
20. Chao et al, *Suppl.* **1,** 296.
21. Radcliffe et al, *Suppl.* **1,** 745.
22. Avé Lallement and Carter, *Suppl.* **3,** 896.
23. Podosek et al, *Geochim. Cosmochim. Acta* **37,** 888–891.
24. Crawford, *Lun. Sci. V, Abstr.*, 142.
25. Meyer and McCallister, *Lun. Sci. IV, Abstr.*, 523.
26. von Engelhardt et al, *Suppl.* **1,** 370.
27. Meyer and McCallister, *Suppl.* **4,** 661.
28. Juan et al, *Lun. Sci. V, Abstr.*, 394.

*Optics*

Plagioclase is usually clear, colorless to white; only rarely is it pale brown.[1] The isotropized plagioclase (i.e., maskelynite) also is usually colorless. Large transparent grains occur in fines samples 61281,8, 67701,26, and 67712,16.[2] A pale green, glassy fragment from a microanorthosite fragment in soil sample 10085 was proven by chemical analysis to be maskelynite (Table 8-1; [14]).[3]

There appears to be good agreement between optical data and chemical composition; for example, glass obtained by melting plagioclase with a chemical composition of $An_{88}$ had $n = 1.564 \pm 0.001$, and plagioclase with a chemical composition of $An_{74}$ had $n = 1.550 \pm 0.001$.[4] In coarse-grained basalt 10062-35, plagioclase ($An_{70\text{-}95}$) had $n\beta = 1{,}563\text{–}1{,}578 \pm 0.002$, and from soil sample 10085-16, plagioclase ($An_{65\text{-}94}$) had $n\beta = 1.536\text{–}1.550 \pm 0.002$.[5] From fines sample 12032,44, calcic bytownite (approximately $An_{89}$) had indices of refraction, determined by (010), of:[6]

$$\alpha' = 1.576 \pm 0.001$$
$$\lambda' = 1.583 \pm 0.001.$$

The average anorthite from anorthosite 15415,22 had[7]

$$\alpha = 1.574 \pm 0.001$$
$$\beta = 1.581 \pm 0.001$$
$$\lambda = 1.585 \pm 0.001$$
$$2V(-) \text{ large}$$
$$r > v \text{ weak.}$$

In lunar bytownites the relations between their optic orientation (determined from Euler *I* angle measurements*) and their chemical composition are consistent with evidence from terrestrial plagioclase. In contrast, most of the anorthite ($An_{90\text{-}92}$) known from the very old basalts of the Moon differ in optic orientation from those of young volcanic rocks of Earth. It has been cautioned, however, that before any firm conclusions can be drawn, optical–chemical data are required on lunar anorthites ($An_{93\text{-}100}$) and on anorthites from metamorphic and plutonic rocks of Earth.[8]

* The Euler *I* angles relate the mutual positions of the three chief vibration directions $[n\alpha]$, $[n\beta]$, and $[n\lambda]$ of the indicatrix to the rectangular cartesian system $X = \perp[001]$ in (010), $Y = (010)$, and $Z = [001]$.[8]

**Table 8-1 Plagioclase analyses**

| | [1] Apollo 12 Double-Core Sample | [2] Luna 20 22001,16 | [3] 15415,22 | [4] 10085–4-10a | [5] Luna 20 22003,1,3a | [6] Luna 20 926-3 | [7] 14310 Core | [8] 14310 Rim |
|---|---|---|---|---|---|---|---|---|
| $SiO_2$ | 47.91 | 42.88 | 43.36 | 44.21 | 44.90 | 44.1 | 45.40 | 45.77 |
| $Al_2O_3$ | 33.43 | 36.08 | 36.04 | 35.68 | 35.60 | 36.2 | 34.93 | 34.35 |
| $TiO_2$ | — | — | 0.04 | 0.02 | 0.01 | — | 0.03 | 0.05 |
| FeO | 0.75 | 0.08 | 0.08 | 0.14 | 0.06 | — | 0.14 | 0.30 |
| MgO | 0.22 | — | 0.07 | 0.10 | 0.16 | 0.09 | 0.24 | 0.14 |
| CaO | 17.15 | 20.38 | 19.34 | 19.28 | 19.40 | 18.9 | 18.82 | 18.45 |
| $Na_2O$ | — | 0.19 | 0.32 | 0.46 | 0.54 | 0.64 | 0.85 | 1.01 |
| $K_2O$ | — | 0.01 | 0.05 | 0.07 | 0.07 | 0.03 | 0.14 | 0.20 |
| Total | 99.46 | 99.62 | 99.30 | 99.96 | 100.76* | 99.96 | 100.55 | 100.27 |

*References*


[1] Sellers et al, *Suppl.* **2,** 670. From a gabbroic fragment in Unit VI. Anorthite ($An_{100}$).

[2] Reid et al, *Geochim. Cosmochim. Acta* **37,** 1014. From a fine-grained soil sample. Anorthite ($An_{98}$).

[3] Stewart et al, *Lun. Sci. III, Rev. Abstr.*, 727. From an anorthosite ($An_{96.5}$).

[4] Agrell et al, *Suppl.* **1,** 109. From a microanorthosite fragment in soil sample. Mean of three analyses. Anorthite ($An_{95.5}$).

[5] Cameron et al, *Geochim. Cosmochim. Acta* **37,** 783. From an anorthositic fragment in soil sample. *Analysis includes MnO = 0.02%. Anorthite ($An_{95}$).

[6] Tarasov et al, *Suppl.* **4,** 340. From an anorthositic fragment in soil sample. Anorthite ($An_{94}$).

[7] Longhi et al, *Suppl.* **3,** 136. From a KREEP basalt. Large zoned crystal. Sodic anorthite (An ~92).

[8] ———, *ibid.* From a KREEP basalt. Large zoned crystal. Sodic anorthite (An ~91).

| | [9] | [10] | [11] | [12] | [13] | [14] | [15] | [16] |
|---|---|---|---|---|---|---|---|---|
| | | | | | Luna 20 | Luna 20 | 10085–4 | |
| | 12032,44 | 12038,63 | 66081,5 | 68415,37 | 808–35 | 22002,2,7d | 10a | 10065 |
| $SiO_2$ | 46.8 | 48.3 | 49.4 | 50.15 | 52.5 | 68.0 | 44.90 | 48.9 |
| $Al_2O_3$ | 33.2 | 31.7 | 31.4 | 31.18 | 28.4 | 19.8 | 34.78 | 32.2 |
| $TiO_2$ | — | — | — | — | 0.13 | — | 0.02 | — |
| FeO | 0.34 | 1.14 | 0.33 | 0.66 | 0.68 | — | 0.21 | 0.46 |
| MgO | — | 0.33 | — | — | 0.11 | 0.03 | 0.05 | 0.30 |
| CaO | 17.7 | 16.7 | 15.8 | 14.50 | 12.9 | 0.58 | 18.92 | 17.2 |
| $Na_2O$ | 1.44 | 1.87 | 2.12 | 2.69 | 3.30 | 11.4 | 0.60 | 1.49 |
| $K_2O$ | 0.01 | 0.16 | 0.23 | 0.19 | 0.74 | 0.13 | 0.08 | 0.10 |
| Total | 99.52* | 100.20 | 99.28 | 99.37 | 98.76 | 99.94 | 99.56 | 100.65 |

*References*


[9] Wenk and Nord, *Suppl.* **2,** 137. From a fines sample. Average of 77 spots on entire specimen. *Analysis includes SrO = 0.03%. Calcic bytownite ($An_{88.8}$).

[10] Keil et al, *Suppl.* **2,** 324. From a coarse-grained basalt. Bytownite ($An_{82.4}$).

[11] Taylor and Carter, *Suppl.* **4,** 294. From a soil sample. Bytownite ($An_{79.3}$).

[12] Helz and Appleman, *Lun. Sci. IV*, *Abstr.*, 353. From a porphyritic basalt. Na-rich plagioclase laths. Sodic bytownite (An ~72).

[13] Tarasov et al, *Suppl.* **4,** 340. From a possible mare basalt fragment in soil sample. Labradorite ($An_{65.3}$).

[14] Cameron et al, *Geochim. Cosmochim. Acta* **37,** 783. Handpicked crystal from an anorthositic fragment in soil sample. Albite ($An_3$).

[15] Agrell et al, *Suppl.* **1,** 109. From a microanorthosite fragment in fines sample. Maskelynite ($An_{94.1}$).

[16] Quaide and Bunch, *Suppl.* **1,** 721. From a microbreccia. Maskelynite ($An_{85.9}$).

In porphyritic basalts 12018, 12021, and 12065, and ophitic basalt 12039, plagioclase is among the phases studied by luminescence petrography. The compositional range of these plagioclases is the same as that for plagioclase from the Apollo 11 igneous rocks. The most calcic regions emit a yellowish luminescence, grading to duller shades in less calcic rims or regions. In the plagioclase from the coarse-grained rocks, the duller luminescing regions show a definite reddish tinge.[9] By resorting to synthesis of large calcic plagioclase crystals with known impurities, the activators responsible for the luminescence in plagioclase, from fine-grained gabbro 10050,37, have been identified as:[10]

| | | | | |
|---|---|---|---|---|
| blue | 460 nm | $Ti^{4+}$ | 300 ppm | strong |
| green | 550 nm | $Fe^{2+}$ | 5000 ppm | medium |
| yellow | 570 nm | $Mn^{2+}$ | 300 ppm | medium |

*References*

1. Meyer and McCallister, *Suppl.* **4,** 662.
2. Housely et al, *Lun. Sci. IV*, *Abstr.*, 381.
3. Agrell et al, *Suppl.* **1,** 109.
4. Bailey et al, *Suppl.* **1,** 184-185.
5. Carter and MacGregor, *Suppl.* **1,** 248, 250.
6. Wenk and Nord, *Suppl.* **2,** 139.
7. Stewart et al, *Lun. Sci. III*, *Rev. Abstr.*, 726–727.
8. Wenk et al, *Suppl.* **3,** 581, 588.
9. Sippel, *Suppl.* **2,** 249–251.
10. Mariano et al, *GSA '73*, *Abstr.*, 726.

*Chemical Composition*

When compared with normal terrestrial plagioclase many lunar plagioclase analyses, show a surplus of Si and a deficiency of Al. This cannot be attributed to inclusions, since no foreign phase has been detected by scanning electron microscopy. If the anorthite content is inferred from the Al/Si ratio rather than from the Ca/Na ratio, the lunar plagioclase corresponds more closely to normal volcanic plagioclase. The difference between $An = 4(Al/Al + Si) - 1$ and $An = Ca/Ca + Na + K$ is caused by substitutions that appear so far to be unique to lunar feldspars. The deficiency in Al + Si (up to 0.06 per formula unit) is compensated for by Fe, Mg and small amounts of Na or Ca occupying tetrahedral sites. The Ca/Na ratio accordingly is increased to balance the electric charge. Vacancy coupled substitution $Ca + \square \rightarrow 2\ Na$ also could cause an increased Ca/Na ratio. (This mechanism has yet to be verified by structure analyses.) Alkali evaporation from plagioclase, if this process occurred, is probably small. These unusual substitutions are an expression of special conditions on the Moon during the crystallization of plagioclase; apparently the important factors are crystallization at high temperatures, rapid cooling, and peculiar magma composition.[1] Separate experiments have shown that the anomalous composition of the lunar plagioclase is not due to selective volatilization of alkalies under vacuum, nor is it necessarily correlated with the cooling rate, as defined by the rock textures. It also is not entirely a disequilibrium feature because it persists unchanged in lunar plagioclase samples that have been reequilibrated and then rapidly quenched.[2]

The excess of $SiO_2$ and deficiency of $Al_2O_3$ for a given CaO content is a striking feature of the mare basalt plagioclase. In basalt 12021 the amount of deviation decreases toward the margins of all the analyzed plagioclase grains. The greatest departure from ideal composition has been noted in hollow and sector-zoned Apollo 12 plagioclase. The deviation is less in zoned plagioclase from basalts 15555 and 70035, and it is least in more sodic hollow plagioclase from basalt 10024.[3]

Unrecrystallized plagioclase of $An_{98}$ occurs in anorthositic samples.[4] Plagioclase as sodic as $An_{60}$ (labradorite) has been found in porphyritic basalts 12052 and 12053;[5] in a light-colored vein in breccia 12013, a still more sodic labradorite ($An_{50-55}$) is dominant.[6] In an Apollo 12 rock fragment consisting largely of a potash feldspar, there are minor amounts of plagioclase in the andesine compositional range.[7] A possible albite occurrence in a Luna 20 soil sample has been mentioned. Thus although the compositional range of all lunar plagioclase is wide, various lunar rock types are characterized by more restricted plagioclase compositions.

Bytownite predominates over anorthite in the mare basalts, and in the Apollo 11 coarse-grained basalts, the bytownites are more calcic than those from fine-grained rocks.[8] The plagioclase in the Apollo 12 rocks is chiefly calcic bytownite, and its composition is even more limited than that from Apollo 11 samples.[9] From Apollo 14 samples, the plagioclase range in the basalts is more limited ($An_{84-94}$) than in the breccias ($An_{76-93}$).[10] In coarse-grained gabbro 14053,6, the plagioclase is relatively homogeneous ($An_{91}$) except near the rim, which is zoned to sodic bytownite and commonly is in contact with potash feldspar, cristobalite, and glass. In an anorthosite rock fragment from breccia 14321,22 the plagioclase zones from a core of $An_{90.8}$ to a rim of $An_{82.4}$. In KREEP basalt 14310, the plagioclase often is in contact with potash feldspar.[11] In larger crystals asymmetric zoning proceeds from a maximum of $An_{96}$ to rims of $An_{90-88}$, and MgO has been observed to increase from 0.13 to 0.23% from one edge to the other edge of the grain instead of outward from the grain centers. The cores of the larger crystals are compositionally similar to the plagioclase in the "genesis rock" anorthosite 15415 ($An_{96-97}$). In 14310 there is no indication, either optically or by means of microprobe examination, of hollow or sector-zoned plagioclase.[12] In an anorthositic fragment from a Luna 20 fines sample, high-Ca plagioclase is the chief rock-forming mineral (Table 8-1; [6]), but in the mare-basalt fragments the amount and textures of the plagioclase are similar to those of Luna 16 basalt fragments, and the compositional range is from labradorite (Table 8-1; [13]) to calcic bytownite.[13] In poikilitic basalt 60315,63, the plagioclase inclusions in pyroxene poikiloblasts are $An_{93-89}$. These crystals have significant Na enrichment at their peripheries and low variable concentrations of Fe and Mg (i.e., FeO = 0.0–0.18%; MgO = 0.12–0.19%). Relicts and interstitial plagioclase grains in this rock have a more variable composition and more calcic cores ($An_{97}$), with MgO = 0.07 to 0.08% and no Fe detected. From vesicular poikiloblastic breccia 77135,18, the plagioclase inclusions in pyroxene are euhedral laths of $An_{91-89}$, and the relict plagioclase grains are more calcic ($An_{96-94}$).[14]

Minor element content of the plagioclase is low but significant with respect to the rock types in which the plagioclase occurs. Iron is the most abundant transition element in the lunar plagioclase with a general FeO range between 0.03 and 1.2%.[15] A most striking feature of the Luna 20 plagioclase is its relatively low iron content,[16] with 95% of the plagioclase having less than 0.3% FeO[17] and some having no detectable iron (Table 8-1; [6], [14]). In several Luna 20 plagioclase crystals that are irregularly zoned in iron content, there are oriented inclusions of needlelike iron crystals. The crystallographic control by the host plagioclase of these iron crystals seems to imply that they are an exsolution phenomenon, although the possibility of simultaneous epitaxial growth of both phases cannot be ruled out entirely. The iron crystals contain only one other element—about 1 to 2% of nickel. If the iron crystals are the result of exsolution, the process must have taken place after the crystallization of the host plagioclase.[18] From anorthosite 14258,28 (1419-7), plagioclase has FeO = 0.05%.[19] On the other hand, the plagioclase for porphyritic basalt 12050 shows a range of FeO from 1.1 to 1.6%,[9] and bytownite from coarse-grained basalt 12038 has a high FeO content, up to 1.92%.[20] An average analysis of plagioclase from this rock is shown in Table 8-1, [10]. An FeO range of 0.18 to 1.8% has been reported for Luna 16 plagioclase.[21] The FeO is believed to be integral to the feldspar, since no inclusions were present in the area where the probe analyses were made.[22] Some of the Fe may be in tetrahedral coordination.[9] The nonsystematic variation of Fe within individual crystals is not understood, but it is suggested that part of the iron content is not ferrous. According to electron spin resonance studies, plagioclase crystals from anorthosite 15415, with FeO of 0.08 to 0.10% had more than 1% of the total iron as $Fe^{3+}$. It substitutes for $Al^{3+}$ in at least two nonequivalent sites.[23] Mössbauer studies of the plagioclase also seem to indicate a significant contribution of $Fe^{3+}$ to the spectra, with the ratio $Fe^{3+}/(Fe^{2+} + Fe^{3+})$ approximately between 0.02 and 0.1. The $Fe^{3+}$ may be present in the form of a complex pattern that is unknown.[15] The distribution of the minor elements in the plagioclase is partly determined by the changing composition of the melt during crystallization. There appears to be also some crystallochemical control of the various element distributions within the plagioclase grains; for example, in grains that show sector zoning, the (001) sector contains a smaller proportion of all minor elements than (010). Minor elements are more abundant in the sodic portions of crystals, with Mg varying the most and Ti the least.[3]

The foregoing data allow the following conclusions:

1. Mare basalts thus far studied have high Fe content (i.e., FeO > 0.5%) with high-Ca plagioclase.
2. KREEP basalts with a wider compositional range ($\sim An_{87}$–$An_{75}$) and low Fe content (i.e., FeO < 0.7%) have more sodic plagioclase.
3. Feldspar-rich igneous rocks have low Fe content (i.e., FeO < 0.4%) and high-Ca plagioclase ($An_{95}$ or more).[24]

Trace element content of some lunar plagioclase is given in Table 8-2.

*References*


1. Wenk and Wilde, *Contrib. Mineral. Petrol.* **41,** 98, 100.
2. Storey and O'Hara, *Lun. Sci. V, Abstr.*, 752.
3. Crawford, *Suppl.* **4,** 709, 713–714.
4. Wood et al, *Suppl.* **1,** 966.
5. Gay et al, *Suppl.* **2,** 382.
6. Drake et al, *EPSL* **9,** 118.
7. Mason et al, *Lun. Sci. Conf.* '71, *Abstr.*, 257.
8. Stewart et al, *Suppl.* **1,** 927.
9. Appleman et al, *Suppl.* **2,** 120–121, 124–125.
10. Czank et al, *Suppl.* **3,** 603.
11. Kushiro et al, *Suppl.* **3,** 121.
12. Trzcienski and Kulick, *Suppl.* **3,** 596–597.
13. Tarasov et al, *Suppl.* **4,** 339.
14. Bence et al, *Suppl.* **4,** 602, 607.
15. Schürmann and Hafner, *Suppl.* **3,** 615, 619.
16. Adams et al, *Geochim. Cosmochim. Acta* **37,** 734.
17. Reid et al, *Geochim. Cosmochim. Acta* **37,** 1015.
18. Bell and Mao, *Geochim. Cosmochim. Acta* **37,** 757–758.

**Table 8-2 Trace Elements in Some Lunar Plagioclase**

| | | [1] | [2] | [3] | [4] | [5] | [6] |
|---|---|---|---|---|---|---|---|
| | | 10084,75 | 14063 | 14321, 184–55 | Apollo 15 | 61222,3 | 65015 |
| Ru | (ppb) | <12– 39 ± 15 | | | | | |
| Os | (ppb) | 1–5.6 | | | | | |
| F | (ppm) | 17–60 | | | | | |
| Br | (ppm) | ≥0.06–0.25 | | | | | |
| U | (ppm) | 0.16 (est.) | | | | | |
| Sc | (ppm) | | 4.0 | | | | |
| Sm | (ppm) | | 6.8 | | | | |
| Eu | (ppm) | | 1.06 | | | | |
| La | (ppm) | | 2.56 | | | | |
| Ce | (ppm) | | 5.0 | | | | |
| Tb | (ppm) | | 0.22 | | | | |
| Dy | (ppm) | | 1.52 | | | | |
| Yb | (ppm) | | 0.86 | | | | |
| Lu | (ppm) | | 0.113 | | | | |
| Hf | (ppm) | | 0.6 | | | | |
| Rb | (ppm) | | | 2.01 | 0.078– 0.686 | 0.136 | 2.050– 6.54 |
| Sr | (ppm) | | | 22.62 | 244.6– 340.8 | 17.22 | 169.6– 185.8 |

*References*

[1] Reed et al, *EPSL* **11,** 356. From a fines sample.

[2] Helmke and Haskin, *Lun. Sci. III*, *Rev. Abstr.*, 367–368. Handpicked crystals from a breccia.

[3] Mark et al, *Lun. Sci. IV*, *Abstr.*, 501. From a basaltic clast in a breccia.

[4] Papanastassiou et al, *EPSL* **17,** 332. From various basaltic and gabbroic samples.

[5] Mark et al, *Lun. Sci. IV*, *Abstr.*, 501. From a basaltic clast in a breccia.

[6] Papanastassiou et al, *EPSL* **17,** 56. From a KREEP-rich polymict rock.

19. Powell and Weiblen, *Suppl.* **3,** 847.
20. Keil et al, *Suppl.* **2,** 323–324.
21. Jakeš et al, *EPSL* **13,** 261.
22. Agrell et al, *Suppl.* **1,** 109.
23. Hafner, *Lun. Sci. IV*, *Abstr.*, 326–327.
24. Steele and Smith, *Geochim. Cosmochim. Acta* **37,** 1076–1077.

*X-Ray Data*

The lunar plagioclases may be uniquely informative of the ordering process in terrestrial feldspars because of their crystalline perfection and freedom from chemical alteration. Although the exact nature of the Al/Si ordering is not known in any lunar plagioclase, simple estimates suggest that many Apollo 11 and 12 plagioclases are neither as highly disordered as has been observed in some heated terrestrial samples of nearly equivalent composition, nor as highly ordered as has been observed in plagioclase with nearly the same composition from terrestrial intrusive rocks.[1]

Some lunar bytownites show "incoherent" two-phase diffraction patterns and are interpreted as being submicroscopically intergrown phases with different An contents.[2] By means of single-crystal X-ray studies, antiperthitic intergrowths have been recognized in some of the plagioclase from breccia 12013,10. The host crystals are in the labradorite–bytownite range and the included phase is believed to be a potash feldspar of some kind. The nature of this included phase has not been established finally,[3] but an earlier report suggested that it was albite.[4]

TEM studies of plagioclase from KREEP basalt 14310 revealed isolated submicroscopic twin lamellae, commonly occurring singly or as two or three parallel lamellae about 1 $\mu$ wide. A few lamellae are as small as 0.2 $\mu$. Twin laws identified are albite and Baveno-*r* (Table 8-3; [4]). Dark-field electron micrographs of the plagioclase show both large and small *b*-antiphase domains, and an exsolution structure in crystals display *b*-split reflections in the diffractogram. Diffuseness of *c* reflections in the X-ray photographs and the inability to resolve *c* domains in the electron micrographs of this anorthite ($An_{94}$) indicate relatively rapid cooling of rock 14310 compared with plutonic rocks.[5] From an extensive study of antiphase domains in lunar (as well as some terrestrial) plagioclase, several conclusions have been reached. There appears to be a small compositional region ($\sim An_{95}$) where at rather high temperatures, both *c* and *b* domains can form in the same crystal. With increase in Na content, *b* domains become more and more crystallographically oriented and sometimes are difficult to distinguish from exsolution structures. This suggests that *b* domains may act as nuclei for exsolution processes. At present it can be concluded only qualitatively that rather rapid cooling from above 1000°C was a rule for basaltic rocks, that anorthosites cooled more slowly, and that anorthites in breccias often have been annealed, therefore displaying domains. This information has been obtained also from observations of textures in thin sections.[6]

The total range of variation in the cell parameters of lunar plagioclase is very limited and apparently shows a dependence on Ca content.[1] Cell constants for some lunar plagioclase are given in Table 8-3.

**Table 8-3 X-Ray Data for Some Lunar Plagioclase**

| X-Ray Data (Å) | [1]<br>15415,22 | [2]<br>Luna 20<br>22003,1,3a | [3]<br>Luna 20<br>926–3 | [4]<br>14310 | [5]<br>12032,44 | [6]<br>Luna 20<br>22002,2,7d |
|---|---|---|---|---|---|---|
| $a$ | 8.179(1) | 8.19 | 8.14 ±0.05 | 8.18 | 8.16 ±0.01 | 8.16 |
| $b$ | 12.879(1) | 12.86 | 12.9 ±0.2 | 12.87 | 12.89 ±0.01 | 12.88 |
| $c$ | 14.179(1) | 14.21 | 14.12 ±0.1 | 14.19 | 14.18 ±0.01 | 7.17 |
| Space Group | $P\bar{1}$ | $P\bar{1}$ | $P\bar{1}$ | $P\bar{1}$ | $I\bar{1}$ | $C\bar{1}$ |

*References*

[1] Stewart et al, *Lun. Sci. III*, *Rev. Abstr.*, 726–727. From an anorthosite. Anorthite ($An_{96.5}$) see Table 8-1; [3].

[2] Cameron et al, *Geochim. Cosmochim. Acta* **37,** 788. From an anorthositic fragment in soil sample. Anorthite ($An_{95}$). See Table 8-1; [5].

[3] Tarasov et al, *Suppl.* **4,** 339. From an anorthositic fragment in soil sample. Anorthite ($An_{94}$). See Table 8-1; [6].

[4] Wenk et al, *Suppl.* **3,** 575. From a KREEP basalt. Anorthite ($An_{93.4}$).

[5] Wenk and Nord, *Suppl.* **2,** 139. From a fines sample. Calcic bytownite ($An_{88.8}$). See Table 8-1; [9].

[6] Cameron et al, *Geochim. Cosmochim. Acta* **37,** 788. Handpicked crystal from an anorthositic fragment in soil sample. Albite ($An_3$). See Table 8-1; [14].

*References*

1. Appleman et al, *Suppl.* **2,** 125, 131.
2. Jagodzinski and Korekawa, *Suppl.* **3,** 555.
3. Bown and Gay, *EPSL* **11,** 23, 26.
4. Bancroft et al, *Lun. Sci. Conf. '71, Abstr.*, 256.
5. Wenk et al, *Suppl.* **3,** 569, 575–576.
6. Wenk et al, *Suppl.* **4,** 920–921.

## Potash Feldspars $KAlSi_3O_8$

*Synonymy*

Alkali feldspar; Agrell et al, *Suppl.* **1,** 125.
Ba-bearing alkali feldspar; Weill et al, *Suppl.* **2,** 414.
Ba, K-rich phase; Anderson and Smith, *Suppl.* **2,** 437.
Barian K-feldspar; Keil et al, *Suppl.* **2,** 325.
Barian sanadine; Brown et al, *Suppl.* **3,** 154.
Barium alkali feldspar; Weill et al, *Suppl.* **2,** 417.
Barium-bearing sanadine; Peckett, *Moon* **3,** 406.
Ba-rich K-feldspar; Lovering et al, *Suppl.* **3,** 288.
Ba-sanadine; Brown et al, *Lun. Sci. III, Rev. Abstr.*, 95.
Celsian orthoclase feldspar; Tzrcienski and Kulick, *Suppl.* **3,** 591.
Hyalophane; Christophe-Michele-Lévy et al, *Suppl.* **3,** 889.
K, Ba feldspar; Meyer et al, *Suppl.* **2,** 396.
K, Ba-rich feldsparlike phase; Anderson and Smith, *Suppl.* **2,** 431.
K-feldspar; Keil et al, *Suppl.* **1,** 579.
$K_2O$-BaO feldspar; Gancarz et al, *EPSL* **12,** 5.
$K_2O$-BaO-rich feldspar; Gancarz et al, *EPSL* **12,** 4.
"K-rich phase"; Albee and Chodos, *Suppl.* **1,** 138.
Orthoclase; Weill et al, *Suppl.* **2,** 413.
Potassic feldspar; Drake et al, *EPSL* **9,** 113.
Potassium-barium feldspar; Dence et al, *Suppl.* **1,** 322.
Potassium feldspar; Anderson and Smith, *Suppl.* **2,** 433.
Sanadine; Pickart and Alperin, *Suppl.* **2,** 2081.

*Occurrence and Form*

Some potash feldspar may be an exsolution phase in the lunar plagioclase. For example, in the KREEP basalts of Apollo 14, the sodic plagioclase (containing more $K_2O$ than the more abundant calcic plagioclase) has what appears to be an exsolved potash feldspar phase. More commonly

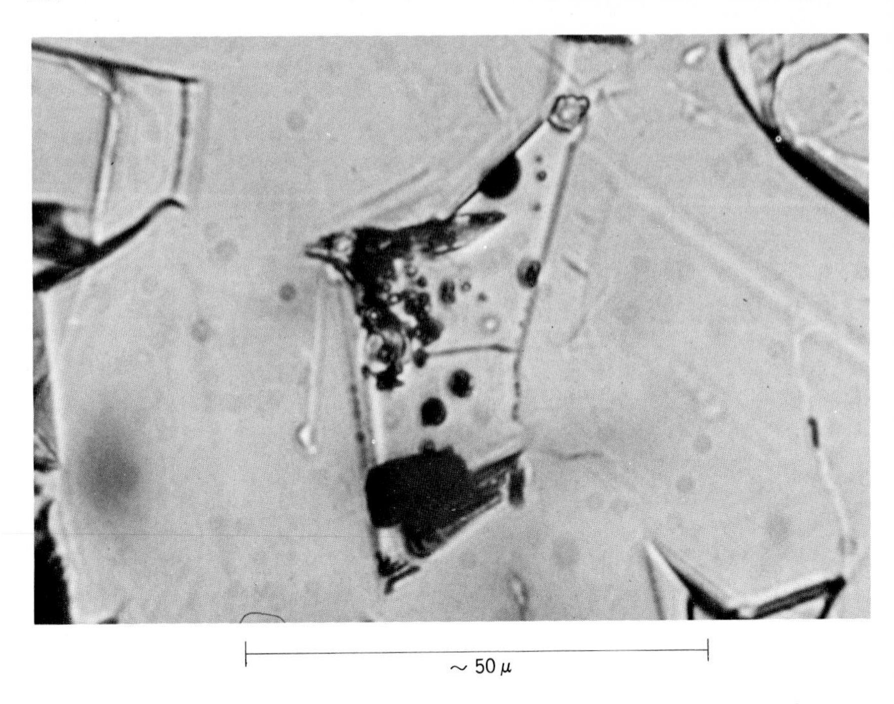


Fig. 8–14 Pool of potassic glass 0.05 mm long, center of photograph. Inclusions are Fe-pigeonite (larger dark grains) and opaque globules. From *Suppl.* **4**, 652. Photograph courtesy of R. Helz, U.S. Geological Survey, Reston, Va.

the potash feldspar is interstitial to the plagioclase and orthopyroxene in these rocks.[1]

The potash feldspar, variously termed orthoclase[2] or sanadine,[3] often contains some BaO. In the igneous rocks the potash feldspar and barian potash feldspar appear as interstitial patches frequently associated with residual glass rich in $SiO_2$ and $K_2O$ (Figure 8–14).[2] In breccia 10046-48-2, a unique angular fragment is composed of 100-$\mu$-long interlocking laths of potash feldspar with a silica mineral (probably tridymite) in the interstices (Table 8-4; [1]).[4] In several Apollo 14 breccias, so-called rhyolitic fragments contain cores of quartz and potash feldspar,[5] and in the glassy matrix of breccia 12057,22, a 1-$mm^2$ clast is an intergrowth of potash feldspar and quartz.[6] It is this type of intergrowth that has given rise to the term "granitic component" discussed previously. In porphyritic basalt 12021, silica-rich patches of the interstitial phases contain rare anhedral crystals of barian K-feldspar up to 50 $\mu$ in diameter,[2] and barian sanadine has been reported from KREEP basalt 14310 as well as basalt 14073.[7]

In breccia 14066, the barian K-feldspar occurs as a euhedral crystal about 350 μ long.[8]

*References*

1. Meyer et al, *Suppl.* **2,** 405.
2. Weill et al, *Suppl.* **2,** 413, 425.
3. Pickart and Alperin, *Suppl.* **2,** 2081.
4. Agrell et al, *Suppl.* **1,** 125.
5. Anderson et al, *Suppl.* **3,** 823.
6. Sclar and Bauer, *Lun. Sci. V, Abstr.*, 688.
7. Brown et al, *Lun. Sci. III, Rev. Abstr.*, 95.
8. Christophe-Michel-Lévy et al, *Suppl.* **3,** 889.

*Optics*

Little optical data have been reported for the potash feldspars. In the "gray mottled" basalts of Apollo 12, what may be a K-rich glass rather than a feldspar was noted to have low birefringence.[1] A similar phase (called simply a "K-rich phase") from LR-1 may be a feldspar but is predominantly isotropic.[2] From KREEP basalt 14310, a barian K-feldspar crystal has a 2V $< 20°$ and therefore has been termed sanadine.[3] In a study of lunar phases by luminescence petrography, a blue luminescence was noted in minute flakes of potash feldspar from some Apollo 12 porphyritic basalts.[4]

*References*

1. Anderson and Smith, *Suppl.* **2,** 435.
2. Albee and Chodos, *Suppl.* **1,** 138.
3. Brown et al, *Suppl.* **3,** 153.
4. Sippel, *Suppl.* **2,** 250.

*Chemical Composition*

Potassium is a very minor constitient in the lunar rocks. A small percentage of this element is contained in some of the plagioclase, but in general the potassium is concentrated in the residual melt. There it appears in the potash feldspars.

Some analyses report no barian content (e.g., Table 8-4; [1]–[3]), but BaO is a common and variable constituent of the potash feldspars. From KREEP basalt 14310, potash feldspar that shows some compositional zoning ($An_{5.7}Ab_{12.8}Or_{81.5}$ to $An_{3.1}Ab_{6.4}Or_{90.5}$) also contains BaO ranging from 3 to 6%).[1] The BaO content ranges inversely with the $K_2O$ content,[2] indicating some solid solution between orthoclase ($KAlSi_3O_8$) and celsian

**Table 8.4 Potash Feldspar Analyses**

| | [1] 10046–48–2 | [2] 14162,41 | [3] 10045–29 | [4] 14305,77 | [5] 14206, 53–20A | [6] 14310 |
|---|---|---|---|---|---|---|
| $SiO_2$ | 64.63 | 65.1 | 64.6 | 62.5 | 62.5 | 60.79 |
| $Al_2O_3$ | 18.15 | 18.7 | 19.9 | 19.5 | 18.4 | 19.24 |
| $TiO_2$ | — | 0.38 | — | 0.30 | — | — |
| FeO | 0.10 | 0.14 | 0.6 | 0.23 | — | 0.24 |
| MgO | 0.02 | — | — | 0.02 | — | 0.01 |
| CaO | 0.33 | 0.6 | 0.72 | <0.1 | 0.8 | 0.58 |
| $Na_2O$ | 2.22 | 1.0 | 0.33 | 0.79 | 0.5 | 0.83 |
| $K_2O$ | 12.55 | 12.7 | 13.1 | 15.2 | 13.7 | 13.56 |
| BaO | — | — | — | 1.0 | 2.6 | 3.84 |
| Total | 98.00 | 98.62 | 99.25 | 99.60* | 98.5 | 99.09 |

*References*

[1] Agrell et al, *Suppl.* **1,** 125. From a breccia. Potash feldspar.

[2] Powell and Weiblen, *Suppl.* **3,** 846. From a norite fragment in soil sample. Potash feldspar.

[3] Keil et al, *Suppl.* **1,** 579. From a fine-grained olivine basalt. Potash feldspar.

[4] Lovering et al, *Suppl.* **3,** 289. From a breccia. Average of two analyses. *Analysis includes $P_2O_5$ = 0.06% and MnO = <0.05%. Potash feldspar.

[5] Anderson et al, *Suppl.* **3,** 823. From a "granitic" fragment in a breccia. Barian K-feldspar.

[6] Trzcienski and Kulick, *Suppl.* **3,** 597. From a KREEP basalt. Barian K-feldspar.

($BaAl_2Si_2O_8$). It has been noted that silica contents often are in excess of stoichiometric (K,Ba) feldspar, and it is possible that this phase is intergrown with free silica.[2] Amounts of BaO exceeding 12% have been found in the barian K-feldspar from basalt 12063,15 (Table 8-4; [9]). The analysis calculated to a formula of

$$(K_{0.57}Na_{0.01}Ca_{0.03}Ba_{0.22}Fe_{0.04})_{0.87}Al_{1.27}Si_{2.75}O_8,$$

and the phase has been termed a celsian-orthoclase ($Cn_{27}$).* If the low total of the analysis is due to partial volatilization of 1.50% $K_2O$, a new formula, based on $K_2O = 10.3\%$, yields

$$(K_{0.66}Na_{0.1}Ca_{0.03}Ba_{0.24}Fe_{0.04})_{0.98}Al_{1.26}Si_{2.73}O_8.$$

* This is an incorrect term, since orthoclase and celsian are structurally different phases.

| | [7] | [8] | [9] | [10] | [11] |
|---|---|---|---|---|---|
| | 14276,13 | 12039,3 | 12063,15 | 10085–LR–1 | |
| $SiO_2$ | 61.44 | 57.2 | 54.45 | 61.0 | 59.6 |
| $Al_2O_3$ | 19.45 | 20.1 | 21.35 | 20.5 | 20.7 |
| $TiO_2$ | — | — | — | — | 0.27 |
| FeO | 0.02 | 0.35 | 1.00 | — | 0.30 |
| MgO | 0.02 | — | — | — | 0.04 |
| CaO | 0.61 | 0.55 | 0.55 | 2.07 | 1.36 |
| $Na_2O$ | 1.73 | 0.81 | 0.13 | 1.24 | 0.89 |
| $K_2O$ | 12.02 | 11.1 | 8.83 | 10.9 | 9.42 |
| BaO | 4.35 | 9.3 | 12.15 | 2.73 | 6.25 |
| Total | 99.64 | 99.41 | 98.46 | 98.44 | 98.90* |

*References*

[7] Gancarz et al, *EPSL* **16,** 313. From a gabbroic anorthosite. Barian K-feldspar.

[8] Keil et al, *Suppl.* **2,** 325. From a coarse-grained basalt. Barian K-feldspar.

[9] Trzcienski and Kulick, *Suppl.* **3,** 594. From a basalt. Barian K-feldspar (termed celsian orthoclase [$Cn_{27}$]).

[10] Albee and Chodos, *Suppl.* **1,** 138. From "gray mottled" basalt in soil sample. (Low-Ba phase.) Shock-isotropized barian K-feldspar or interstitial residual glass?

[11] ———, *ibid.* From a "gray mottled" basalt in soil sample. (High-Ba phase.) *Analysis includes $Cr_2O_3$ = 0.04% and MnO = 0.03%. Shock-isotropized barian K-feldspar or interstitial residual glass?

This calculates to $Cn_{24}$.[3] Careful X-ray study is needed to determine whether there is a structural difference between this high Ba-bearing feldspar and the other lunar barian K-feldspars before a new phase can be assumed. A euhedral crystal of barian K-feldspar, from spinel-bearing breccia 14066, has been termed hyalophane (Figure 8–15). The formula of this feldspar is given as $K_{0.765}Na_{0.082}Ca_{0.027}Ba_{0.126}Al_{1.153}Si_{2.84}O_8$.[4] This calculates to approximately 7.5% BaO. Hyalophane (with BaO = 6.29%) has been reported also from soil sample 15532,11.[5]

There is considerable compositional variation in the "K-rich phases" from LR-1 (Table 8-4; [10], [11]). Both the high-Ba and low-Ba analyses approach the stoichiometry of feldspars, but it has been suggested that these phases may be interstitial residual glass rather than shock-isotropized feldspar.[6]

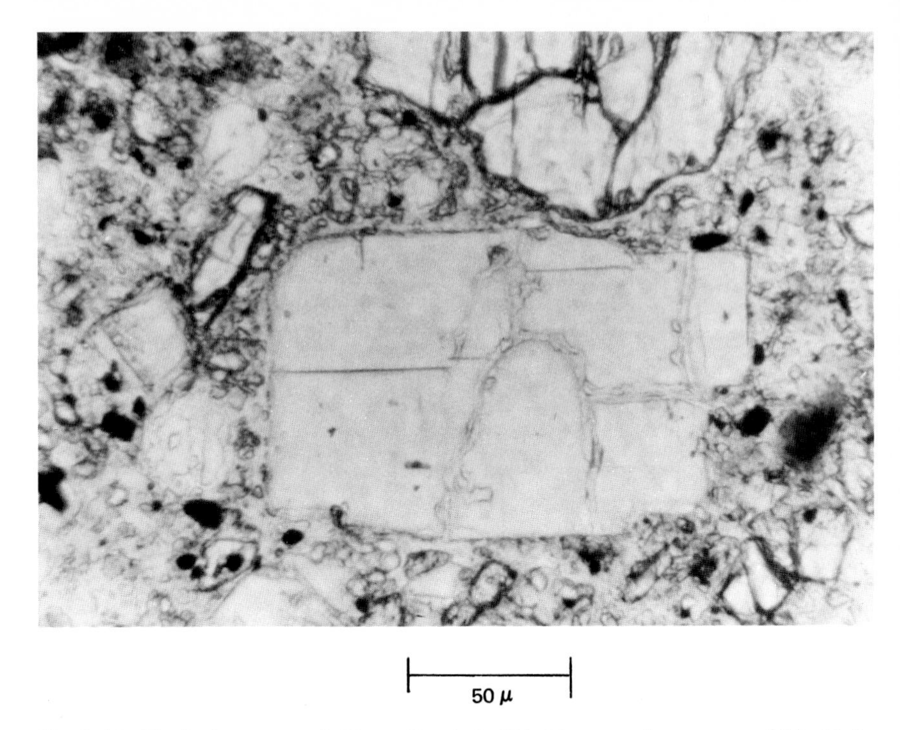


Fig. 8–15 Hyalophane crystal. From *Suppl.* **3,** 891. Photograph courtesy of M. Christophe-Michel-Lévy, Université de Paris, France.

## *References*


1. Kushiro et al, *Suppl.* **3,** 121.
2. Meyer et al, *Suppl.* **2,** 405.
3. Trzcienski and Kulick, *Suppl.* **3,** 591, 594.
4. Christophe-Michel-Lévy et al, *Suppl.* **3,** 889.
5. Weiblen and Roedder, *Suppl.* **4,** 687.
6. Albee and Chodos, *Suppl.* **1,** 138.

# CHAPTER NINE | SILICATES: PYROXENES AND PYROXFERROITE

In the lunar crystalline rocks and many of the breccias, the pyroxenes are the most abundant group of minerals.[1] Most of them are clinopyroxenes, but orthopyroxenes, predominantly the magnesian varieties, also occur. The lunar pyroxenes fall within the diopside ($MgCaSi_2O_6$)-hedenbergite ($MgFeSi_2O_6$)-ferrosilite ($Fe_2Si_2O_6$)-enstatite ($Mg_2Si_2O_6$) compositions, and their chemical variations and nomenclature are indicated in the pyroxene quadrilateral (Figure 9–1). Although no lunar pyroxene approaches wollastonite ($CaSiO_3$), this phase is a significant component of the augites, and up to 46 mole % of $CaSiO_3$ has been reported.[2]

Chemical trends in the lunar pyroxenes can be correlated with bulk composition of the host rock, paragenetic sequence, emplacement history, and variations in oxygen fugacity.[3] There is continuous chemical zoning within many individual grains, and with rapid crystallization exsolution of pigeonite in augite (and augite in pigeonite) has occurred. Sectoral zoning often is pronounced.

Augite and subcalcic augite are dominant in the Apollo 11 rocks,[4] and in both the fine-grained and coarser-grained igneous rocks, the augite-pigeonite series makes up almost 50% of the rocks' volume.[5]

Because of their chemical variability, exsolution textures, epitaxy with augite and orthopyroxene, domain structure, range of intracrystalline cation ordering, and relative abundance, lunar pigeonites have played a significant role in a reconstruction of the crystallization histories of mare basalts. For example, pyroxene vitrophyre 15597 is believed to have arrived at the lunar surface in an essentially undifferentiated state; probably it was separated from the main magma body and quenched to low temperature in or on the lunar regolith. In this rock, pigeonite and augite nucleated in a rapid metastable fashion before final solidification and compositional zoning occurred, accompanied by overgrowth of augite.

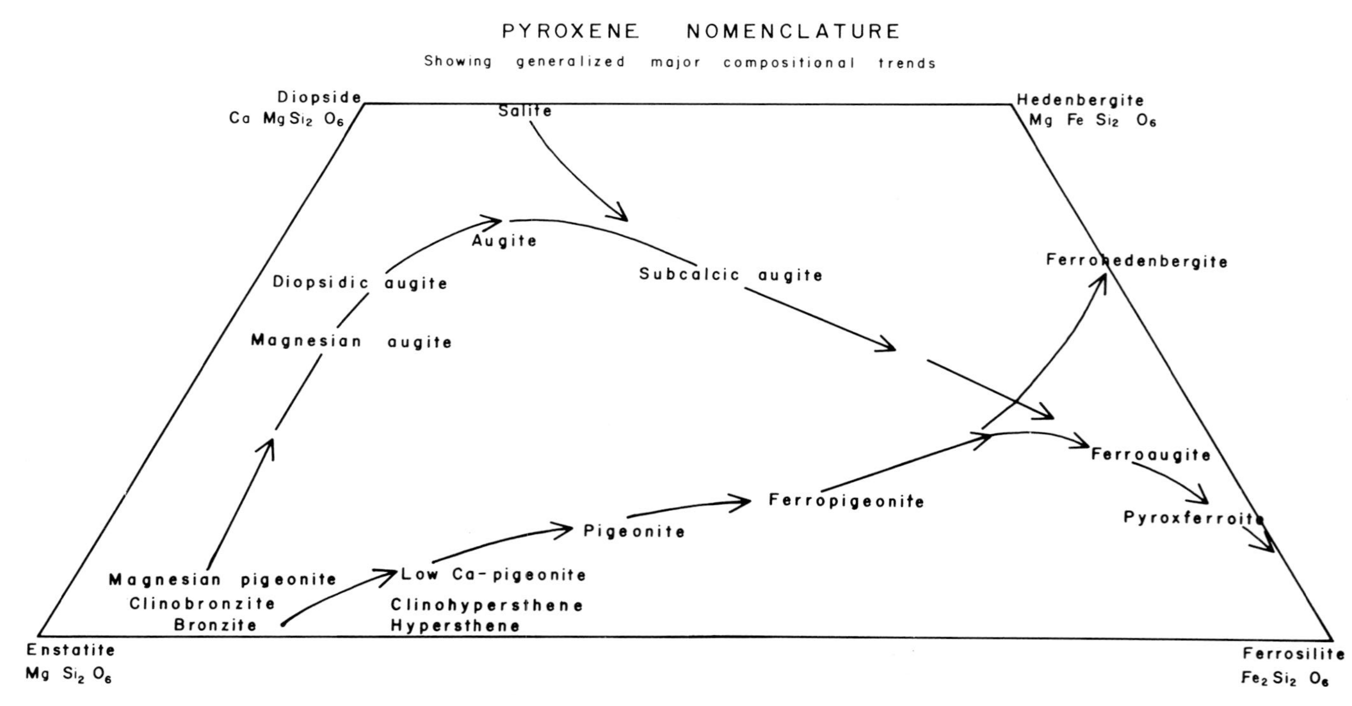


Fig. 9–1 Pyroxene quadrilateral with nomenclature and generalized major compositional trends of lunar pyroxenes.

The crystallographic features of the pigeonites may represent one extreme of pigeonite cooling history on the Moon.[6]

The pyroxenes in KREEP basalt 14310 fall into three major groups: bronzitic orthopyroxenes, twinned magnesian pigeonites, and untwinned Fe-rich pigeonites. The crystallographic features of the pyroxenes, such as twinned magnesian pigeonite overgrowths on bronzite with common (100) are distinct from those of mare pyroxenes and suggest that such features may be useful in characterizing lunar rocks and their crystallization trends. The mode of exsolution and the chemical trends of these pyroxenes indicate that the crystallization was much more rapid than that of plutonic or intrusive rocks but was not as metastable as that of many mare rocks.[7] A strong correlation of KREEP glass with orthopyroxene is well established. Orthopyroxenes have not been reported in the indigenous Apollo 11 or 12 basalts and are considered to be indicators of an exotic KREEP component in their soils.[8] Cation distribution and crystal structures of the Apollo 14 orthopyroxenes are similar to those of terrestrial pyroxenes of volcanic origin.[7]

Compositions of single crystals of pyroxene as well as plagioclase in the Apollo 15 brown glass microbreccias (constituting about 90% of the soil fragments) suggest that the Imbrium ejecta below the regolith are composed of comminuted KREEP basalt, recrystallized noritic rocks, and anorthositic rocks.[9]

A new pyroxenoid mineral structurally identical with terrestrial pyroxmangite, but containing only minor Mn, was named pyroxferroite for its chemical composition.[10] At a moderately late crystallization stage pyroxenes no longer crystallized, but pyroxferroite appeared.[11] It is believed that this metastable triclinic phase grew (instead of a more stable assemblage) at a composition that is less stable in the monoclinic than in the triclinic modification because there was already a pyroxene substrate on which to nucleate.[12] In late stages of crystallization of the mesostasis, where the pyroxferroite is unstable, a pyroxene close to hedenbergite (i.e., ferrohedenbergite) appeared.[11]

Two main trends are indicated in plots of pyroxene compositions from a number of lunar rock types (Figure 9–2).

1. Highly magnesian clinopyroxenes and orthopyroxenes that occur in anorthositic and noritic crystalline rocks and poikiloblastic breccias.
2. More ferrous clinopyroxenes from mare basalts, olivine and feldspathic microgabbros, and feldspathic peridotites.

*References*

1. Dence et al, *Suppl.* **1,** 319.
2. Hargraves and Hollister, *Sci.* **175,** 430.

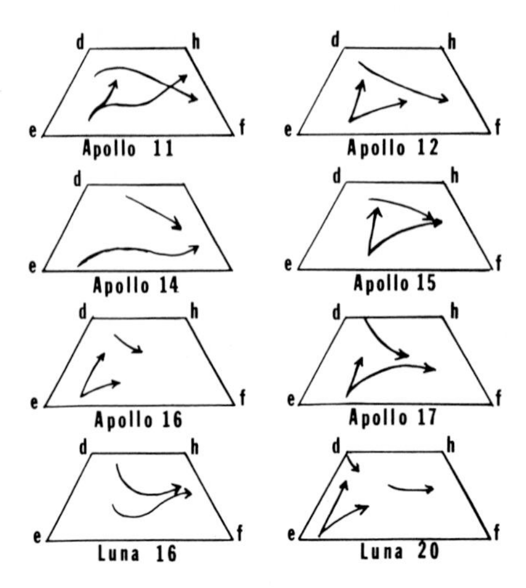


Fig. 9–2 Generalized compositional trends for pyroxenes from the various lunar missions.

3. Papike and Bence, *Suppl.* **3,** 461.
4. James and Jackson, *J. Geophys. Res.* **75,** 5802.
5. Schmitt et al, *Suppl.* **1,** 29, 31.
6. Brown and Wechsler, *Suppl.* **4,** 887–888, 894.
7. Takeda and Ridley, *Suppl.* **3,** 425, 429.
8. Meyer et al, *Suppl.* **2,** 404.
9. Cameron et al, *EPSL* **19,** 20.
10. Chao et al, *Suppl.* **1,** 78.
11. Agrell et al, *Suppl.* **1,** 101.
12. Hollister and Hargraves, *Suppl.* **1,** 549.

## Clinopyroxenes: Diopside ($CaMgSi_2O_6$)-Hedenbergite ($MgFeSi_2O_6$)-Ferrosilite ($Fe_2Si_2O_6$) Series

*Synonymies*

AUGITE Ca(Mg, Fe) $Si_2O_6$

Al-poor ferroaugite; Boyd and Smith, *J. Petrol.* **12,** 456.
Al-rich augite; Boyd and Smith, *J. Petrol.* **12,** 442.
Calcic augite; Stewart et al, *Lun. Sci. III*, *Rev. Abstr.*, 727.
Calcic clinopyroxene; Meyer, *Geochim. Cosmochim. Acta* **37,** 945.
Calcic pyroxene (in part); Dence et al, *Suppl.* **2,** 285.
Diopsidic augite; Hargraves and Hollister, *Sci.* **175,** 430.
Fe-augite; Dowty et al, *Suppl.* **3,** 484.

Fe-poor augite; Weigand and Hollister, *Suppl.* **3,** 476.
Fe-rich augite; Boyd and Smith, *J. Petrol.* **12,** 441.
Fe-rich subcalcic augite: Carter et al, *Suppl.* **2,** 780.
Ferroaugite; Frondel et al., *Suppl.* **1,** 445.
Mg-augite; Dowty et al, *Suppl.* **3,** 484.
Mg-rich augite; Boyd and Smith, *J. Petrol.* **12,** 441.
Subcalcic augite; Frondel et al, *Suppl.* **1,** 445.
Subcalcic clinopyroxene; Smith et al, *Suppl.* **1,** 920.
Subcalcic ferroaugite; Kushiro and Nakamura, *Suppl.* **1,** 609.
Ti-rich augite; Agrell et al, *Suppl.* **1,** 102.
Titanaugite; Keil et al, *Sci.* **167,** 597.

DIOPSIDE $CaMgSi_2O_6$

Fe-bearing diopside; Walter et al, *Lun. Sci. Conf. '72, Abstr.*, 685.
Magnesian diopside; Essene et al, *Suppl.* **1,** 393.
Salite; D. Walker, Department of Geological Sciences, Harvard University, personal communication.
Subcalcic diopside; Essene et al, *Suppl.* **1,** 391.

FERROSILITE $Fe_2Si_2O_6$

Fe-rich pyroxene; Albee et al, *EPSL* **13,** 357.

HEDENBERGITE $CaFeSi_2O_6$

Ferrohedenbergite; Essene et al, *Suppl.* **1,** 391.
Hedenbergitic clinopyroxene; Kushiro et al, *Suppl.* **3,** 118.
Low-calcium ferropyroxene; Brown et al, *Suppl.* **3,** 142.

PIGEONITE $m(MgFeSi_2O_6) + n(Ca_2Si_2O_6)$

Calcic pigeonite; Carter et al, *Suppl.* **2,** 780.
Ferropigeonite; James and Jackson, *J. Geophys. Res.* **75,** 5795.
Iron-rich pigeonite; Weill et al, *Suppl.* **2,** 416.
Low-calcium pigeonite; Gay et al, *Suppl.* **3,** 352.
Magnesian pigeonite; Takeda and Ridley; *Lun. Sci. Conf. '72, Abstr.*, 652.
Mg-pigeonite; Dowty et al, *Suppl.* **3,** 484.
Mg-rich pigeonite; Takeda and Ridley, *Lun. Sci. Conf. '72, Abstr.*, 653.
Subcalcic ferropigeonite; Brown et al, *Lun. Sci. Conf. '72, Abstr.*, 87.

CLINOBRONZITE $MgFeSi_2O_6$

Ghose et al, *Geochim. Cosmochim. Acta* **37,** 833.

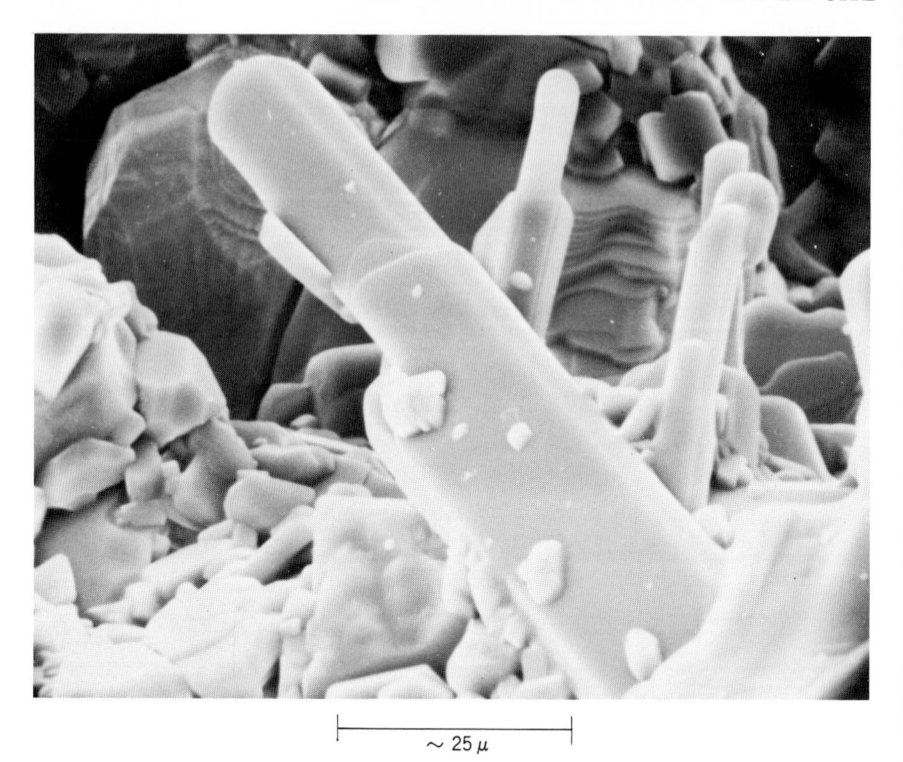


Fig. 9–3 SEM photograph of low-Ca prismatic pyroxene crystals in a vug (from a breccia fragment in fines sample 14161). Pyroxene needles smaller than $2\mu$ in diameter are rounded and have hemispherical terminations; larger crystals have regular faces. From *Suppl.* **3,** 743. NASA photograph S-71-58419, courtesy of U. Clanton, Johnson Space Center, Houston.

CLINOENSTATITE $Mg_2Si_2O_6$

Takeda and Ridley, *Suppl.* **3,** 425.

CLINOHYPERSTHENE $FeMgSi_2O_6$

Ghose et al, *Geochim. Cosmochim. Acta* **37,** 834.

*Occurrence and Form*

Clinopyroxenes occur as equant grains up to several hundred microns in diameter, and as smaller elongated grains in complex intergrowth with plagioclase.[1] In fine-grained rocks they are usually anhedral, but some clinopyroxenes are also in euhedral prisms. In the coarse-grained rocks

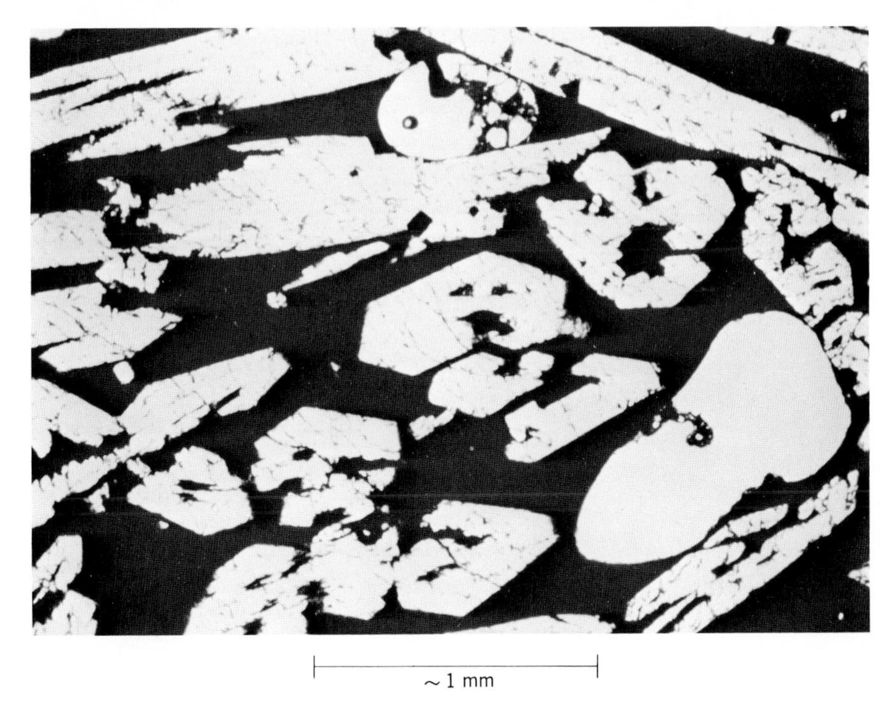


Fig. 9–4 In basalt 15485,3, elongate euhedral clinopyroxene phenocrysts are zoned from hollow cores (now filled with groundmass) through pigeonite to augite rims. NASA photograph S-71-51742, courtesy of R. Laughan, Johnson Space Center, Houston.

they are subhedral to ophitic.[2] A few euhedral crystals of pyroxene found in the fines and in vugs of the microgabbros of Apollo 11 were short prismatic in habit with (100), (010), small (110), and rounded terminal faces of (001) and (0*kl*).[3] Elongate prisms of pyroxene occur in a vug of an Apollo 14 breccia fragment (Figure 9–3).[3a] Twins on (100) were noted both in euhedral crystals and in grains in thin sections.[3] In medium-grained picritic basalt 12002, phenocrysts of pyroxene, as much as 3 mm long, are lathlike, complexly zoned, and intergrown with olivine. The clinopyroxene also forms bundles of phenocrysts radiating from a common nucleus.[4] In mare basalts samples 15604,4 and 15604,5, euhedral skeletal clinopyroxenes occur in a matrix of dark brown glass. The pyroxenes are also in fine intergrowth with plagioclase and opaques.[5] In fragments of pyroxene basalts from Apollo 15 rake samples, highly systematically zoned hollow pyroxene phenocrysts occur in a groundmass of pyroxene, plagioclase, and ilmenite laths (Figures 9–4 and 9–5). In pyroxene vitrophyre 15597, zoned hollow pyroxene crystals are in a glassy matrix. In other rock fragments of olivine basalts,

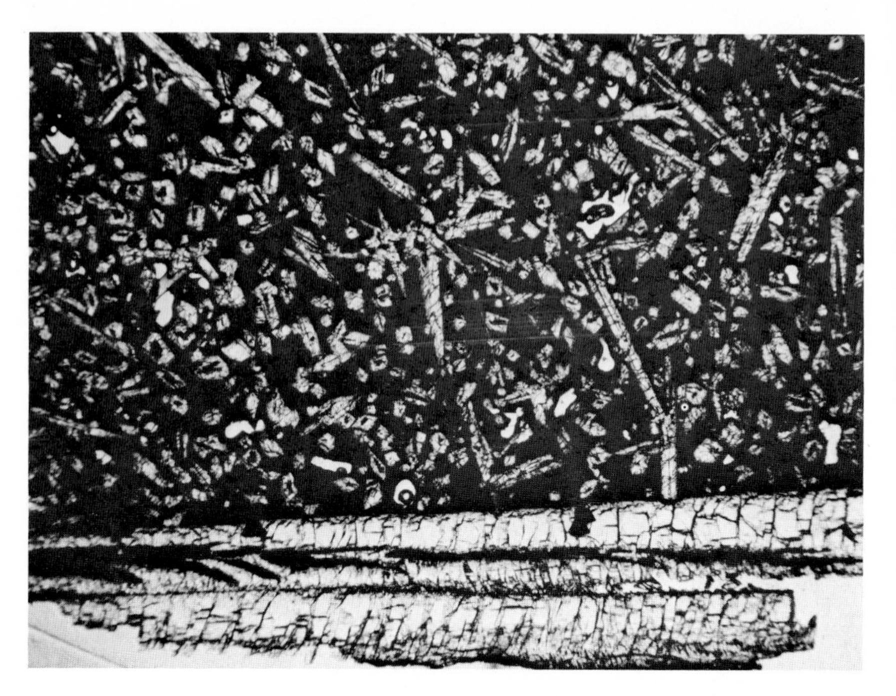

Fig. 9–5 Zoned hollow pyroxene crystals in glass matrix of pyroxene vitrophyre 15597. NASA photograph S-71-52209, courtesy of R. Laughan, Johnson Space Center, Houston.

the groundmass pyroxenes are almost equant and frequently poikilitically enclosed in somewhat larger plagioclase laths.[6] In numerous poikilitic rocks fragments from Apollo 16 rake samples, large pyroxene crystals up to 3 mm long (termed oikocrysts) enclose abundant smaller crystals (termed chadocrysts) of mostly feldspar and some mafic silicates.[7] Inclusions of ilmenite in pyroxene have been observed (Figure 9–6).

Continuous zoning (also termed normal zoning) such as seen in pyroxenes from the mare basalts, is due to depletion of the melt in Mg by the growth of an abundant mineral in what is essentially a closed system.[8] The remarkable zoning of the subcalcic clinopyroxenes in the microgabbros is related to the fractionation of the magma.[9] In anorthositic gabbro 68415, individual clinopyroxene crystals are only weakly zoned and vary compositionally from grain to grain. They range from Mg-rich pigeonite, rimming olivine, to ferroaugite molded with clear glass in the angular interstices of fine plagioclase laths.[10] The smaller ($\sim$ 0.2 mm) anhedral to irregular clinopyroxene grains, interstitial to the plagioclase framework, are chiefly Mg rich and do not display extensive zoning to Fe-rich compositions.[11]

Fig. 9–6 Ilmenite inclusions in a pyroxene grain (center of picture). Photograph courtesy of C. Frondel, Harvard University, Cambridge, Mass.

In breccia 67955, pyroxene poikiloblasts are more than 0.05 mm thick and are conspicuously homogeneous compared with the extremely zoned mare pyroxenes.[12] In brecciated anorthositic gabbro 77017, augite and pigeonite have a zoning pattern suggesting that each oikocryst grew as a chemically isolated system. This is consistent with their present isolation from each other; nowhere are pigeonite and augite oikocrysts observed in contact.[13]

Hourglass or sector zoning indicates crystallographic control of zoning superimposed on the magma effect[9] (i.e., different compositions develop behind crystallographically distinct growth faces[14]). The sector zoning in

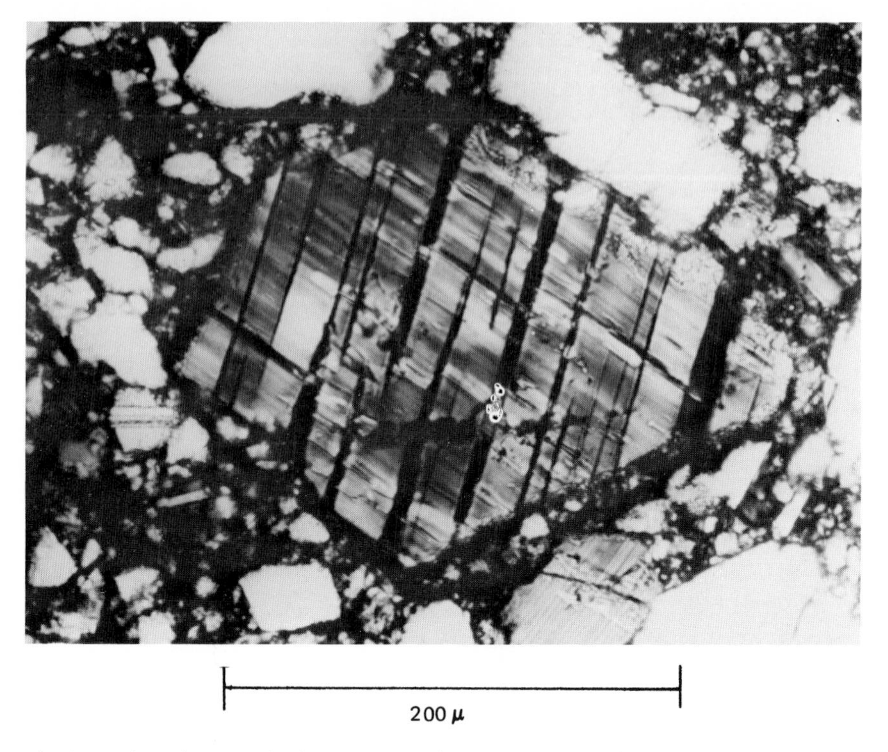


Fig. 9–7 Pigeonite crystal with coarse augite exsolution lamellae parallel to (001) and finer augite lamellae parallel to (100). From *Suppl.* **4,** 511. Photograph courtesy of M. Brown, Department of Geological Sciences, Durham University, England.

the pyroxenes, such as in medium-grained basalt 10058, is believed to be due to rapid growth, possibly at a supercooled condition, with the preservation of different surface compositions.[8] In coarse-grained basalt 10044, the pyroxene grains show pronounced sector zoning,[15] and compared with pyroxenes in terrestrial rocks, they are remarkably free of inclusions of oxides and sulfides.[16]

Many clinopyroxenes in the Apollo 11 microbreccias show multiple narrow lamellae parallel to (001) (Figure 9–7). Polysynthetic twinning on (100) occurs, but rarely. The multiple lamellar twinning is believed to be shock induced.[17] In many clinopyroxenes, very fine ($\sim$ 60–100 Å thick) lamellae, alternately augite and pigeonite, are oriented for the most part on (001) and to a lesser extent on (100).[18] The bands often appear straight-edged (Figure 9–8) and their boundaries can be seen to be exceptionally sharp even when viewed at the highest magnifications. This lamellar

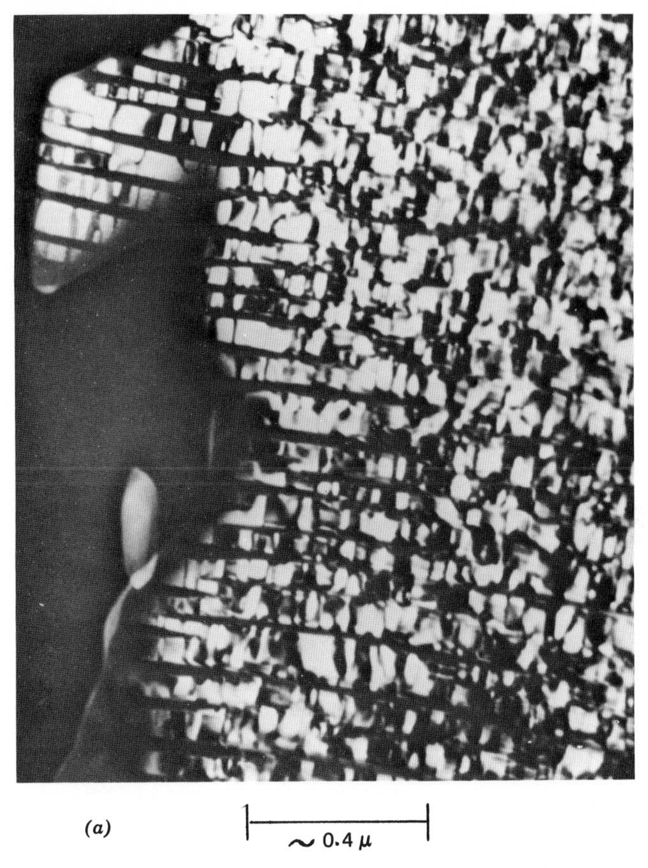


Fig. 9.8 (*a*) Strain-free grain of pigeonite with exsolved augite, from feldspathic basalt 14321,42. From *Suppl.* **3,** 413. (*b*) Dark field micrograph showing exsolution lamellae in pyroxene from olivine diabase 10029-1. From *Suppl.* **1,** 743. (*c*) Stacking faults and augite lamellae in host pigeonite in ophitic basalt 12038. From *Suppl.* **2,** 75. Photographs courtesy of J. S. Lally, U.S. Steel Corp., Research Laboratory, Monroeville, Pa., and G. Nord, Jr., Department of Metallurgy, Case-Western Reserve University, Cleveland.

structure is interpreted as an exsolution phenomenon occurring during the cooling history subsequent to crystallization.[19] In basalt 12052, exsolution textures have been observed in the pyroxenes on a finer scale than those found in Apollo 11 rocks. This indicates a faster cooling rate.[20] In medium-grained picritic basalt 12002, in addition to normal and sector zoning, two types of epitaxial overgrowth were observed—an augite rim on pigeonite (with a sharp compositional break) and an overgrowth of edge pyroxene ranging from ferropigeonite to ferroaugite.[4]

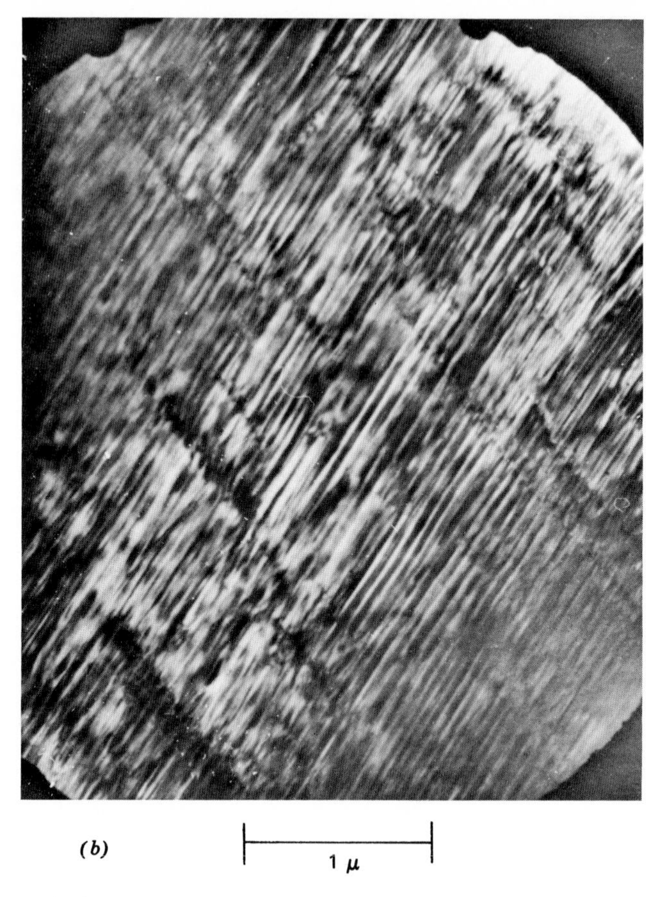


Fig. 9–8 (*continued*).

Augitic pyroxenes crystallized later than olivine but earlier than or simultaneously with plagioclase.[21] In breccia 67955, poikilitic diopside, as well as hypersthene, envelop olivine and plagioclase.[12] In coarse-grained basalt 10047, subcalcic diopside forms cores in single-zoned clinopyroxene grains.[22] In anorthosite 15415,18, diopsidic augite occurs as small triangular crystals up to 250 $\mu$,[23] or as thin septa interstitial to the larger plagioclase grains and as equant inclusions in the plagioclase.[24] From fine-grained basalt 70215,149 (as well as from some other Apollo 17 rocks) an early crystallizing clinopyroxene, close to salite in composition, has been identified.[25] In fine-grained noritic hornfels 76055,10, augite (together with armalcolite, nickel-iron, troilite, and pink spinel) is only a minor constituent.[26] Magnesium-rich augite ocurs in the inner mantles of phenocrysts

Fig. 9.8 (*continued*).

in coarse-grained basalt 12021.[27] Grains of epitaxial augites and subcalcic augites on pigeonite cores are typically subhedral and form ophitic, subophitic, or granular textures with plagioclase.[28] Subcalcic augite is prominent in the coarse-grained basalts of Apollo 11 and has been found also in Luny Rock 1.[29] From a Luna 16 fragment (638/2) the majority of the pyroxenes are subcalcic augite or ferroaugite.[30] In a fine-grained Luna 20 soil sample, the dominant clinopyroxenes are calcic, with pigeonite less common than orthopyroxenes.[31] In thin sections of Apollo 11 basalts 10017-20, 10022-41, 10024-23, 10072-33, and 10072-49, augite (or subcalcic ferroaugite) predominates over pigeonite, and both clinopyroxenes are closely associated.[32] Titanaugite has been observed intergrown with ilmenite,[3] and an hourglass-shaped, sectorially zoned variety of clinopyroxene, typical of titanaugite, is found in the coarsest Apollo 11 basalts. These basalts also contain an abundance of subcalcic augite zoning to a ferroaugite.[33] In thin section FQM-151 from a Luna 16 sample, a clinopyroxene close to ferrosilite occurs in a 70-$\mu$ cluster of grains, possibly parts of a single crystal, associated with patches of K-rich mesostasis.[34] In medium-grained ophitic basalts of Apollo 11, the Fe-rich clinopyroxene ferrohedenbergite, a late-stage phase intergrown with silica and fayalite, occurs as a decomposition of pyroxferroite.[35] Minor amounts of ferrohedenbergite were found with andesine and quartz in an Apollo 12 rock fragment consisting largely of potash feldspar.[36]

With only a few exceptions, pigeonites are absent from the clinopyroxenes of the Apollo 11 samples. In porphyritic basalts 12021 and 12052, primary pigeonite is common and typically is in epitaxial contact with augite. Most clinopyroxenes from both rocks have hollow cores and are composed of central pigeonites with complex augite–high-Fe–pigeonite rims.[37] Some of the zoning of these pigeonites is oscillatory.[38] The pigeonite phenocrysts, mantled by zoned Ca-rich pyroxenes, are up to several centimeters long and 3 to 4 mm in diameter. In Apollo 12 rocks, magnesian pigeonite occurs as cores in pigeonite,[27] and it has been reported also from ophitic basalt, anorthosite, and gabbroic anorthosite fragments in the coarse fines of 14257.[39] In basalt 14053 it forms large prisms up to 2 mm long.[40] Unusual multiple twinning in two pigeonite crystals was observed in KREEP basalt 14310. The crystals are in subophitic relationship with plagioclase, and the twin axis in both is $\perp(\bar{1}\bar{2}2)$. Pyroxenes twinned on this axis are rare in terrestrial rocks. This twinning may be a primary feature associated with rapid growth.[41] In basaltic vitrophyre 15597, pigeonitic clinopyroxenes make up about 60% of the rock's volume and occur in acicular crystals, together with very minor chromian spinel in a dark brown quench glass. Most of the crystals are less than 100 $\mu$, and many contain glass cores rimmed by epitaxial (100) augite and/or overgrown by (100) augite rims. About half the crystals are twinned on (100), and both twinned and untwinned crystals have pigeonitic cores.[42] The pigeonites typically are zoned from magnesian pigeonite cores to ferroaugite mantles.[43] Pigeonite often occurs in augite as subrounded or irregularly elongated inclusions. The irregular intergrowth of elongated augite and pigeonite is also common, as well as the intergrowths of needlelike subcalcic augite and plagioclase, which are indicative of rapid growth of the two phases, but the very fine intergrowths of augite and pigeonite lamellae (in the order of 5 $\mu$ or less) are attributed to rapid exsolution of metastable subcalcic augite.[44] The fine-scale (100) twinning may have resulted from inversion from an orthopyroxene phase.[40] In anorthositic gabbro 68415, an extremely Fe-rich pigeonite occurs as separate grains in patches of glass of granitic composition. This rock also contains both augite and calcic pigeonite, which is predominant, and there is exsolved pigeonite in the strongly zoned augite. This coexisting pyroxene composition is characteristic of igneous rather than metamorphic assemblages.[45] A very complex pigeonite grain from Luna 20 sample K-4 shows coarse augite exsolution lamellae paralleling (001) and (100), a finer set of second generation augite lamellae parallel to (001), and rods of exsolved chrome spinel nearly parallel and perpendicular to (001). This large-scale exsolution of augite and spinel, together with a partial phase transformation of clinobronzite to bronzite in the pigeonite, indicates a slow cooling. This pyroxene may have been derived from the original lunar crust.[46] It has been suggested that given sufficient time, lunar pigeonites invert to orthopyroxenes in very anhydrous environments.[47]

Clinohypersthene occurs as exsolution lamellae, rarely seen under the microscope but detected by X-ray diffraction or electron microscopy.[48] From porphyritic basalt 12065, an investigation of an epitaxial crystallization sequence of pyroxenes revealed an early phase that possibly was clinohypersthene.[49] Lamellae in a complex pigeonite from Luna 20 sample K-4 are termed clinohypersthene on the basis of their composition.[50]

*References*


1. Weill et al, *Suppl.* **1,** 939, 944.
2. Dence et al, *Suppl.* **1,** 319.
3. Frondel et al, *Suppl.* **1,** 454–455, 458–459.

3a. McKay et al., *Lun. Sci. III, Rev. Abstr.*, 531.

4. Grove et al, *Suppl.* **4,** 995–996, 998.
5. Powell et al, *Lun. Sci. IV, Abstr.*, 597–598.
6. Dowty et al, *Lun. Sci. IV, Abstr.*, 181.
7. Simonds et al, *Suppl.* **4,** 614.
8. Hollister and Hargraves, *Suppl.* **1,** 547–548.
9. Smith et al, *Suppl.* **1,** 920.
10. Walker et al, *Suppl.* **4,** 1014.
11. Gancarz et al, **16,** 315–316.
12. Hollister, *Suppl.* **4,** 636.
13. Helz and Appleman, *Lun. Sci. V*, Abstr., 322.
14. Hollister et al, *Suppl.* **2,** 533.
15. Albee and Chodos, *Suppl.* **1,** 147.
16. Bailey et al, *Suppl.* 1, 179.
17. Sclar, *Suppl.* **1,** 851.
18. Radcliffe et al, *Suppl.* **1,** 731.
19. Fernandez-Moran et al, *Suppl.* **1,** 411, 414.
20. Champness and Lorimer, *Contrib. Mineral. Petrol.* **33,** 171.
21. Mason et al, *Suppl.* **1,** 656

22. Essene et al, *Suppl.* **1,** 391.
23. Hargraves and Hollister, *Sci.* **175,** 430–432.
24. James, *Sci.* **175,** 434
25. Walker, D., Department of Geological Sciences, Harvard University, personal communication.
26. Chao, *Suppl.* **4,** 720.
27. Boyd and Smith, *J. Petrol.* **12,** 439, 441–442.
28. Keil et al, *Suppl.* **1,** 567–568.
29. Papanastassiou et al, *EPSL* **8,** 4.
30. Grieve et al, *EPSL* **13,** 237.
31. Meyer, *Lun. Sci. IV, Abstr.*, 520.
32. Kushiro and Nakamura, *Suppl.* **1,** 609.
33. Agrell et al, *Suppl.* **1,** 100.
34. Albee et al, *EPSL* **13,** 357.
35. James and Jackson, *J. Geophys. Res.* **75,** 5799.
36. Mason et al, *Lun. Sci. Conf. '71, Abstr.*, 257.
37. Bence et al, *Suppl.* **2,** 560–561.
38. Dence et al, *Suppl.* **2,** 285.
39. Klein et al, *Lun. Sci. Conf. '72, Abstr.*, 408.
40. Melson et al, *Lun. Sci. Conf. '72, Abstr.*, 474.
41. Johnston and Gibb, *Mineral. Mag.* **39,** 248–249.
42. Brown et al, *Lun. Sci. IV, Abstr.*, 91.
43. Brown and Wechsler, *Suppl.* **4,** 887, 891–893.
44. Kushiro and Nakamura, *Suppl.* **1,** 609.
45. Helz and Appleman, *Suppl.* **4,** 652.
46. Ghose and Wan, *Suppl.* **4,** 901, 905.
47. Papike and Bence, *GSA '71, Abstr.*, 667.
48. Mason et al, *Sci.* **167,** 657.
49. Hollister et al, *Lun. Sci. Conf. '71, Abstr.*, 249.
50. Ghose et al, *Geochim. Cosmochim. Acta* **37,** 831.

### *Optics*

Within the clinopyroxenes there are complex patterns of color varying generally from very light pink to dark reddish brown[1] and even deep purple, although colorless, yellow, and some greenish yellow grains have been observed. As a rule, pleochroism is very weak.

The clinopyroxenes in the Apollo 11 fine-grained rocks show a large variation in 2V. Early diopsidic augite, with a 2V = 48°, frequently is zoned with 2V = 42 to 20°. The optic plane is parallel to (010).[2] From basalt 10044, clinopyroxenes with Mg-augite cores and ferroaugite rims are weakly pleochroic from pale pink to pale pinkish brown. They are zoned in birefringence as well as color and have a range of 2V from 37 to 49°.[3]

Universal stage measurements on larger diopsidic augite from anorthosite 15415,18 yielded a 2V of 50°.[4] From olivine-pyroxene vitrophyre 15474,4, skeletal prismatic "totem pole" clinopyroxenes are pink and pleochroic.[5] Most augite grains are pink or tan, becoming darker with increasing Ti content. Titanaugite is dark purple and markedly pleochroic[6] or dark brown and intergrown with ilmenite.[7] However, from poikilitic basalt 10017 the augite is nearly colorless to very pale brown and has extinction $Z \wedge C$ = 37 to 39°.[8] Ferroaugite is yellowish brown,[7] and with increase in the ferrosilite content the grains appear darker. As the composition of the clinopyroxene reaches 81 mole % of ferrosilite, the normally clear grains suddenly become virtually opaque.[9] In Luna 16 sample FQM-151, the clinopyroxene in the mesostasis is high in the ferrosilite component and brown rather than yellow, like the compositionally close pyroxferroite.[10] From ophitic basalt 12051, ferrohedenbergite has a 2V = 45 to 50°.[11] An intense red pyroxene (containing some $Ti^{3+}$) has been observed in mare basalt 74275.[12]

Pigeonite is generally yellow-green or nearly colorless in the cores; it has lower birefringence and often is surrounded by red-brown rims. It is weakly pleochroic, with a small 2V (14–16°), and sometimes has a crossed axial plane ($\alpha \parallel [010]$).[13] From gabbroic anorthosite 68416,77, euhedral (0.4 × 0.3 mm) pigeonite grains are pale yellow with 2V = 21°.[14] From fine-grained Luna 20 sample 22001,17, the major clinopyroxene is a slightly brownish pink calcic variety, and colorless pigeonite is less common.[15]

There are several types of optically discernible disorientation in the pyroxenes: undulatory extinction, discontinuous changes in extinction with subboundaries separating zones of homogeneous extinction, and irregular blocky or patchy extinction. It is suggested that these disorientations would have occurred if the pyroxenes had grown rapidly from a melt, perhaps during quenching.[16] Also clearly visible are "sectorial" extinction and apparent warping or bending of the (110) cleavage planes. These effects are caused by lattice rotation of two or more blocky, sometimes pie-shaped "domains" within a crystal.[1] (See under X-Ray Data.)

*References*

1. Ross et al, *Suppl.* **1,** 839.
2. Agrell et al, *Suppl.* **1,** 101.
3. Bailey et al, *Suppl.* **1,** 179.
4. Hargraves and Hollister, *Sci.* **175,** 430.
5. Powell et al, *Lun. Sci. IV, Abstr.,* 598.
6. Keil et al, *Suppl.* **1,** 567–568.
7. Frondel et al, *Suppl.* **1,** 454, 458.
8. French et al, *Suppl.* **1,** 436.

9. Ware and Lovering, *Sci.* **167,** 518.
10. Albee et al, *EPSL* **13,** 357.
11. Gay et al, *Suppl.* **2,** 379.
12. Sung et al, *Lun. Sci. V, Abstr.*, 758.
13. Christie et al, *Suppl.* **2,** 70,72.
14. Juan et al, *Lun. Sci. IV, Abstr.*, 421.
15. Meyer, *Geochim. Cosmochim. Acta* **37,** 945
16. Carter et al, *Suppl.* **1,** 271.

### *Chemical Composition*

Normal chemical zoning in the clinopyroxenes is from more Mg- and Ca-rich varieties to the Fe-rich varieties. The most magnesian diopsides, from coarse-grained basalt 10047, have as much as 6% $Al_2O_3$. With a slight iron enrichment, the $Al_2O_3$ content drops to 2.3%.[1] In the diopsidic augite of anorthosite 15415,18, a content of up to 46 mole % of $CaSiO_3$ has been reported (Table 9-1; [2]).[2] From Apollo 16 rake samples, plots of the pyroxene compositions in the poikilitic rock fragments fall into two main groups: high-Mg augite and high-Mg, low-Al hypersthene and/or pigeonite. The large pyroxene poikiloblasts are only slightly zoned, with cores 2 to 4 mole % richer in an enstatitic component than the rims.[3] The clinopyroxenes from breccia 67955 are chiefly diopside, so-named to emphasize their high Mg and Ca content.[4] Augites from breccia 14303 contain more Mg than the most magnesian augites reported from Apollo 11 samples.[5]

Although the pyroxene in the Apollo 11 larger igneous rock samples is mostly augite, spot analyses spread over much of the pyroxene quadrangle. Most analyses from the centers of the crystals fall in a marked cluster (an average augite, Table 9-1; [6]), with the more iron-rich points coming from the grain margins. Zoning involves an increase in Fe with decrease in Ca, Mg, Al, and Ti.[6] It has been noted that augite contains appreciable Ti and Al, indicating a possible $CaTiAl_2O_6$ component.[1] In the various augite crystals examined from coarse-grained basalt 10050, the $TiO_2$ and $Al_2O_3$ contents vary from about 1.5% in the pale red areas to about 3% in the darker red areas. There is a coupled substitution of $R^{2+} + 2Si^{4+}$ by $Ti^{4+} + 2Al^{3+}$.[7] Pink augite phenocrysts from olivine-pyroxene vitrophyre 15474,4 are unusually rich in $TiO_2$ (3.1%) and $Al_2O_3$ (6.8%).[8] Very few augites contain less than 1.5% $TiO_2$ ,and their Cr content is higher than in terrestrial analogs.[9] Pyroxenes with more than 2% $TiO_2$ have been referred to as titanaugites (Table 9-1; [4]).[10] One titanaugite contained 7% $TiO_2$ as well as 7.9% $Al_2O_3$.[11] A $TiO_2$ content of 9.3% has been reported. Some trivalent titanium has been found in pyroxenes from mare basalts 70017, 71055, and 74275. The data, obtained from electron microprobe analysis, absorption spectroscopy, and Mössbauer spectroscopy, confirm earlier evidence cited for traces of $Ti^{3+}$ in some Apollo 11 and 15 pyroxenes.[12] Even though they are strongly zoned, the Apollo 11 clinopyroxenes show less variation in their trace element content than do the plagioclases;[13] but pyroxenes that connot be distinguished in terms of major components Ca, Mg, and Fe may be clearly distinguished on the basis of relative Cr, Ti, and Al contents.[4] Pyroxenes from the soil samples yield an average composition identical to the average pyroxenes from the igneous rocks.[6] In fine-grained basalt 10022-41 there is a large compositional variation in augite-subcalcic ferroaugite, with a trend toward ferrosilite (Table 9-1; [6]–[10]); the most iron-rich clinopyroxene is close to ferrosilite in composition (Table 9-1; [15]).[14]

Pigeonite is usually of the magnesian variety and, except for Mn, generally contains lower concentrations of Al, Cr, and Ti than augite.[9] The pigeonites from breccia 14303 form a fairly discrete group,[4] and the diopsidic augites and pigeonites in Luna 20 clinopyroxenes are even more magnesian than the Apollo 14 pyroxenes, which they otherwise resemble.[15] In pyroxene-phyric basalt 15116, the sequence of crystallization is most clear from magnesian pigeonite to magnesian augite to a more iron-rich composition. Other rock types display a random spread of the plotted analyses throughout the pyroxene quadrilateral.[16] In feldspathic basalt 68415, the narrow compositional gap between pigeonite and augite is consistent with a relatively high temperature and possibly rapid crystallization history.[17] Pyroxene vitrophyre 15597 shows no evidence of orthopyroxene cores in the twinned pigeonites, although their average inner core composition is well within the range of lunar orthopyroxenes.[18] In medium and coarse-grained Apollo 11 basalts 10003, 10047, and 10050, a chemically inhomogeneous subcalcic augite appears to be the only primary pyroxene, and pigeonite exsolved from it later.[7] However, pigeonite in fine-grained basalt 10022-28 is the product of primary crystallization from a melt. In this sample it occurs as cores surrounded by augite. The pigeonite is quite homogeneous, whereas the augite is complexly zoned. The compositional break between the pigeonite and augite is sharp.[19] In olivine-cristobalite basalt 15555, the pyroxene trend is unusual. Intergrown pigeonite and augite zone to a common focus at subcalcic ferropigeonite with a composition of $Ca_{22}Mg_{22}Fe_{56}$.[20] In recrystallized polymict breccia 22006 of Luna 20, pigeonite is more abundant than augite, but both are quite homogeneous and have the same Fe/Mg. This homogeneity plus a granoblastic texture suggests that the rock has been metamorphosed sufficiently to equilibrate the Fe and Mg within these minerals. The two pyroxenes are probably in equilibrium with each other.[21]

**Table 9-1 Clinopyroxene and Pyroxferroite Analyses**

| | [1] 14162,31 (1407–11) | [2] 15415,18 | [3] 14166, 6E2 | [4] 10045–35 | [5] 70215, 149 | [6] 10085–42 | [7] 12019 | [8] Luna 16 G 38/2 | [9] 12065,8 | [10] 14053 | [11] 12021 |
|---|---|---|---|---|---|---|---|---|---|---|---|
| $SiO_2$ | 47.3 | 53.42 | 53.9 | 48.00 | 42.74 | 49.35 | 47.35 | 49.48 | 46.0 | 46.0 | 47.0 |
| $Al_2O_3$ | 3.2 | 0.68 | 1.6 | 4.93 | 8.77 | 2.39 | 7.44 | 1.99 | 1.40 | 0.87 | 0.26 |
| $Cr_2O_3$ | 1.31 | 0.19 | 0.5 | 0.70 | 0.64 | — | 1.15 | 0.34 | 0.04 | 0.08 | 0.04 |
| $TiO_2$ | 3.2 | 0.36 | 0.9 | 5.91 | 6.57 | 1.94 | 2.48 | 1.54 | 1.43 | 0.96 | 0.37 |
| FeO | 5.6 | 9.70 | 2.4 | 8.49 | 9.28 | 13.46 | 14.87 | 22.52 | 34.8 | 39.0 | 41.7 |
| MgO | 24.7 | 13.83 | 17.4 | 13.27 | 10.71 | 12.82 | 10.82 | 12.63 | 2.23 | 1.90 | 3.8 |
| MnO | 0.18 | 0.24 | 0.06 | 0.09 | 0.19 | 0.28 | 0.41 | 0.47 | 0.46 | 0.65 | 0.76 |
| CaO | 14.8 | 22.14 | 23.9 | 19.88 | 20.58 | 17.82 | 15.89 | 11.41 | 13.5 | 9.87 | 5.4 |
| $Na_2O$ | 0.4 | 0.03 | 0.14 | — | 0.12 | — | — | — | — | 0.03 | <0.01 |
| Total | 100.69 | 100.59 | 100.80 | 101.27 | 99.60 | 98.06 | 100.39 | 99.38 | 99.86 | 99.36 | 99.33 |

*References*


[1] Powell and Weiblen, *Suppl.* **3,** 844. From a basaltic particle in a soil sample. Fe-poor augite (magnesian augite).

[2] Hargraves and Hollister, *Sci.* **175,** 431. From an anorthosite. Diopsidic augite.

[3] Steele and Smith, *Nature* **240,** 5. From a fines sample. Diopside.

[4] Agrell et al, *Suppl.* **1,** 102. From a basalt. Titanaugite.

[5] Walker, D., Department of Geological Sciences, Harvard University. Personal communication. From a fine-grained basalt. Salite.

[6] Agrell et al, *Suppl.* **1,** 102. From a coarse-grained basalt fragment in fines sample. Augite.

[7] Brown et al, *Suppl.* **2,** 590. From a basalt. Al-rich augite.

[8] Grieve et al, *EPSL* **13,** 236. From a basaltic fragment in soil sample. Subcalcic augite.

[9] Kushiro et al, *Suppl.* **2,** 483. From a porphyritic basalt. Subcalcic ferroaugite.

[10] ———, *Suppl.* **3,** 118. From a coarse-grained dolerite. Ferroaugite.

[11] Smith and Finger, *Carnegie Inst. Yearb.* '70, 134. From a porphyritic basalt. Magnesian pyroxferroite.

| | [12] | [13] | [14] | [15] | [16] | [17] | [18] | [19] | [20] | [21] | [22] |
|---|---|---|---|---|---|---|---|---|---|---|---|
| | | | | | | | | 14310,22 | | | |
| | | | | | Luna 20 | Luna 20 | Luna 20 | | | 10085 | |
| | 10058–23 | 10058–23 | 15076/12 | 10022–4 | 22003,2 | 22002,2,5b | K–4 | Inner Rim | Outer Rim | 12–34 | 15555/39 |
| $SiO_2$ | 45.86 | 45.4 | 44.46 | 44.1 | 55.4 | 53.5 | 53.83 | 50.8 | 47.1 | 45.0 | 45.52 |
| $Al_2O_3$ | 0.74 | 0.49 | 1.00 | 1.43 | 1.03 | 0.99 | 0.49 | 1.35 | 0.95 | 1.47 | 2.62 |
| $Cr_2O_3$ | 0.10 | 0.03 | 0.08 | 0.06 | 0.37 | 0.43 | 0.35 | 0.39 | 0.08 | — | — |
| $TiO_2$ | 0.59 | 0.51 | 0.89 | 1.06 | 0.45 | 0.57 | 0.37 | 0.89 | 0.74 | 1.51 | 2.59 |
| FeO | 44.19 | 44.7 | 47.54 | 45.8 | 9.5 | 19.7 | 23.02 | 19.8 | 34.1 | 30.4 | 29.31 |
| MgO | 1.38 | — | 0.20 | 1.72 | 30.9 | 23.0 | 21.69 | 20.3 | 9.17 | 3.54 | 0.40 |
| MnO | 0.64 | 0.65 | 0.51 | 0.59 | — | — | 0.42 | 0.33 | 0.47 | 0.49 | 0.37 |
| CaO | 6.21 | 6.17 | 4.44 | 3.52 | 1.94 | 2.1 | 0.89 | 5.07 | 5.74 | 17.8 | 18.54 |
| $Na_2O$ | — | — | — | 0.03 | — | 0.03 | 0.01 | 0.02 | 0.04 | 0.05 | — |
| Total | 99.71 | 97.95 | 99.32 | 98.31 | 99.59 | 100.32 | 101.07 | 98.95 | 98.39 | 100.29* | 99.35 |

*References*


[12] Agrell et al, *Suppl.* **1,** 103. From a coarse-grained diabasic rock. Calcian pyroxferroite.

[13] Brown et al, *Suppl.* **1,** 203. From a coarse-grained diabasic rock. Mg-free pyroxferroite.

[14] Brown et al, *Suppl.* **3,** 142. From a basalt. Subcalcic pyroxferroite (Fe-rich pyroxferroite).

[15] Kushiro and Nakamura, *Suppl.* **1,** 611. From a fine-grained basalt. Ferrosilite.

[16] Prinz et al, *Geochim. Cosmochim. Acta* **37,** 988. From a spinel-troctolite fragment in soil sample. Magnesian pigeonite.

[17] Cameron et al, *Geochim. Cosmochim. Acta* **37,** 781. From a soil sample. Rim on zoned crystal. Low-Ca pigeonite.

[18] Ghose et al, *Geochim. Cosmochim. Acta* **37,** 833. From a soil sample. Pigeonite (clinohypersthene).

[19] Kushiro et al, *Suppl.* **3,** 118. From a KREEP basalt. Inner part of zoned clinopyroxene mantle on an orthopyroxene core. Pigeonite.

[20] ———, *ibid.* Outer part of zoned clinopyroxene mantle on an orthopyroxene core. Ferropigeonite.

[21] James and Jackson, *J. Geophys. Res.* **75,** 5805. From a fine-grained basalt fragment in fines sample. *Analysis contains $K_2O$ = 0.03%. Ferrohedenbergite.

[22] Brown et al, *Suppl.* **3,** 142. From a basalt. Ferrohedenbergite.

Pyroxenes in microgabbro 12064 are zoned from a pigeonite core through augite to ferroaugite, and there is a strong increase in Ca content in the ferrohedenbergite region toward hedenbergite. Hedenbergite found interstitially in fine-grained granular fayalite had a composition of $Wo_{42}En_3$-$Fs_{55}$.[22] Most lunar clinopyroxenes that have been termed hedenbergite are compositionally ferrohedenbergite. So-called hedenbergite from a fine-grained basalt fragment in fines sample 10085,12-34 is a decomposition of pyroxferroite, and analysis shows it to be ferrohedenbergite (Table 9-1; [21]).[23] Some pyroxene from basalt 15555/39 zones to almost Mg-free ferrohedenbergite rather than pyroxferroite (Table 9-1; [22]).[24]

From a basalt clast in polymict breccia 14321,184-55, a pyroxene contained Rb = 0.290 ppm, K = 142.0 ppm, and $^{86}Sr$ = 1.50 ppm.[25]

No evidence for $Fe^{3+}$ or $Cr^{2+}$ in the pyroxenes has been found either by Mössbauer or absorption spectroscopy.[26]

*References*

1. Essene et al, *Suppl.* **1,** 393.
2. Hargraves and Hollister, *Sci.* **175,** 430.
3. Simonds et al, *Suppl.* **4,** 623–624.
4. Hollister, *Suppl.* **4,** 633, 638.
5. Weigand and Hollister, *Suppl.* **3,** 473.
6. Reid et al, *Suppl.* **1,** 753, 757, 759.
7. Ross et al, *Suppl.* **1,** 839, 848.
8. Powell et al, *Lun. Sci. IV, Abstr.*, 598.
9. Keil et al, *Suppl.* **1,** 567–568, 572.
10. ———, *Sci.* **167,** 597.
11. Frondel et al, *Suppl.* **1,** 454.
12. Sung et al, *Lun. Sci. V, Abstr.*, 758–759.
13. Anderson et al, *Suppl.* **1,** 159.
14. Kushiro and Nakamura, *Suppl.* **1,** 609.
15. Reid et al, *Geochim. Cosmochim. Acta* **37,** 1018.
16. Dowty et al, *Suppl.* **4,** 432.
17. Walker et al, *Suppl.* **4,** 1014.
18. Brown and Wechsler, *Suppl.* **4,** 893.
19. Weill et al, *Suppl.* **1,** 944.
20. Brown et al, *Lun. Sci. Conf. '72, Abstr.*, 87.
21. Podosek et al, *Geochim. Cosmochim. Acta* **37,** 888, 890.
22. Klein et al, *Suppl.* **2,** 277.
23. James and Jackson, *J. Geophys. Res.* **75,** 5805.
24. Brown et al, *Suppl.* **3,** 142.
25. Mark et al, *Lun. Sci IV, Abstr.*, 501.
26. Burns et al, *Suppl.* **4,** 988.

*X-Ray Data*

Exsolution lamellae not observed optically have been revealed on a submicroscopic scale through X-ray studies[1] and transmission electron microscopy. Augites (e.g., in medium-grained basalt 14053), with space group $C2/c$, have pigeonite exsolved on (001), and the pigeonites, with space group $P2_1/c$, have augite exsolved on (001) and (100).[2]

Observed pie-shaped domains in the clinopyroxenes are shown in precession photographs to be offset from one another by rotation about a common *b* axis.[3] These domains presumably originated during the $C2/c \rightarrow P2_1/c$ transition in pigeonite, and the diffuseness of the *b* reflections is a measure of the size of the domains.* Provided the chemical composition is the same, quick quenching causes small domains to form (diffuse *b* reflections), and slow cooling produces larger domains (sharp *b* reflections). Domain size, then, can serve as an indicator of the rock's cooling history. These domains have been observed directly through transmission electron microscopy. In pigeonite from medium-grained porphyritic basalt 12021, the *b*-type reflections are sharp, but from finer-grained porphyritic basalt 12053, the *b*-type reflections in the pigeonite are diffuse. These observations have led to the suggestion that following crystallization at approximately 1200°C, 12021 cooled slowly at a fairly uniform rate comparable to that of terrestrial rocks, and 12053, following crystallization, was quenched rapidly at about 1200°C.[4]

Cell constants of some of the lunar clinopyroxenes are given in Table 9-2.

*References*

1. Agrell et al, *Suppl.* **1,** 102.
2. Papike and Bence, *GSA '71, Abstr.*, 667.
3. Ross et al, *Suppl.* **1,** 839.
4. Ghose et al, *Suppl.* **3,** 516, 529.

## Orthopyroxenes: $MgSiO_3$-$FeSiO_3$ Series

*Synonymies*

BRONZITE (Mg, Fe)$SiO_3$

Aluminous bronzite; Ridley et al, *Suppl.* **3,** 162.
Bronzitic pyroxene; Ridley et al, *GSA '71, Abstr.*, 682.
Chromium-rich bronzite; Ghose et al, *Geochim. Cosmochim. Acta* **37,** 831.

* The b-type reflections are $h + k$ = odd.

**Table 9-2 X-Ray Data for Some Lunar Pyroxenes and Pyroxferroite**

| | [1] | [2] | [3] | [4] | [5] | [6] | [7] | [8] |
|---|---|---|---|---|---|---|---|---|
| | | | 12052 | | | | | |
| X-Ray Data (Å) | 14053 | 15597 | Rim | Core | 15597 | Luna 20, K-4 | | |
| $a_0$ | 9.77 | 9.74 | 9.726 ± 0.002 | 9.688 ± 0.003 | 9.70 | 9.66 | 9.69 | 9.77 |
| $b_0$ | 8.96 | 8.89 | 8.909 ± 0.003 | 8.890 ± 0.003 | 8.89 | 8.90 | 8.90 | 8.90 |
| $c_0$ | 5.26 | 5.27 | 5.268 ± 0.001 | 5.238 ± 0.002 | 5.24 | 5.20 | 5.25 | 5.25 |
| Space group | $C2/c$ | $C2/c$ | $C2/c$ | $P2_1/c$ | $P2_1/c$ | $P2_1/c$ | $C2/c$ | $C2/c$ |

*References*


[1] Papike and Bence, *GSA '71*, *Abstr.*, 667. From a coarse-grained basalt. Augite.

[2] Brown and Wechsler, *Suppl.* **4,** 889. From a basaltic vitrophyre. Augite.

[3] Brett et al, *Suppl.* **2,** 307. From a basalt. Augite.

[4] ———, *ibid.* From a basalt. Pigeonite.

[5] Brown and Wechsler, *Suppl.* **4,** 889. From a basaltic vitrophyre. Untwinned crystal. Pigeonite.

[6] Ghose et al, *Geochim, Cosmochim. Acta* **37,** 832. From a soil sample. Zone, in a complex pigeonite, called clinohypersthene because of composition.

[7] ———, *ibid*, From a soil sample. (001) Augite zone.

[8] ———, *ibid.* From a soil sample. (100 )Augite 1 zone.

[9] ———, *ibid.* From a soil sample. (100) Augite 2 zone.

| X-Ray Data (Å) | [9] Luna 20, K-4 | [10] 14310,90 | [11] 76535 | [12] Luna 20, K-1 | [13] Luna 20, K-4 | [14] 12033,97 –2B | [15] 10047 | [16] 12021 |
|---|---|---|---|---|---|---|---|---|
| $a_0$ | 9.47(?) | 18.301(3) | 18.27 ± 0.01 | 18.29 | 18.30 | 18.30 | 6.6213 ± 0.0005 | 6.619(6) |
| $b_0$ | 8.90 | 8.869(2) | 8.83 ± 0.01 | 8.85 | 8.90 | 8.884 | 7.5506 ± 0.0009 | 7.55(2) |
| $c_0$ | 5.25 | 5.215(1) | 5.20 ± 0.01 | 5.22 | 5.20 | 5.212 | 17.3806 ± 0.0016 | 17.35 (6) |
| Space group | $C2/c$ | $Pbca$ | $P2_1ca$ | $Pbca$ | $Pbca$ | $Pbca$ | $P\bar{1}$ | $P\bar{1}$ |

*References*


[10] Takeda and Ridley, *Suppl.* **13,** 424. From a KREEP basalt. Bronzite.

[11] Smyth, Low orthopyroxene from a deep crystal rock: A new pyroxene polymorph of space group $P2_1ca$; *Geophys. Res. Let.* **1,** 27. From a troctolitic granulite. Low orthopyroxene bronzite.

[12] Ghose et al, *Geochim. Cosmochim. Acta* **37,** 832. From a soil sample. Chromium-rich bronzite.

[13] ———, *ibid.* From a soil sample. Inverted (100) hypersthene.

[14] Meyer et al, *Suppl.* **2,** 404. From a glassy KREEP breccia. Hypersthene not inverted from pigeonite.

[15] Burnham, *Suppl.* **2,** 49. From a microgabbro. Mean of five independent determinations. Calcian pyroxferroite.

[16] Smith and Finger, *Carnegie Inst. Yearb.* '70, 134. From a porphyritic basalt. Magnesian pyroxferroite.

High-alumina orthopyroxene, Brown et al, *Suppl.* **3,** 142.
Low-alumina bronzite; Ridley et al, *Suppl.* **3,** 162.
Low orthopyroxene; Smyth, *Geophys. Res. Lett.* **1,** 27.
Mg-bronzite; Weigand and Hollister, *Suppl.* **3,** 473.
Mg-rich bronzite; Weigand and Hollister, *Suppl.* **3,** 471.

### Enstatite $MgSiO_3$

Aluminous enstatite; Anderson, *J. Geol.* **81,** 219.
Low-alumina orthopyroxene; Brown et al, *Suppl.* **3,** 142.
Magnesian orthopyroxene; Gay et al, *Suppl.* **2,** 387.
Mg-orthopyroxene; Ridley et al, *Lun. Sci. Conf. '72, Abstr.*, 571.
Mg-rich orthopyroxene; Ridley et al, *GSA '71, Abstr.*, 683.

### Hypersthene (Fe, Mg) $SiO_3$

Eulite; Pickart and Alperin, *Suppl.* **2,** 2082.
Protohypersthene; Gay et al, *Suppl.* **3,** 355.

#### *Occurrence and Form*

Rare orthopyroxenes have not been reported from the basalts in Apollo 11 samples; rather, they occur chiefly in the microbreccias and regolith samples (e.g., in the fragments of anorthositic and noritic rocks),[1] and a few orthopyroxenes occur as exsolution bands in clinopyroxene.[2] Small orthopyroxene grains from the "gray mottled" basalts of Apollo 12 are poikilitically enclosed in plagioclase grains.[3] In Apollo 14 samples, the opthopyroxenes commonly are undeformed and contain lamellae of clinopyroxene up to 5 $\mu$ wide.[4] Orthopyroxenes overgrown by clinopyroxene and sharing (100) are characteristic of KREEP basalt 14310.[5] Some orthopyroxenes in the Luna 20 samples show exsolved fine augite lamellae in twinned orientation $\parallel$ (100).[6] The abundance of orthopyroxenes in the Luna 20 samples indicates that a bronzite-bearing rock (e.g., norite or gabbroic anorthosite) is a prominent constituent of the lunar highlands in this area.[7] In thin section of anorthositic gabbro 15459,38, the orthopyroxenes (up to 1 mm in diameter) contain lamellae and irregular masses of augite up to 10 $\mu$ wide.[8] In recrystallized polymict breccia 60315, anhedral poikiloblasts up to 2 mm across are composed, in general, of slightly zoned orthopyroxene enclosing 2 to 30 $\mu$ grains of plagioclase, olivine, and rare metal (Figure 9–9).[9] One theory about the formation of the pyroxene poikiloblasts is that they grew largely by solid-state recrystallization, in the presence of a

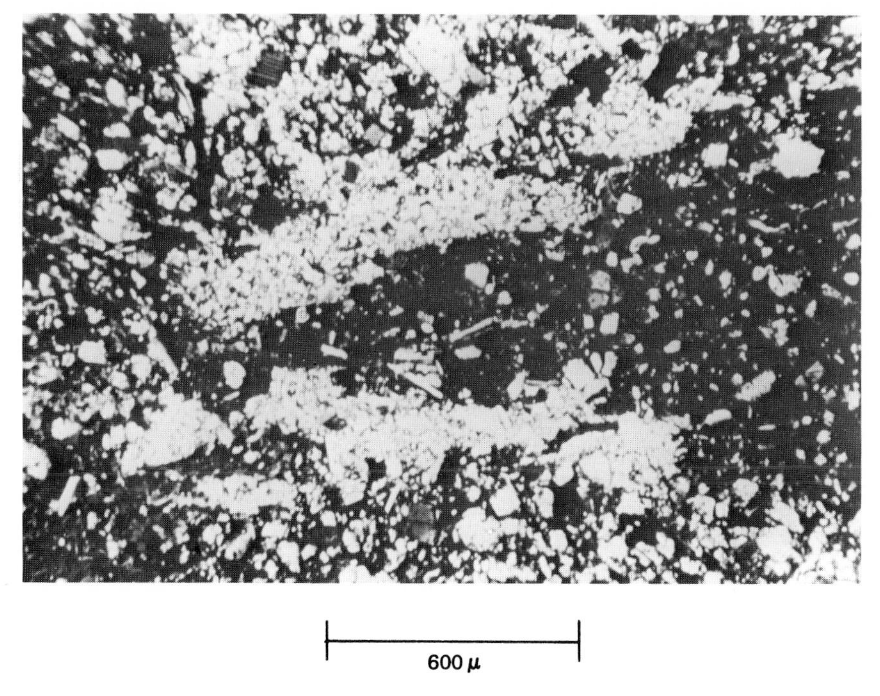


Fig. 9–9 Orthopyroxene poikiloblast with augite rim including olivine, plagioclase, and opaques; from breccia 60315. From *Suppl.* **4,** 1019. Photograph courtesy of D. Walker, Harvard University, Cambridge, Mass.

small amount of liquid, during high-grade metamorphism brought about by a large meteorite impact and the effects of an insulating ejecta blanket. From a fragment in coarse-fined 68503,16 there are zoned orthopyroxene poikiloblasts larger than 1 mm, with rare rims of Mg-augite.[10] In micronorite 76055,10, orthopyroxene porphyroblasts contain inclusions of Ni-Fe and pink Mg-spinel.[11]

Small grains of enstatite are found in less than 20-μ-sized plagioclase-rich fragments in breccias and fines of 200 to 300 μ size.[12] An enstatitic orthopyroxene may have been the earliest pyroxene to crystallize in some Apollo 12 rocks. It appears to have resorbed margins and to be enclosed by magnesian pigeonite.[13] In peridotite 15445,10, an aluminous enstatite occurs as interlocking grains with olivine, spinel, plagioclase, and rutile (?).[14]

Bronzite has been found in fines sample 12070 and in soil fragments of anorthositic gabbro.[15] In KREEP basalt 14310 bronzite cores occur in pigeonite (Figure 9–10).[5] In a thin section of breccia 14315,55, a chondrule 1 mm in diameter was composed mostly of euhedral to subhedral bronzite

Fig. 9–10 Large medium-gray grain is pigeonite with a bronzite core. In KREEP basalt 14310,30. From *Suppl.* **3,** 135. Photograph courtesy of D. Walker, Harvard University, Cambridge, Mass.

plus minor olivine and glass.[16] A norite component in a soil sample from Station 8 of the Apollo 17 mission has minor but conspicuous coarse (1–2 mm) single crystals of unshocked orthopyroxene of $En_{77\text{-}81}$ composition. Some of the crystals have attached plagioclase. Some have minor diopside-rich veins, for which exsolution is not responsible because the diopside is too Mg-rich and the veins are irregular. Fluid (liquid or vapor) intrusion, with possible fractional crystallization, produced these veins.[17]

Hypersthene occurs as exsolution lamellae in augite[18] and also as cores in grains showing epitaxial zoning of clinopyroxene. In porphyritic basalt 12065, hypersthene was the first pyroxene to crystallize (Figure 9–11).[19] Three grains of highly shocked hypersthene from microbreccia 10061-37, found together with maskelynite, may be of meteoritic origin.[18] From Luna 20, rare primary hypersthene is found in mare-type basalt fragments, occurring as minute cores in pigeonite-augite crystals.[7] Anorthosite 60025 contains a relatively Fe-rich hypersthene forming cores in larger crushed subequant grains with augite rims or twinned augite. In smaller grains there are thin rods and stringers of hypersthene 50 to 200 $\mu$ long in cores of clean, intact crystals and in the granulated matrix.[20]

Fig. 9–11 Epitaxial zoning of clinopyroxene on core hypersthene in prophyritic basalt 12065. (Large, long crystal in lower right of picture.) NASA photograph S-70-20963, courtesy of R. Laughan, Johnson Space Center, Houston.

## *References*


1. Agrell et al, *Sci.* **167,** 584.
2. Sclar, *Sci.* **167,** 676.
3. Anderson and Smith, *Suppl.* **2,** 433.
4. LSPET, *Sci.* **173,** 684.
5. Takeda and Ridley, *Suppl.* **3,** 426.
6. Ghose and Wan, *Suppl.* **4,** 901.
7. Ghose et al, *Geochim. Cosmochim. Acta* **37,** 834–835.
8. Takeda, *Suppl.* **4,** 876.
9. Bence et al, *Lun. Sci. IV, Abstr.*, 60.
10. ———, *Suppl.* **4,** 598.
11. Chao, *Suppl.* **4,** 719, 721.
12. Dence et al, *Suppl.* **1,** 324–325.

13. Gay et al, *Suppl.* **2,** 387.
14. Anderson, *J. Geol.* **81,** 219.
15. Brown et al, *Suppl.* **2,** 589–590, 596.
16. Juan et al, *Suppl.* **3,** 692.
17. Steele et al, *Lun. Sci. V, Abstr.*, 727–730.
18. Keil et al, *Suppl.* **1,** 567, 573.
19. Hollister et al, *Suppl.* **2,** 529, 531.
20. Walker et al, *Suppl.* **4,** 1018.

### *Optics*

Enstatite from a Luna 20 sample (22001,17-A16) is colorless.[1] The enstatite-bronzites in the noritic and gabbroic fragments of Luna 20 are optically negative.[2] The bronzite from gabbroic anorthosite 68416,77 has $2V_x = 74°$.[3] In highly shocked Luny Rock 1 from Apollo 11, hypersthene has a large negative 2V and parallel extinction in most grains.[4] In clinopyroxenes from anorthositic gabbro 15473,3,2, there are conspicuous exsolution lamellae of hypersthene about 3 $\mu$ wide. These grains appear dark and somewhat opaque because of the presence of abundant micron-size inclusions of (possibly) ilmenite.[5]

### *References*

1. Meyer, *Geochim. Cosmochim. Acta* **37,** 945–946.
2. Tarasov et al, *Suppl.* **4,** 335.
3. Juan et al, *Lun. Sci. IV, Abstr.*, 421.
4. Albee and Chodos, *Suppl.* **1,** 138.
5. Cameron et al, *EPSL* **19,** 14.

### *Chemical Composition*

Most lunar orthopyroxenes are predominantly magnesian. In the KREEP component of Apollo 11 and 12 soils, they range compositionally from $Fs_{10}$ to $Fs_{30}$, with a CaO content between 1 and 2%.[1] In KREEP basalt 14310, bronzitic orthopyroxene cores zone to ferropigeonite with exsolved subcalcic ferroaugite.[2] In this sample there are also aluminous bronzites that zone to low-alumina bronzites (i.e., $Al_2O_3 \sim 1\%$).[3] Aluminous enstatite from peridotite clast 15445,10 has more $Al_2O_3$ than most analyzed lunar pyroxenes, and some of the grains also have up to 1.61% $TiO_2$. (Table 9-3; [2]).[4] In breccia 14303, Mg-bronzites form a unique group with virtually no variation in chemical composition (Table 9-3; [3]).[5] From a Luna 16 sample, a single crystal of orthopyroxene enclosed in a large plagioclase crystal, had a composition of $Wo_{5.8}En_{71.6}Fs_{22.6}$.[6] In KREEP basalts 15024,4, 15264,4, and 15404,3, and 5, and anorthositic

**Table 9-3 Orthopyroxene Analyses**

| | [1] Luna 20, 22001,17 A 16 | [2] 15445,10 | [3] 14303–53 B | [4] 76535 | [5] Luna 20, 928–1 | [6] 14161/20 | [7] 14310/20 | [8] Luna 20, K–1 | [9] 12033,2B | [10] 10061–3 |
|---|---|---|---|---|---|---|---|---|---|---|
| $SiO_2$ | 41.3 | 55.0 | 56.2 | 55.9 | 56.6 | 55.17 | 52.84 | 54.86 | 54.4 | 52.5 |
| $Al_2O_3$ | 0.09 | 5.1 | 0.68 | 1.5 | 1.49 | 1.25 | 3.21 | 0.90 | 0.4 | 0.65 |
| $Cr_2O_3$ | 0.17 | 0.41 | 0.52 | 0.6 | 0.44 | 0.57 | 0.60 | 0.62 | 0.3 | — |
| $TiO_2$ | 0.01 | 0.78 | 0.38 | 0.4 | 0.04 | 0.31 | 0.73 | 0.42 | 0.4 | 0.75 |
| FeO | 11.6 | 5.8 | 9.19 | 7.4 | 9.03 | 10.42 | 11.65 | 15.00 | 17.9 | 23.1 |
| MgO | 47.5 | 33.6 | 33.5 | 32.7 | 31.3 | 30.84 | 27.35 | 26.33 | 25.7 | 20.2 |
| MnO | 0.12 | — | 0.17 | 0.1 | 0.09 | 0.13 | 0.25 | 0.25 | — | 0.47 |
| CaO | 0.29 | — | 0.79 | 1.5 | 1.42 | 0.74 | 3.20 | 1.94 | 1.4 | 1.86 |
| $Na_2O$ | — | — | 0.01 | — | — | — | — | 0.02 | — | 0.03 |
| Total | 101.08 | 100.69 | 101.44 | 100.1 | 100.41 | 99.43 | 99.83 | 100.34 | 100.5 | 99.59* |

*References*


[1] Meyer, *Geochim. Cosmochim. Acta* **37,** 947. From a soil sample. Enstatite.

[2] Anderson, *J. Geol.* **81,** 221. From a periodtite clast. Aluminous enstatite.

[3] Weigand and Hollister, *Suppl.* **3,** 476. From a breccia. Average of nine analyses. Mg-bronzite.

[4] Smyth, *Geophys. Res. Lett.* **1,** 29. From a troctolitic granulite. Low-orthopyroxene bronzite.

[5] Tarasov et al, *Suppl.* **4,** 337. From an anorthositic fragment in soil sample. Bronzite.

[6] Brown et al, *Suppl.* **3,** 142. From a fines sample. Bronzite.

[7] ———, *ibid*, From a KREEP basalt. Aluminous bronzite.

[8] Ghose et al, *Geochim. Cosmochim. Acta* **37**, 833. From a soil sample. Hypersthene (also termed a chromium-rich bronzite).

[9] Meyer, *Suppl.* **2,** 398. From a KREEP glass fragment. Hypersthene.

[10] Keil et al, *Suppl*, **1,** 564. From a microbreccia. *Analysis includes $K_2O$ = 0.03%. Hypersthene.

gabbro 15434,4, and 7, skeletal orthopyroxene phenocrysts show fractionation trends from hypersthene cores ($Wo_3En_{75-82}Fs_{15-22}$) to intermediate pigeonite ($Wo_{10}En_{52}Fs_{38}$).[7] Small grains (< 0.13 mm) of orthopyroxene from anorthositic breccia 60215,13 are more Mg-rich than the orthopyroxenes from anorthosite 15415 but less magnesian than Luna 20 orthopyroxenes.[8] Of the analyzed Luna 20 pyroxenes, 41% are orthopyroxene,[9] and they are the most magnesian-rich yet reported from lunar samples[10] ranging between $En_{88}$ and $En_{68}$ (Table 9-3; [1]).[11]

*References*

1. Meyer et al, *Suppl.* **2,** 404–405.
2. Ridley et al, *GSA '71, Abstr.*, 682.
3. ———, *Suppl.* **3,** 162.
4. Anderson, *J. Geol.* **81,** 219.
5. Weigand and Hollister, *Suppl.* **3,** 473.
6. Bence e al, *EPSL* **13,** 304.
7. Powell et al, *Lun. Sci. IV, Abstr.*, 597–598.
8. Meyer and McCallister, *Suppl.* **4,** 662–663.
9. Reid et al, *Geochim. Cosmochim. Acta* **37,** 1018.
10. Meyer, *Geochim. Cosmochim. Acta* **37,** 946.
11. Meyer, *Lun. Sci. IV, Abstr.*, 520, 522.

*X-Ray Data*

X-Ray study has confirmed the presence in the lunar rocks of some orthopyroxenes that are compositionally very close to clinohypersthene or low-Ca pigeonite. The presence of magnesian orthopyroxene in KREEP basalt 14310 was confirmed by precession photographs of single pyroxene crystals devoid of augite exsolution.[1] X-Ray diffraction studies on orthopyroxene with coarse augite lamellae, from breccia 15459, have aided in presenting further evidence that slowly cooled rocks exist in the lunar highlands.[2] It is believed that troctolitic granulite 76535 also cooled slowly because the bronzite crystals from it have a postulated, but previously unreported, space group $P2_1ca$. This space group resulted from an inversion of *Pbca* as a displacive, solid-state ordering reaction indicative of extremely long cooling times.[3] From a pigeonite (K-4) of a Luna 20 soil sample, detailed crystallographic and chemical studies have revealed a most complex zoning. Augite exsolution lamellae on (001) of the pigeonite host are optically visible, in addition to exsolved rods and platelets of a chrome-rich spinel. With precession photographs five major phases could be determined:

1. A clinohypersthene.
2. Augite lamellae parallel to (001) of the clinohypersthene host.

3. Augite lamellae parallel to (100).
4. Hypersthene lamellae parallel to (100).
5. Chrome spinel.

X-Ray data for phases 1 to 4 are given in Table 9-2 ([6]–[9], [13]).[4]
Cell constants for some orthopyroxenes are given in Table 9-2.

*References*

1. Ridley et al, *Lun. Sci. Conf. '72, Abstr.*, 570–571.
2. Takeda, *Suppl.* **4,** 884.
3. Smyth, *Geophys. Res. Lett.* **1,** 27–29.
4. Ghose et al, *Geochim. Cosmochim. Acta* **37,** 834–835.

PROTOHYPERSTHENE

An early orthopyroxene of composition $Wo_7En_{67}Fs_{26}$, protohypersthene disappeared by reaction with the liquid after the appearance of pigeonite.[1] It is not abundant, but it has been reported from porphyritic basalt 12020 and ophitic basalt 12040.[2] It has been suggested, however, that the core pigeonite in porphyritic basalt 12065 originally was not protohypersthene but "high-clinopyroxene" with *C*2/*c* symmetry.[3]

*References*

1. Biggar et al, *Lun. Sci. Conf. '71, Abstr.*, 214.
2. ———, *Suppl.* **2,** 620.
3. Kushiro et al, *Suppl.* **2,** 485.

EULITE

From chips of 12071,6, the iron-rich pyroxene (0.17 $MgDiO_3$,0.83 $FeSiO_3$) showing possible magnetic ordering, may be the orthopyroxene eulite.[1]

*Reference*

1. Pickart and Alperin; *Suppl.* **2,** 2082.

## Pyroxferroite: Close to $(Fe_{0.83}Ca_{0.13}Mg_{0.02}Mn_{0.02})SiO_3$

*Synonymy*

CaFe pyroxenoid; Frondel et al, *Sci.* **167,** 681.
Calcian pyroxferroite; Chao, *Suppl.* **1,** 67.
Fe-rich "pyroxene"; Weill et al, *Sci.* **167,** 635.
Fe-rich pyroxmangite; Anderson et al, *Sci.* **167,** 587.

Fe-rich pyroxferroite; Boyd and Smith, *J. Petrol.* **12,** 442.
Ferrosilite III; Agrell et al, *Sci.* **167,** 585.
Ferrous-rich pyroxferroite; Brown, *J. Geophys. Res.* **75,** 64, 88.
Iron-wollastonite; Agrell et al, *Sci.* **167,** 585.
Low-calcium ferropyroxene; Brown et al, *Suppl.* **3,** 142.
Magnesian pyroxferroite; Smith and Finger, *Carnegie Inst. Yearb.* '70, 133.
Mg-free pyroxferroite; Brown et al, *Suppl.* **1,** 203.
Mg-rich pyroxferroite; Boyd and Smith, *J. Petrol.* **12,** 442.
Phase A (Douglas); Douglas et al, *Sci.* **167,** 595.
"Pyroxmangite"; Keil et al, *Sci.* **167,** 597.
"Subcalcic pyroxferroite"; Brown et al, *Lun. Sci. III, Rev. Abstr.*, 95.

*Occurrence and Form*

Pyroxferroite is found in gabbros or microgabbros as anhedral or, more rarely, euhedral grains, and appears to be a late-stage mineral together with tridymite and/or cristobalite (Figure 9–12).[1] In porphyritic basalt 12021, the pyroxferroite forms unusually large crystals (up to 1 mm in diameter) intergrown with subcalcic augite, tridymite, and calcic plagioclase.[2] In medium-grained basalt 10003-44, a mosaic texture of interlocking grains is well displayed. Pyroxferroite occurs also intergrown with or mantling clinopyroxene. In some grains the mantling is sharp, with the boundary clearly visible under the microscope by a change in the birefringence (Figure 9–13). Other grains exhibit a gradual change with no compositional break, but there is a gradual increase in the triclinicity and partial birefringence as the pyroxenoid takes over and grades outward toward the still more iron-rich margins.[3] Cleavage in pyroxferroite is not as pronounced as in terrestrial pyroxmangite. The one good cleavage is (010) and the poor one (001).[1] In thin section pyroxferroite tends to split parallel to the cleavage, imparting a fibrous appearance to the grains.[3]

*Optics*

The point at which triclinicity takes over appears to be just about 2V = 38°, with a small extinction angle of 1° to the (010) cleavage in a section perpendicular to [001]. The extinction on (010) increases toward the grain margins, where $Z \wedge C = 47°$ and 2V = 31°. There is little dispersion of the optic plane but considerable dispersion of the *XY* and *YZ* planes.[3]

In thin section the mineral is almost colorless or yellow, or faintly pleochroic yellow to yellow-orange. The range of refraction indices is:[1]

$$n_\alpha = 1.748 \pm 0.003 - 1.756$$
$$n_\beta = 1.750 \pm 0.003 - 1.758$$
$$n_\gamma = 1.765\text{–}1.768 \pm 0.003$$

Fig. 9–12 Pyroxferroite (light grains in central and lower portions of right half of picture). Other minerals are ilmenite (black), plagioclase (long lath), clinopyroxene (medium gray, upper and lower parts of picture), and cristobalite (dark gray, mosaic texture). From coarse-grained basalt 10047. From *Suppl.* **1,** 68. Photograph courtesy of A. G. Plant, Geological Survey of Canada, Ottawa.

*Chemical Composition*

The composition of pyroxferroite is close to ferrosilite but a little more Ca-rich. From microgabbro 10047 it is approximately $(Fe_{0.83}Ca_{0.13}Mg_{0.02}Mn_{0.02})SiO_3$,[4] and the major elements vary only over narrow ranges.[1] The composition where the change to triclinic optics has been observed is $Ca_{27}Mg_{33}Fe_{40}$, with MgO slightly less than 5.90%.[3]

Apollo 11 material is a calcian pyroxferroite with a composition closest to $FeSiO_3$ of all natural *Siebenerketten* materials (Table 9-1; [12]).[4] In the mesostasis of coarse-grained diabase 10058-23, the pyroxferroite is zoned to a Mg-free variety (Table 9-1; [13]).[5] In porphyritic basalt 12021, the Mg-for-Fe substitution in pyroxferroite ranges from $(Ca_{0.13}Mg_{0.03}Fe_{0.84})SiO_3$ to $(Ca_{0.12}Mg_{0.13}Fe_{0.75})SiO_3$.[4] Some of the unusually large crystals in this sample have been referred to as magnesian pyroxferroite (Table 9-1; [11]).[2] The subcalcic pyroxferroite, rimming zoned clinopyroxene grains in basalt 15076/12, could be called an Fe-rich pyroxferroite (Table 9-1; [14]).[6]

Fig. 9–13 Pyroxferroite (Py) has an irregular but sharp boundary against zoned ferroaugite (Fe-aug.) in porphyritic basalt 12021,2. From *Suppl.* **2,** 291. Photograph courtesy of A. G. Plant, Geological Survey of Canada, Ottawa.

It is assumed in the analyses of pyroxferroite that the titanium is in the quadrivalent state, although it has been suggested that the titanium may be trivalent.[1]

*X-Ray Data*

Pyroxferroite is a triclinic pyroxenoid, with space group $P\bar{1}$ and a S*iebenerketten* structure (i.e., the single silicate chains have a repeat period of seven silicon tetrahedra). Its specific gravity, calculated from its cell contents, is 3.82. The measured specific gravity is 3.76.[4] Cell constants and X-ray powder data appear in Tables 9-2 and 9-4, respectively.

**Table 9-4 X-Ray Power Data for Pyroxferroite (Apollo 11 Mean*)**

| *d* (Å) | *I* | *hkl* |
|---|---|---|
| 6.87 | 20 | 010 |
| 6.55 | 25 | 100 |
| 5.56 | 5? | 011? |
| 4.93 | 5? | $1\bar{1}2$? |
| 4.68 | 40 | 110 |
| 4.20 | 15 | $\bar{1}11$ |
| 3.76 | 8 | $0\bar{2}2$ |
| 3.68 | 6 | $0\bar{2}1$, $0\bar{2}3$ |
| 3.53 | 8 | 013 |
| 3.44 | 15 | 020 |
| 3.38 | 5 | $1\bar{2}2$ |
| 3.32 | 10 | $1\bar{2}3$ |
| 3.28 | 12 | 200, 201 |
| 3.19 | 5 | 113 |
| 3.14 | 15 | 202, $\bar{2}01$ |
| 3.09 | 45 | 021, $2\bar{1}1$, $2\bar{1}2$ |
| 3.01 | 25 | 120, $0\bar{2}5$ |
| 2.973 | 20 | $2\bar{1}3$ |
| 2.934 | 100 | 014, 210 |
| 2.838 | 10 | 211 |
| 2.805 | 8 | $\bar{2}11$, 121 |
| 2.674 | 60 | $0\bar{2}6$ |
| 2.621 | 30 | $\bar{1}14$ |
| 2.579 | 35 | $2\bar{2}2$ |
| 2.509 | 5 | $0\bar{3}3$ |
| 2.490 | 20 | $0\bar{3}2$ |
| 2.446 | 15 | 214 |
| 2.409 | 10 | $2\bar{2}0$ |
| 2.377 | 8 | $1\bar{3}2$ |
| 2.311 | 8 | $2\bar{1}6$ |
| 2.284 | 8 | $13\bar{3}$, 030 |
| 2.234 | 8 | $0\bar{3}6$ |
| 2.217 | 8 | $2\bar{2}6$, $22\bar{4}$ |
| 2.186 | 25 | 024 |
| 2.156 | 40 | 206, $3\bar{1}2$ |
| +35 more to 1.022 | 5 | |

* Combination of data by five Apollo 11 Principal Investigative Teams. FeK and CrK radiation.

*Reference.* Chao et al, *Suppl.* **1,** 72–73.

### Iron Wollastonite and Ferrosilite III

Electron diffraction examination of crystals from coarse-grained basalt 10044/43 revealed about equal parts of unmixed phases with similar structures that may represent iron-wollastonite and ferrosilite III.[8] Ferrosilite III has a *Neunerketten* structure, whereas pyroxferroite has a *Siebenerketten* structure.[9] However, until further X-ray investigation of these crystals proves otherwise, the suggestion that both phases may be pyroxferroite is entertained.

### *References*

1. Chao et al, *Suppl.* **1,** 65–78.
2. Smith and Finger, *Carnegie. Inst. Yearb.* '*70*, 133–134.
3. Agrell et al, *Suppl.* **1,** 103–107.
4. Burnham, *Suppl.* **2,** 47.
5. Brown et al, *Suppl.* **1,** 203.
6. ———, *Lun. Sci. III, Rev. Abstr.*, 95, 97.
7. Agrell et al, *Sci.* **167,** 585.
8. Burnham, *Sci.* **154,** 513.

# CHAPTER TEN | SILICATES: OLIVINES

## Forsterite-Fayalite Series

Olivine, together with Mg-pyroxene, is generally among the earliest phases to crystallize from lunar basalts.[1] Most of the olivines have a high forsterite component.[2] These minerals commenced and ceased to crystallize at an early stage.[3] In several anorthositic rock fragments, some only moderately magnesian olivine may have crystallized at an intermediate stage, and fayalite is a late-crystallizing olivine found in the residuum. Hence there are early, middle, and late olivines, but usually not all in the same rock.[4]

*References*


1. Jakeš, *Lun. Sci. V*, *Abstr.*, 381.
2. James and Jackson, *J. Geophys. Res.* **75,** 5799–5800.
3. Agrell et al, *Suppl.* **1,** 98.
4. Smith et al, *Suppl.* **1,** 919.


### FORSTERITE $Mg_2SiO_4$

*Synonymy*

Chrysolite; Kim et al, *Suppl.* **2,** 750.
Forsteritic olivine; Brown et al, *Suppl.* **1,** 203.
Iron-poor olivine; Adams et al, *Geochim. Cosmochim. Acta* **37,** 736.
Magnesian olivine; Brown et al, *Suppl.* **1,** 202.

*Occurrence and Form*

Forsteritic olivine occurs in anhedral, subhedral,[1] and sometimes euhedral grains (Figure 10-1). The grains are usually no longer than 0.5 mm,[2] but

Fig. 10–1 Olivine crystal of late deposition standing on pyroxene. From a vug in basalt 15555,105. Photograph courtesy of J. Jedwab, Université Libre de Bruxelles, Belgium.

olivines up to 3 mm long have been noted.[3] Some olivines occur with abundant cristobalite, as in fine-grained olivine basalt 10045; and compared with terrestrial rocks, this coexistence is anomalous. These olivines may have occasional rims of augite that are compositionally gradational and indicative of a slight reaction between olivine and the liquid.[4] From fine-grained olivine basalt 10022-41, some olivine crystals that are in contact with both augite and pigeonite are rounded and may have reacted with the liquid to form the pyroxenes.[5] Dendritic olivine microphenocrysts

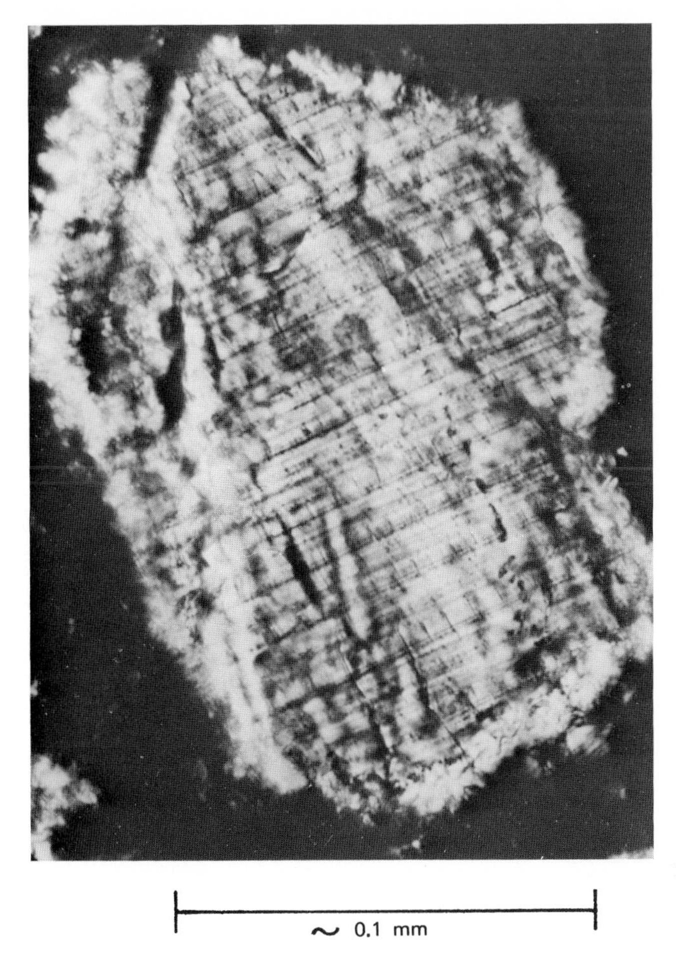


Fig. 10–2 Shock lamellae parallel to {110} in olivine. From *Suppl.* **3,** 902. Photograph courtesy of H. Avé Lallement, Department of Geology, Rice University, Houston.

occur in a glassy portion of breccia 10061 that may be part of a quenched picritic lava.[6] In numerous samples of Apollo 11 crystalline rocks, breccias, and soil, it was observed that a growth phenomenon has caused the trapping of large melt inclusions in most of the larger olivine crystals and even in some of the very tiny crystals. These inclusions, which are oriented parallel to olivine (100), are almost all ilmenite, although they once may have been armalcolite.[7] Only a few olivine grains in Apollo 11 samples show any development of planar or lamellar features that are considered to be indicative of shock (Figure 10-2).[8] Within a diaplectic plagioclase glass fragment from what is probably a gabbroic anorthosite, olivine

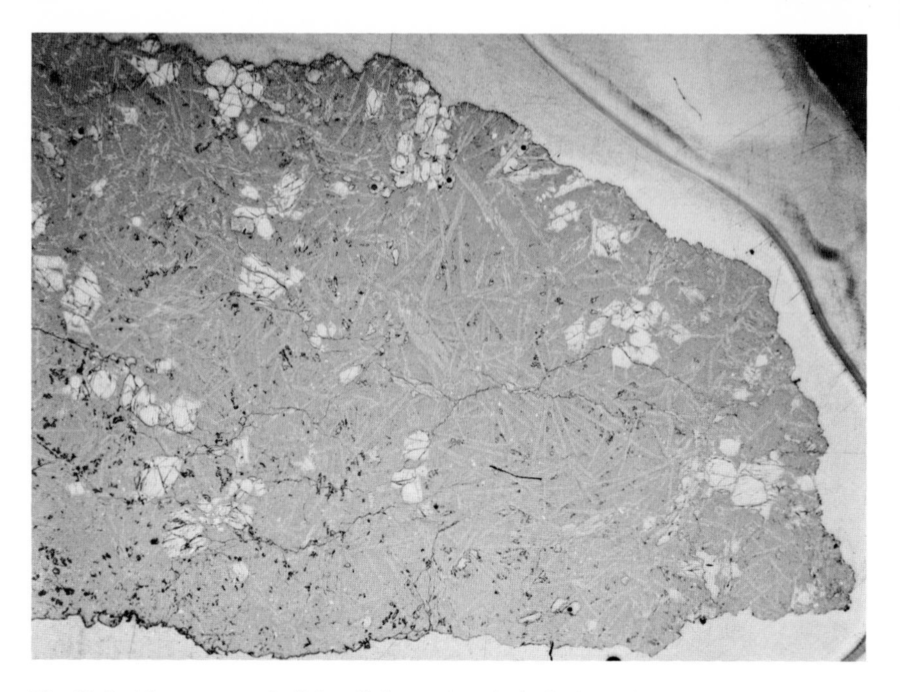

Fig. 10–3 Phenocrysts of olivine (light gray) and plagioclase (elongate medium gray) in porphyritic olivine basalt 12009. NASA photograph S-70-25427, courtesy of R. Laughan, Johnson Space Center, Houston.

grains exhibited planar deformation structures—for example, very thin lamellae parallel to {100}, {011}, {001}, and {130}. The last are characteristic of dynamic deformation.[9]

From porphyritic olivine basalts 12004, 12008, 12009, and 12022 (Figure 10–3), olivine phenocrysts are subhedral, stoutly columnar, about 0.3 mm in size, and make up from 9 to 16% of the rocks' volume. Subhedral to euhedral grains of olivine up to 2 mm in size, occurring in a matrix of anhedral plagioclase grains, compose about 35% of granular basalt 12035.[10] Olivines in Mg-rich magmas tend to crystallize rapidly in skeletal forms. Hollow olivines or olivines with inclusions of groundmass are also typical products of rapid crystallization (Figure 10–4). Excellent examples of skeletal olivines are found in basalt 12009,[11] and in picritic basalt 12002, zoned olivine phenocrysts up to 1.5 mm in diameter are subhedral skeletal to anhedral (Figure 10–5).[12]

In an unusual lunar breccia 14068, an olivine-rich groundmass has conspicuous olivine, more or less skeletal in habit, with hollow cores where the crystal outline is relatively blocky.[13] Forsteritic olivine appears

Fig. 10–4 Hollow diamond-shaped olivine crystals intergrown with hollow plagioclase laths in microtroctolite 62295. Photograph courtesy of D. Walker, Harvard University, Cambridge, Mass.

in subophitic basalt 14072 as large, subrounded phenocrysts up to 2.0 mm across, as inclusions in large pyroxenes, and as anhedral members of the general framework. In addition, an olivine of intermediate composition occurs as part of the late-stage assemblage, with mosaic-textured cristobalite and a spongy network of native iron.[14] Some Apollo 14 breccias contain chondrules that have the texture and mineralogy of common types of meteoritic chondrules (i.e., euhedral olivine plus pyroxene crystals in a brown transparent glass: Figure 10–6). In fines sample 14259,33, a barred

Fig. 10–5 Rounded subequant olivine phenocrysts (white) in picritic basalt 12002,7. NASA photograph S-70-31576, courtesy of R. Laughan, Johnson Space Center, Houston.

olivine-glass chondrule has been noted. There are also olivine-rich cores surrounded by recrystallized aureoles or diffusion haloes of pyroxenes and opaques. It is believed that these chondrules were formed on the Moon by mechanisms related to large impacts.[15]

Some Apollo 15 rake samples have yielded pyroxene basalt fragments containing skeletal olivine phenocrysts. Other fragments of olivine basalts have zoned microphenocrysts of olivine ($\sim$ 0.5 mm) in a fine groundmass of pyroxene, plagioclase, ilmenite, and a few smaller olivine crystals. The outer margins of the phenocrysts are euhedral but embayed; thus the overall outline tends to be ameboid. Zoning is not very pronounced in olivine microgabbro fragments.[16] Lunar peridotite clast 15445,10 contains interlocking irregular masses of olivine, spinel, pyroxene, plagioclase, and rutile (?). Olivine grains up to 0.1 mm in maximum dimension, with the pyroxene, make up 80% of the volume of the rock.[17]

In addition to the large olivine grains (up to 3 mm), olivine relicts are scattered throughout a number of Apollo 16 poikilitic rocks.[3] In breccia 67955, coarse olivines are inclusions in poikilitic hypersthene and diopside.

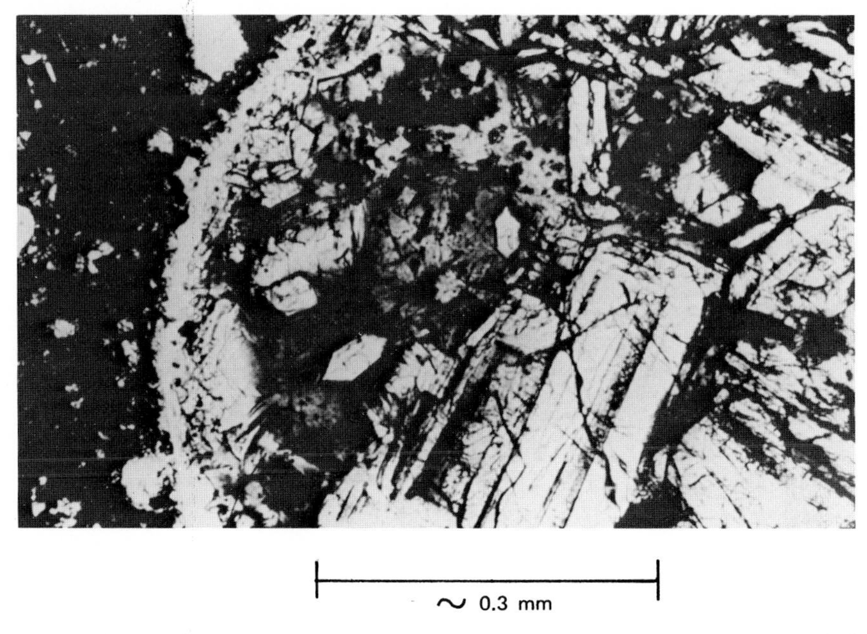


Fig. 10–6 Orthopyroxene-olivine-glass lunar chondrule from breccia 14313. Euhedral olivine (small) and pyroxene (large) crystals occur in a matrix of light brown transparent glass. From *Suppl.* **3,** 675. Photograph courtesy of E. King, Department of Geology, University of Houston, Texas.

This distinctive texture is believed to result from high-temperature crystallization in the presence of a fluid phase, possibly a silicate melt.[18] In anorthositic gabbro 68415,8, olivine has been observed throughout the thin section as small anhedral to irregular grains interstitial to plagioclase laths, as cores in pyroxene or rimmed by pyroxene, and also mantled by plagioclase without a pyroxene rim. The textures suggest that the crystallization of olivine preceded that of the pyroxene and that it reacted with with the melt to form pyroxene rims except where it was mantled by plagioclase.[19] The olivine from xenoclastic basalt 60335,75E is slightly more magnesian and in places is rimmed by orthopyroxene, which in turn is rimmed by pigeonite.[20] In spinel troctolite, the rims of olivine grains contain colorless octahedra of a spinel.[21]

In coarse-grained anorthositic gabbro 77071, olivine occurs as euhedral to subhedral crystals surrounded by single-crystal mantles of pyroxene, and as relatively small grains enclosed in plagioclase.[22] In 76055,10, a rock described as an impact-generated melt-rock,[23] there are abundant olivine xenocrysts that have, in general, no reaction rims; instead rounded outlines appear where the xenocrysts are enveloped by orthopyroxene.[24] From Apollo 17 deep-trench samples, several 1-mm fragments of high-Mg olivine have a cataclastic texture and minute oriented exsolution products, indicating a metamorphic event.[25] Fragments from 72415 to 72418, of a highly crushed dunite, inferred to be a product of primary lunar differentiation, contain rounded large (up to 10 mm) crystals of olivine in a granular matrix. The matrix is also predominantly olivine and has formed by crushing without recrystallizing. Both matrix and clasts have inclusions of plagioclase, Cr-spinel, high- and low-Ca pyroxenes, and Fe metal.[26]

*References*

1. Frondel et al, *Suppl.* **1,** 463.
2. Dence et al, *Suppl.* **1,** 323.
3. Simonds et al, *Suppl.* **4,** 614.
4. Brown et al, *Suppl.* **1,** 203.
5. Kushiro and Nakamura, *Suppl.* **1,** 617.
6. Smith et al, *Suppl.* **1,** 906.
7. Roedder and Weiblen, *Suppl.* **1,** 804.
8. Sclar, *Suppl.* **1,** 854.
9. von Engelhardt et al, *Suppl.* **1,** 369–370.
10. Butler, *Geochim. Cosmochim. Acta* **36,** 773–774.
11. Drever et al, *Suppl.* **3,** 172.
12. Grove et al, *Suppl.* **4,** 995.
13. Helz, *Suppl.* **3,** 866.
14. Longhi et al, *Suppl.* **3,** 135.
15. King et al, *Suppl.* **3,** 674.
16. Dowty et al, *Lun. Sci. IV, Abstr.,* 181.
17. Anderson, *J. Geol.* **81,** 219.
18. Hollister, *Suppl.* **4,** 633.
19. Gancarz et al, *EPSL* **16,** 316.
20. Walker et al, *Lun. Sci. IV, Abstr.,* 752.
21. ———, *Lun. Sci. IV, preprint.*
22. Sclar, *Lun. Sci. V, Abstr.,* 688.
23. *Lunar Sample Information Catalog, Apollo* 17, 295.
24. Chao, *Suppl.* **4,** 721.
25. Steele et al, *Lun. Sci. V, Abstr.,* 728.
26. Albee et al, *Lun. Sci. V, Abstr.,* 3–5.

*Optics*

The olivines range from colorless, as from gabbroic anorthosite 68416,77, to lemon-yellow[2] or yellow-green, as reported on transparent and unaltered grains from Apollo 11 fines,[3] and pale greenish yellow for xenocrysts of microbreccia 77215.[4]

Some refractive indices are:

from Apollo 11 fines:[3]

$n_\alpha$ = 1.710
$n_\beta$ = 1.730
$n_\gamma$ = 1.750
with 2V ~ 100°

from gabbroic fragments in another fines' sample 10085-16, $n_\beta$ had a range of 1.692 to 1.744 ±0.002 for compositions of:[5]

$Fo_{55}$ to $Fo_{78}$

from ophitic olivine basalt 10062-35, $n_\beta$ had a range of 1.69 to 1.71 ±0.02 for compositions of:

$Fo_{70}$ to $Fo_{76}$

and from fine-grained basalt 10045, the $2V_x$ = 81° to 83°, indicating some slight zoning of:[6]

$Fo_{68}$ to $Fo_{74}$.

In general, discrete olivine fragments and grains from microbreccias have a high degree of crystalline perfection, as shown by their great transparency in particles even as large as 1 mm.[7]

In addition to containing fractures and faults, many olivine crystals in the Apollo 14 breccias show undulatory extinction, and a few crystals contain kink bands. The olivines commonly display a mosaic structure and are partially recrystallized to mosaics of new strain-free crystals along their borders. Many of the crystals have recrystallized totally, making the host grain unrecognizable.[8] From dunite fragments 72415 to 72418, large olivine crystals have undulatory extinction and planar partings resulting in a mosaic of rhomb-shaped domains bounded by partings or strain bands.[9] Rare shocked olivine from coarse fines 10083-12-53 shows weak extinction variations and faint lamellae that may be due to deformation.[10] One large, homogeneous grain in a partly shocked anorthosite fragment of microbreccia 10019-22 has fine mosaic texture, probably because of moderate shock.[11]

*References*

1. Juan et al, *Lun. Sci. IV, Abstr.*, 421.
2. Kim et al, *Suppl.* **2,** 750.

3. Frondel et al, *Suppl.* **1,** 463.
4. Chao and Minkin, *Lun. Sci. V, Abstr.*, 110.
5. Carter and MacGregor, *Suppl.* **1,** 248, 250.
6. Agrell et al, *Suppl.* **1,** 98.
7. Sclar, *Suppl.* **1,** 853.
8. Avé Lallement and Carter, *Suppl.* **3,** 901–903.
9. Albee et al, *Lun. Sci. V, Abstr.*, **3.**
10. Chao et al, *Suppl.* **1,** 300.
11. Keil et al, *Suppl.* **1,** 581.

*Chemical Composition*

Most lunar olivines are highly magnesian, often of quite homogeneous composition, and they are higher in Ca and Cr than terrestrial olivines. They contain very little Ni, minor Mn, Ti, and Al, and the Cr may be divalent.[1] Olivines from highland rocks have lower Cr than those from mare basalts, but there is no well-defined trend with respect to Mg/(Mg + Fe). In mare basalts the highest Cr content is in the olivine range $Fo_{65}$ to $Fo_{80}$.[2] From anorthositic gabbro 68415,8, although there is little variation in composition within grains or between grains of olivine, the minor elements show some differences; for example,

$$\begin{aligned} MnO &= 0.23\text{–}0.33\% \\ Cr_2O_3 &= 0.07\text{–}0.17\% \\ TiO_2 &= 0.05\text{–}0.09\% \\ CaO &= 0.02\text{–}0.29\% \end{aligned}$$

However, there appears to be no correlation of these minor element occurrences with the Mg/Fe, which is the same as that of the most magnesian pyroxene observed in the thin section.[3] The olivine from lunar peridotite clast 15445,10 has a very high Mg/Fe with Fo = 91, and the CaO content is notably low (Table 10-1; [2]). Four analyzed grains had from 0.00 to 0.05% CaO. Lunar olivines with comparable low CaO are rare,[4] although they have been noted (e.g., Table 10-1; [3], [5]).

Many olivines approach the forsterite end member. From an Apollo 17 deep-trench plagioclase-rich basalt fragment, the olivine composition is unusually Mg-rich $Fo_{94\text{-}75}$, with an average of $Fo_{90}$.[5] Crystalline (possibly effusive) rock 62295 contains two 0.8-mm olivine phenocrysts of composition $Fo_{89\text{-}93}$, with skeletal overgrowths of olivine of $Fo_{82\text{-}85}$.[6] In the dark lithic portion of basalt 61016-215, the Mg-rich olivine has a very limited composition, and more than half the analyzed grains are $Fo_{89\text{-}93}$.[7] From partially melted breccia 67435,14, the olivine has a composition of $Fo_{89}$ (Table 10-1; [1]).[8] Olivines from a variety of rock types in the Luna 20 fragments are very forsteritic, and samples of high-feldspathic highland material were even more Mg-rich than those from non-highland sites. Although the olivines range in composition from $Fo_{40}$ (an intermediate olivine) to $Fo_{92}$, more than 50% of them have an Fo content greater than 80%.[9] The individual olivine grains are completely homogeneous and may be divided into two classes on the basis of Fe/(Fe + Mg). One group has a composition of $Fo_{80}$ or more, and the other group has an Fo content of less than 72%.[10] From fines sample 12070,98, olivine with composition $Fo_{82}$ was called chrysolite,[11] and many lunar olivines have approximately this composition (e.g., from Luna 20 soil sample 22007:[12] Table 10-1; [6]).

Less forsteritic olivine ($Fo_{63}$), from coarse-grained gabbroic igneous rock 14053, has been found as inclusions in pigeonite crystals,[13] and from 68415,37, possibly a high-grade metamorphic rock, the olivine ranges from $Fo_{70}$ to $Fo_{67}$ with no evident zoning.[14] In basalt 15555 the olivine, commonly enclosed in plagioclase, is more Fe-rich than that of the other mare basalts, yet it is still largely magnesian (e.g., $Fo_{58}$), and in basalt 15256 large olivine clasts have an approximate composition of $Fo_{65}$. The average olivine composition from fines sample 15471 is $Fo_{61}$, and fines sample 15271 yielded an intermediate olivine ($Fo_{56\text{-}60}$) as well as a highly forsteritic one ($Fo_{88\text{-}90}$).[15] In a poikiloblastic diabasic igneous rock 60315,63, anhedral olivines included in orthopyroxenes are uniform in composition with a maximum range of $Fo_{74\text{-}77}$. One more Mg-rich large olivine relict ($Fo_{78\text{-}80}$) is unequilibrated with the pyroxene in the rock. Unzoned discrete olivine grains in poikiloblastic rock 68503,16,6 have a compositional range of $Fo_{64\text{-}76}$, where they occur as inclusions in plagioclase, but the olivine inclusions in the orthopyroxene are restricted to $Fo_{70}$.[16] In clasts of breccia 67955, tiny (up to 0.03 mm) perfect spheres of olivine, occurring as inclusions in plagioclase, have a composition of $Fo_{78}$. Larger grains, included in clinopyroxene, are just slightly more Fe-rich ($Fo_{77}$). All the olivines are low in Cr, Ca, and Ti, suggesting that they did not grow under surface lava conditions.[17]

Zoning in forsteritic olivine is usually slight. In the unmetamorphosed intersertal basalts of Apollo 11, zoned phenocrysts range from $Fo_{63}$ to $Fo_{68}$, with some smaller cores of an intermediate olivine of $Fo_{41}$.[18] There is a wide range of olivine compositions from fine-grained olivine basalt 10022-41, between $Fo_{71}$ and $Fo_{41}$, even within a single-zoned crystal.[19] Fragments of mare basalts from Apollo 15 rake samples yielded the following compositional information on the olivines:

1. From pyroxene-phyric basalts 15125 and 15666, large skeletal phenocrysts are predominantly magnesian, with some zoning to a more fayalitic composition.

**Table 10-1 Chemical Analyses of Olivines (Forsterite-Fayalite Series)**

| | [1] | [2] | [3] | [4] | [5] | [6] | [7] | [8] |
|---|---|---|---|---|---|---|---|---|
| | | | 14002, | Luna 20 | Luna 20 | Luna 20 | | |
| | 67435,14 | 15445,10 | 7E–1–8 | 509–10 | 22002,2,5a | 22007 | 60215/13 | 10020–40 |
| $SiO_2$ | 41.2 | 42.4 | 40.8 | 40.3 | 39.9 | 39.41 | 39.1 | 37.5 |
| $Al_2O_3$ | 0.38 | — | 0.01 | 0.01 | — | 0.23 | 0.26 | 0.05 |
| $Cr_2O_3$ | 0.01 | 0.01 | 0.18 | — | 0.05 | 0.22 | 0.20 | 0.21 |
| $TiO_2$ | 0.03 | — | 0.02 | 0.16 | — | 0.16 | 0.03 | 0.09 |
| FeO | 7.5 | 8.3 | 11.1 | 13.2 | 15.6 | 17.14 | 19.0 | 25.5 |
| MgO | 51.3 | 50.7 | 48.7 | 47.0 | 45.2 | 43.25 | 41.3 | 36.5 |
| MnO | 0.08 | — | 0.01 | — | 0.15 | $<0.01$ | 0.14 | 0.30 |
| CaO | 0.24 | 0.05 | 0.05 | 0.15 | 0.05 | 0.27 | 0.34 | 0.33 |
| NiO | — | 0.03 | 0.03 | — | — | $<0.01$ | — | — |
| Total | 100.74 | 101.49 | 100.90 | 100.82 | 100.95 | 100.68 | 100.42* | 100.48 |

*References*

[1] Prinz et al, *Sci.* **179,** 75. From a microbreccia. Forsterite.

[2] Anderson, *J. Geol.* **81,** 221. From a peridotite clast. Forsterite.

[3] Steele and Smith, *Nature* **240**, 5. From a fines sample, Forsterite.

[4] Crawford and Weigand, *Geochim. Cosmochim. Acta* **37,** 820. From an anorthositic noritic troctolite fragment in soil sample. Forsterite.

[5] Cameron et al, *Geochim. Cosmochim. Acta* **37,** 782. From a fine-grained soil sample. Forsterite.

[6] Podosek et al, *Geochim. Cosmochim. Acta* **37,** 889. From a recrystallized vesicular rock-fragment in soil sample. Forsterite (chrysolite).

[7] Meyer and McCallister, *Lun. Sci. IV*, *Abstr.*, 524. From a "troctolitic" clast in an anorthositic cataclasite. *Analysis includes $Na_2O = 0.05\%$. Forsterite.

[8] Haggerty et al, *Suppl.* **1,** 516. From a fine-grained olivine basalt. Forsterite.

| | [9] 68415,37 | [10] Luna 20 22006 | [11] 69941,13 | [12] Luna 20 22002,2,4c | [13] 12018,49 | [14] Luna 20 528–34 | [15] 10058 |
|---|---|---|---|---|---|---|---|
| $SiO_2$ | 37.17 | 36.22 | 34.4 | 32.7 | 30.15 | 30.1 | 29.2 |
| $Al_2O_3$ | 0.02 | 0.19 | 0.06 | 0.30 | — | 0.10 | — |
| $Cr_2O_3$ | — | 0.19 | 0.06 | 0.04 | — | 0.07 | 0.03 |
| $TiO_2$ | 0.03 | 0.09 | 0.12 | 0.14 | — | 0.23 | 0.06 |
| FeO | 28.40 | 30.49 | 41.0 | 55.7 | 64.3 | 67.5 | 68.5 |
| MgO | 34.17 | 32.89 | 22.9 | 8.81 | 2.57 | 0.25 | — |
| MnO | 0.33 | 0.30 | 0.40 | 0.73 | — | 0.85 | 1.01 |
| CaO | 0.30 | 0.25 | 0.31 | 0.51 | 0.61 | 0.81 | 0.40 |
| NiO | — | — | — | — | — | — | — |
| Total | 100.42* | 100.63* | 99.25 | 98.93 | 97.63 | 99.91 | 99.2 |

*References*


[9] Helz and Appleman, *Lun. Sci. IV, Abstr.*, 353. From an anorthositic gabbro. *Analysis includes traces of $K_2O$ and $Na_2O$.

[10] Podosek et al, *Geochim. Cosmochim. Acta* **37,** 889. From a recrystallized anorthositic norite fragment in soil sample. *Analysis includes $Na_2O = 0.01\%$. Forsterite.

[11] Taylor and Carter, *Suppl.* **4,** 294. From a soil sample. Intermediate olivine.

[12] Cameron et al, *Geochim. Cosmochim. Acta* **37,** 782. From a fine-grained soil sample. Fayalite.

[13] El Goresy et al, *Suppl.* **2,** 234. From a porphyritic basalt. Average of two analyses. A reaction between ilmenite and enclosing pyroxene. Fayalite.

[14] Crawford and Weigand, *Geochim. Cosmochim. Acta* **37**, 817. From a mare-basalt fragment in soil sample. Fayalite.

[15] Brown et al, *Suppl.* **1,** 203. From a medium-grained cristobalite basalt. Mg-free fayalite.

2. In the olivine-phyric basalts, the predominantly magnesian olivine (with maximum frequency $\sim Fo_{60}$) is entirely in phenocrysts. A few late-stage fayalites occur in this more coarsely crystallized rock.

3. In feldspathic peridotite the olivine is mainly magnesian but with an Fo content no greater than that in the other rocks.[20]

In vuggy breccia 77115, part A, the olivine xenocrysts have cores of $Fo_{69-89}$, and rims of $Fo_{68-78}$, converging toward the composition of olivine in the matrix, $Fo_{67-71}$.[21]

*References*

1. Haggerty et al, *Suppl.* **1,** 535.
2. Jakeš, *Lun. Sci. V, Abstr.*, 381, 383.
3. Gancarz et al, *EPSL* **16,** 316.
4. Anderson, *J. Geol.* **81,** 223.
5. Steele et al, *Lun. Sci. V, Abstr.*, 728.
6. Agrell et al, *Lun. Sci. IV, Abstr.*, 15.
7. Drake, *Lun. Sci. V, Abstr.*, 177.
8. Prinz et al, *Sci.* **179,** 75.
9. Reid et al, *Geochim. Cosmochim. Acta* **37,** 1021.
10. Taylor et al, *Geochim. Cosmochim. Acta* **37,** 1095.
11. Kim et al, *Suppl.* **2,** 750.
12. Podosek et al, *Geochim. Cosmochim. Acta* **37,** 889.
13. Kushiro et al, *Suppl.* **3,** 121.
14. Helz and Appleman, *Lun. Sci. IV, Abstr.*, 352.
15. Mason et al, *Suppl.* **3,** 791–792.
16. Bence et al, *Suppl.* **4,** 602, 606.
17. Hollister, *Suppl.* **4,** 636.
18. James and Jackson, *J. Geophys. Res.* **75,** 5814.
19. Kushiro and Nakamura, *Suppl.* **1,** 617.
20. Dowty et al, *Suppl.* **4,** 436–437.
21. Chao et al, *Lun. Sci. V, Abstr.*, 109.

*X-Ray Data*

Single-crystal X-ray study of olivines from Apollo 11 fines yielded diffraction patterns typical of forsteritic members, with no sign of actual or incipient alteration or exsolution. According to X-ray examination, inclusions observed optically were not oriented with respect to the olivine host.[1] In some Apollo 11 olivines, X-ray photographs of single grains display elongated streaks demonstrating shock-induced mosaicism.[2] Olivine from mare basalt 15555 exhibits no subsolidus transformations or exsolution reactions, but transmission electron microscopy reveals the presence of some dislocation loops and arrays believed to have formed between 1000 and 1250°C.[3]

From Luna 20 sample 927-2, an anorthositic rock fragment, small (5–20 μ) grains of a low-iron olivine ($Fo_{\sim 87}$) have:[4]

$$a_0 = 4.78 \pm 0.05 \text{ Å}$$
$$b_0 = 10.32 \pm 0.05 \text{ Å}$$
$$c_0 = 6.09 \pm 0.05 \text{ Å}$$

An olivine ($Fo_{84}$, Table 10-1; [5]) from Luna 20 fine-grained soil sample 22002,2,5a has:[5]

$$a_0 = 4.80 \text{ Å}$$
$$b_0 = 10.35 \text{ Å}$$
$$c_0 = 6.05 \text{ Å}$$

Olivine from fine-grained olivine basalt 10020-40 (average composition $Fo_{72} \pm 3$ mole %: Table 10-1; [8]) has:[6]

$$a_0 = 4.755 \pm 0.002 \text{ Å}$$
$$b_0 = 10.284 \pm 0.002 \text{ Å}$$
$$c_0 = 6.0168 \pm 0.0009 \text{ Å}$$

The space group for the olivines is *Pbnm*.

*References*

1. Gay et al, *Suppl.* **1,** 488.
2. von Engelhardt et al, *Suppl.* **1,** 370.
3. Nord et al, *Suppl.* **4,** 955.
4. Tarasov et al, *Suppl.* **4,** 335.
5. Cameron et al, *Geochim. Cosmochim. Acta* **37,** 788.
6. Haggerty et al, *Suppl.* **1,** 535–536.

## FAYALITE $Fe_2SiO_4$

*Synonymy*

Fayalitic olivine; Busche et al, *Am. Mineral.* **57,** 1741.
Iron-rich olivine; Adams et al, *Geochim. Cosmochim. Acta* **37,** 736.

Fig. 10–7 Fayalite crystals (one above and two below the horizontal plagioclase lath) in mesostasis of basalt 10058. From *Suppl.* **1,** 208. Photograph courtesy of M. Brown, Department of Geological Sciences, Durham University, England.

## *Occurrence and Form*

Fayalite is found in trace amounts in the mesostasis of some of the crystalline rocks (Figure 10–7).[1] It is seen as small grains associated with cristobalite and pyroxferroite.[2] In coarse-grained basalt 10044 it occurs in narrow bands between pyroxferroite and ilmenite,[3] and also as small subhedra of fayalite and as fayalite rims on troilite. In the residuum of some microgabbros, fayalite is sympletically intergrown with another phase of composition $CaFeSi_2O_6$ (Figure 10–8), and may result from decomposition of a pyroxene or pyroxferroite,[4] or it may have precipitated directly from a residual melt.[5] In basaltic rocks 12018 and 12063, fayalite not associated with glass appears as a reaction rim between ilmenite and the enclosing pyroxene.[6] The texture of coarse-grained dolerite 14053 is unique because of the breakdown of fayalite due to subsolidus reduction processes. The composition of this mineral is $Fa_{86\text{-}96}$, and it formed during end stages of crystallization and subsequently was broken down to a spongy mass of pure Fe metal plus tridymite and $SiO_2$-rich glass (Figure 10–9).[7] This

Fig. 10–8 Fine-grained intergrowth of fayalite, ferrohedenbergite, and K-rich glass in microgabbro 12064,8. Most of the transparent material is fayalite; ferrohedenbergite is the slightly darker area to the right of middle of picture. Glass blebs occur inside fayalite and between fayalite and ferrohedenbergite. From *Suppl.* **2,** 280. Photograph courtesy of C. Frondel, Harvard University, Cambridge, Mass.

indicates extremely low oxygen fugacity, lower than present during the formation of Apollo 11 or 12 rocks.[8] Locally, within the spongy texture, there are all gradations between fresh native Fe and native Fe that has been converted to troilite. This is interpreted as evidence for a very late-stage period of sulfurization.[7] In a mare-basalt particle from an Apollo 15 soil sample, sieve-textured intergrowths of fayalite and siliceous glass, high in K, occur in the interstices.[9]

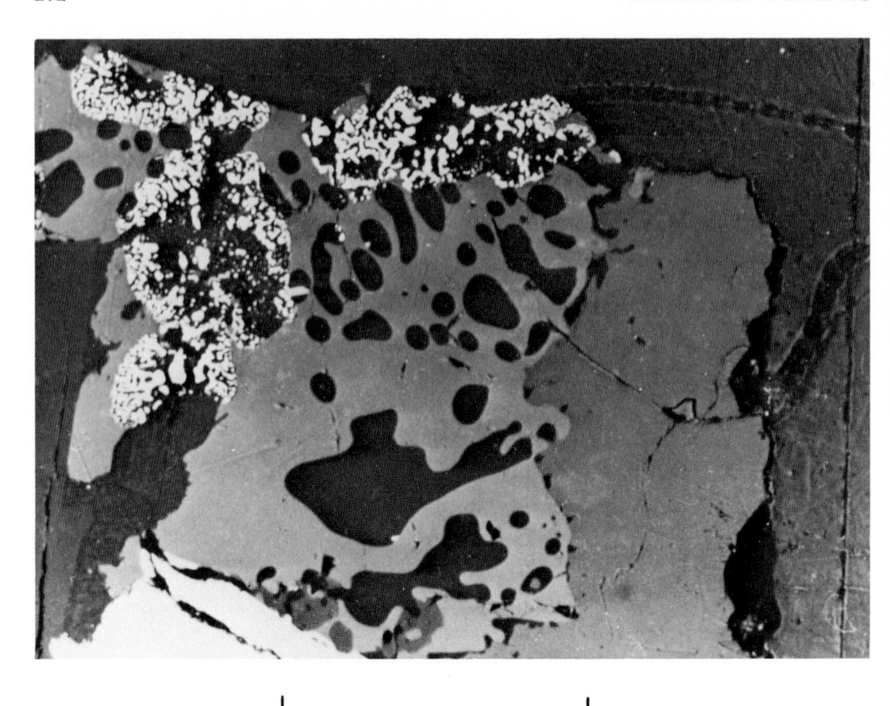


Fig. 10–9 Spongy texture of native Fe plus tridymite and $SiO_2$-rich glass resulting from the breakdown of fayalite (light gray), occurring in a sieve texture. From *Suppl.* **3,** 344. Photograph courtesy of A. El Goresy, Max-Planck-Institut für Kernphysik, Heidelberg, W. Germany.

### *Chemical Composition*

Most of the lunar fayalite is close to $Fe_2SiO_4$ and generally rich in Mn; it has minor Ca and Mg.[10] Its composition apparently is independent of its mode of origin (Table 10-1; [12]–[15]).[7]

### *References*


1. Brown et al, *Suppl.* **1,** 203.
2. Dence et al, *Suppl.* **1,** 323.
3. Fuchs et al, *Suppl.* **1,** 478.
4. Smith et al, *Suppl.* **1,** 919.
5. James and Jackson, *J. Geophys. Res.* **75,** 5795.
6. El Goresy et al, *Suppl.* **2,** 232.
7. ———, *Suppl.* **3,** 343.
8. ———, *Lun. Sci. Conf.* '72, *Abstr.*, 204.
9. Powell et al, *Lun. Sci. IV*, *Abstr.*, 597.
10. Essene et al, *Suppl.* **1,** 395.

# CHAPTER ELEVEN | MINOR SILICATES

## Tranquillityite, Zircon, Titanite, Thorite, Amphiboles, Micas, Garnets, Melilites, and Unidentified Silicates

TRANQUILLITYITE $Fe_8^{2+}(Zr + Y)_2Ti_3Si_3O_{24}$

*Synonymy*

Fe, Ti, Zr silicate containing rare earths and Y; Dence et al, *Suppl.* **1,** 324.
Fe-Ti-Zr silicate without Nb and REE; Gancarz et al, *EPSL* **12,** 15.
Iron-titanium-zirconium mineral (1); Simpson and Bowie, *Suppl.* **2,** 212.
Mineral A; Ramdohr and El Goresy, *Sci.* **167,** 617.
Phase A; Lovering and Wark, *Suppl.* **2,** 153.
Phase C; Douglas et al, *Sci.* **167,** 596.
Phase X (Brett); *Lunar Sample Information Catalog, Apollo 16*, 355.
Ti-Fe-Zr-Y silicate; Cameron, *Suppl.* **1,** 221.
U-Th-rich phase; Gancarz et al, *EPSL* **12,** 14.
Yttrium-zirconium silicate; Cameron, *Suppl.* **1,** 237.
Zr-Ti-Fe-Ca-Y phase; Burnett et al; *Suppl.* **2,** 1509.

Tranquillityite is the Moon's own mineral, a unique phase without a terrestrial counterpart. It is a new silicate, first recognized in Apollo 11 basalts[1] and subsequently identified from rocks (chiefly basalts) of all other lunar landing sites.

The simultaneous discovery of this phase by six separate investigative teams precipitated one of the many exciting discussions of the first Lunar Science Conference at Houston in January 1970. At that time none of the groups of investigators had sufficient data to characterize the new phase. It was agreed that to avoid duplication of effort, the results would be pooled

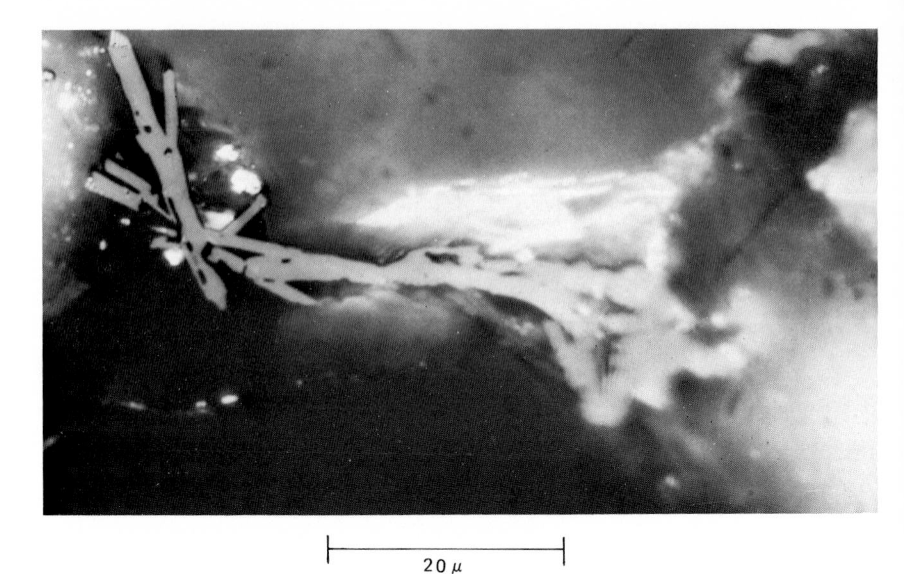


Fig. 11–1 Cluster of tranquillityite laths in fine-grained basalt 10017. Oil immersion; 45° polarizers. From *Suppl.* **2,** 162. Photograph courtesy of P. R. Simpson, Institute of Geological Sciences, London.

under the chairmanship of J. F. Lovering. The high degree of uniformity of the chemical analyses as well as other observations left no doubt that tranquillityite was a new and distinct mineral. The name, which refers to the Sea of Tranquillity, the landing site of Apollo 11, has been approved by the IMA Commission on Mineral Names.[2]

### *References*

1. Cameron, *Suppl.* **1,** 240.
2. Lovering et al, *Suppl.* **2,** 40.

### *Occurrence and Form*

Tranquillityite, a rare accessory mineral, has been found chiefly in the coarse-grained lunar basalts.[1] It occurs as thin laths or sheaves of laths with overall dimensions usually less than 100 μ (Figure 11-1).[2] When clusters or little bundles of crystals, with the longest crystals sharply bent, were first observed, they were presumed to be pseudotetragonal orthorhombic.[3] Later X-ray study has shown that the mineral is hexagonal[2] (see below).

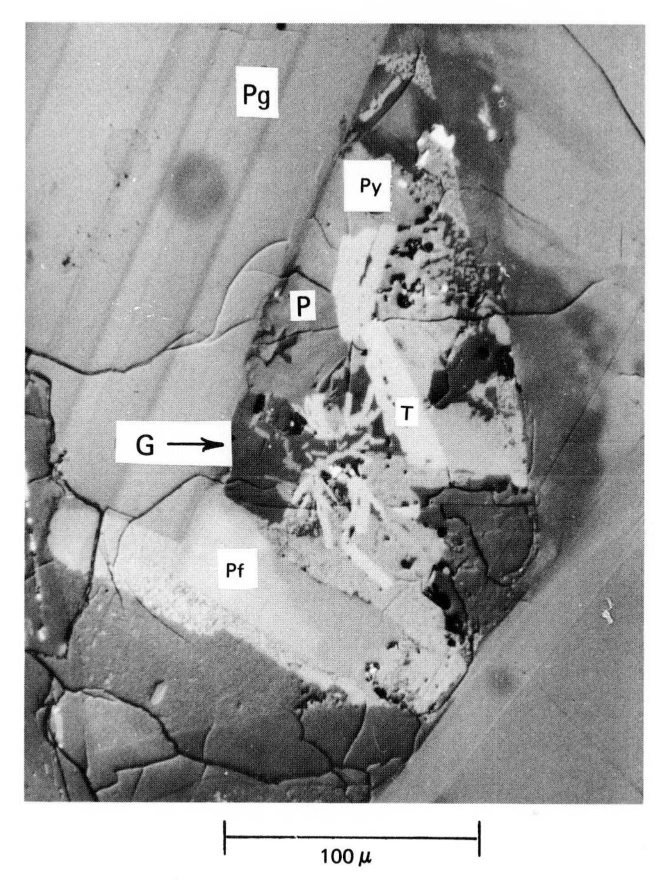


Fig. 11–2 Assemblage of tranquillityite sheaves (T) in interstitial association with pyroxferroite (Pf), glass (G), and phosphate phases (P) between plagioclase (Pg), and clinopyroxene in basalt 10047,20A. From *Suppl.* **2,** 41. Photograph courtesy of J. F. Lovering, School of Geology, University of Melbourne, Parkville, Victoria, Australia.

In the residuum of the crystalline rocks, Zr-rich areas have been noted, with rare P-rich areas.[4] These, as well as areas of high U concentration, are sites for tranquillityite. Tranquillityite is a late-stage crystallization product of the lunar basalts and is associated with interstitial phases such as troilite plus metal, pyroxferroite, tridymite, cristobalite, alkali feldspar, and felsic glass (Figure 11–2).[1]

Small red flakes of tranquillityite are scattered throughout a thin section of coarse-grained basalt 10047.[5] From ophitic basalt 12038,67, a radiate group of 15 μ-crystals of tranquillityite occurs in pyroxene at its contact with ilmenite. Smaller acicular crystals of tranquillityite are joined to an anhedral mass of the same phase.[6] Tranquillityite is relatively abundant in the mesostasis of nonmare basalt 14276,13. Here it occurs as plates 1 to 5 $\mu$ across.[7] Tranquillityite has been observed in the glassy residuum of KREEP basalt 14310,[8] and more than 12 crystals ($< 10\ \mu$) were found in an interstitial cristobalite crystal in basalt 15475.[9] Tranquillityite is abundant also in basalt 75035. In coarse-grained basalts 70017 and 70035 and medium-grained basalt 71055, tranquillityite occurs with baddeleyite, armalcolite, Zr-armalcolite, ilmenite, troilite, and rutile.[10] In fines samples 74220, 74242, 74243, fine-grained basalt 70215, and coarse-grained basalt 79155, minor tranquillityite occurs with minor baddeleyite and zirconolite in an assemblage that includes armalcolite, ulvöspinel, ilmenite, rutile, aluminian-titanian-chromite, and cobaltian iron. In coarse-grained basalt 75055, a tranquillityite breakdown to baddeleyite plus ilmenite and (possibly) pyroxene was observed.[11]

*References*


1. Cameron, *Suppl.* **1,** 240.
2. Lovering et al, *Suppl.* **2,** 40, 42–44.
3. Ramdohr and El Goresy, *Sci.* **167,** 617.
4. Weill et al, *Suppl.* **1,** 947.
5. Dence et al, *Suppl.* **1,** 324.
6. Simpson and Bowie, *Suppl.* **2,** 216.
7. Gancarz et al, *EPSL* **16,** 322.
8. Longhi et al, *Suppl.* **3,** 137.
9. Brown et al, *Suppl.* **3,** 149.
10. ———, *Lun. Sci. V, Abstr.*, 90.
11. El Goresy et al, *Lun. Sci. V, Abstr.*, 209–210.


*Optics*

Tranquillityite is gray, semi-opaque,[1] or deep foxy red in thin crystals in strong transmitted light. Its deep color makes the optical properties difficult to determine, but reportedly tranquillityite is isotropic or weakly anisotropic. It is not pleochroic, and it is optically homogeneous and free of exsolution or alteration products.[2] Originally mistaken for kennedyite because of optical similarities, tranquillityite has a reflectivity similar to that of anosovite or ludwigite[3] (namely, $R = 12.9 \pm 0.2\%$ at 546 nm in air), and from this value an average index of refraction of 2.11 to 2.13 has been calculated.[1]

One 0.1-mm lamellar lath, seen in gabbroic anorthosite 68415,8, is semi-opaque and gray with strong red internal reflections. It is called phase

X[4] but most probably is tranquillityite. (This phase X is not to be confused with phase X [Brown], which is a synonym for Zr-armalcolite[5].)

*References*

1. Cameron, *Suppl.* **1,** 237.
2. Lovering et al, *Suppl.* **2,** 40.
3. Ramdohr and El Goresy, *Sci.* **167,** 617.
4. Brett, in *Lunar Sample Information Catalog, Apollo 16*, 355.
5. Brown et al, *Suppl.* **3,** 149.

*Chemical Composition*

Chemical analyses of tranquillityite show it to be of nearly constant composition.[1] An average analysis for Apollo 11 samples is given in Table 11-1, [2]. Analyses of samples from other Apollo missions reveal some variation in the Si, Ti, and Zr contents (within relatively narrow limits) and generally constant Fe content. Various investigators have suggested the presence of different though related phases on the basis of higher Al content (e.g., Table 11-1; [5]),[2] the absence of Y, or the absence of Nb and REE.[3] Analyses show that some tranquillityite contains as much as 5% $Y_2O_3$ (Table 11-1; [1], [9]), whereas in other samples Y is lacking or appears only in trace amounts (Table 11-1; [3]–[5]). Despite these differences, almost all analyses of tranquillityite permit calculations yielding the formula

$$Fe^{2+}(Zr + Y)_2\,Ti_3\,Si_3\,O_{24}.$$

The REE, when present, are very minor constituents. It has been suggested that Nb should be reported as $Nb_2O_5$ rather than $Nb_2O_3$.[4] Uranium has been reported from some tranquillityite, with an approximate range of 40 to 1000 ppm U. It has been noted with interest that in both Apollo 11 and Apollo 12 basalts the same suites of U-enriched minerals occur (i.e., tranquillityite, zirconolite, baddeleyite, whitlockite, and apatite). Considering the distance between the two landing sites, differences in ages (Apollo 11 basalts ~ 3.7 billion years, Apollo 12 basalts ~ 3.3 billion years[5]), rock chemistries, and textures, the similarity of the late-stage mineral suites appears to imply similar chemical fractionation trends in the final crystallization stages of the mare basalts.[6]

From fines sample 14001,7,3 and KREEP basalt 14310,6, abundant, 5-$\mu$ opaque grains were observed intimately associated with other opaque minerals. Analyses were made on very small hard grains that developed curved surfaces when polished and yielded results with summations too low to permit calculation of a rational formula. It has been suggested that

**Table 11-1 Analyses for Tranquillityite and an Unidentified Fe-Ti-Zr Silicate**

| | [1] | [2] 10047 | | [3] | [4] | [5] | [6] | [7] | [8] | [9] | [10 |
|---|---|---|---|---|---|---|---|---|---|---|---|
| | 12051,59 | *a* | *b* | 14310/20 | 15475/125 | 12038,67 | 14276,13 | 12039,3 | 14310 | 10047 | 10017,50 |
| $SiO_2$ | 13.1 | 13.77 | 2.827 | 14.0 | 14.4 | 14.7 | 14.99 | 15.10 | 15.2 | 17.0 | 31.68 |
| $TiO_2$ | 17.7 | 20.08 | 3.099 | 21.9 | 21.8 | 19.0 | 22.55 | 20.13 | 23.0 | 18.0 | 19.49 |
| $Al_2O_3$ | 1.1 | 0.86 | 0.210 | 1.2 | 1.4 | 4.75 | 1.06 | 1.62 | 1.76 | 0.5 | 1.12 |
| $Cr_2O_3$ | — | 0.10 | 0.012 | 0.2 | — | — | 0.04 | — | — | — | — |
| FeO | 42.0 | 42.63 | 7.321 | 42.1 | 43.5 | 39.85 | 42.74 | 42.80 | 41.4 | 41.0 | 36.99 |
| MnO | 0.16 | 0.35 | 0.062 | 0.3 | 0.3 | — | 0.31 | 0.23 | — | — | — |
| MgO | <0.02 | 0.06 | 0.185 | 1.3 | 0.6 | — | 1.13 | — | 1.33 | <0.1 | — |
| CaO | 1.5 | 1.15 | 0.259 | 1.1 | 1.4 | 1.85 | 1.39 | 1.51 | 1.31 | 1.0 | 1.13 |
| $Na_2O$ | <0.01 | <0.02 | — | — | — | — | 0.01 | <0.02 | — | <0.1 | — |
| $K_2O$ | — | 0.03 | — | — | — | — | 0.01 | — | — | 0.5 | — |
| $ZrO_2$ | 17.5 | 16.95 | 1.704 | 14.5 | 14.0 | 19.45 | 12.24 | 17.10 | 12.6 | 16.0 | 8.57 |
| $HfO_2$ | — | 0.12 | 0.012 | — | — | — | 0.64 | — | 0.45 | — | — |
| $Y_2O_3$ | 5.4 | 2.72 | 0.296 | 0.5 | 0.9 | — | 0.44 | 1.36 | 1.60 | 5.0 | — |
| $Nb_2O_3$ | 0.33 | — | — | — | — | — | <0.01† | — | — | — | — |
| $Pr_2O_3$ | 0.12 | — | — | — | — | — | 0.03 | 0.27 | — | — | — |
| $Nd_2O_3$ | — | — | — | — | — | — | <0.01 | — | — | — | — |
| $Gd_2O_3$ | — | — | — | — | — | — | 0.11 | — | — | — | — |
| Total | 98.91 | 98.82 | 15.987 | 97.1 | 98.3 | 99.6 | 98.38* | 100.12 | 98.65 | 100.0* | 98.98 |

*References*

[1] Keil et al, *Suppl.* **2,** 43. From an ophitic basalt.

[2] Lovering et al, El Goresy and Ramdohr, *Suppl.* **2,** 42. From a coarse-grained basalt. Average of six analyses. **a* = wt% of oxide; *b* = cations calculated for 24 oxygens, permitting the following assignment:

$$
\begin{aligned}
\text{Fe, Ti, Mg, Mn, Ca} &= 7.975 \\
\text{Zr, Hf, Y} &= 2.012 \\
\text{Al, Cr, Ti} &= 3.000 \\
\text{Al, Si} &= 3.000 \\
\hline
&\phantom{=}\ 15.987
\end{aligned}
$$

yielding formula $Fe_8^{2+}(Zr + Y)_2Ti_3Si_3O_{24}$.

[3] Brown et al, *Suppl.* **3,** 151. From a KREEP basalt. Average of analyses on six crystals.

[4] ———, *ibid.* From a mare basalt.

[5] Simpson and Bowie, *Suppl.* **2,** 212. From an ophitic basalt. Average of two analyses.

[6] Gancarz et al, *EPSL* **16,** 322. From a nonmare basalt. Average of two analyses. *Analysis includes:

| | |
|---|---|
| $P_2O_5 = 0.15\%$ | $La_2O_3 = <0.01\%$ |
| $UO_2 = 0.15\%$ | $Dy_2O_3 = <0.01\%$ |
| $ThO_2 = 0.06\%$ | $PbO = 0.03\%$ |
| $Ce_2O_3 = 0.30\%$ | † Nb reported as $Nb_2O_5$ |

[7] Keil et al, *Suppl.* **2,** 43. From an ophitic basalt. Average of three analyses.

[8] El Goresy et al, *Suppl.* **3,** 339. From a KREEP basalt. Average of three analyses.

[9] Douglas and Plant, *Suppl.* **2,** 42. From a coarse-grained basalt. *Analysis includes $RE_2O_3 = 1.0\%$

[10] Rice and Bowie, *Suppl.* **2,** 163. From a fine-grained basalt. Unidentified Fe-Ti-Zr silicate.

the analyses represent two separate phases, since in one group $SiO_2 = 7\%$ with $ZrO_2 = 23.5\%$, and in the other group $SiO_2 = 13.7\%$ with $ZrO_2 = 34.4\%$.[3] However, unless better analyses are obtained, or X-ray studies can indicate differences in the phases, it can be assumed that the analyses probably represent tranquillityite.

*References*

1. Brown et al, *Suppl.* **3,** 149.
2. Simpson and Bowie, *Suppl.* **2,** 217.
3. Gancarz et al, *EPSL* **12,** 14–15.
4. Erlank et al, *Lun. Sci. Conf.* '72, *Abstr.*, 216.
5. *Apollo* 17 *Prelim. Sci. Rep.*, NASA SP-330, **29**–28.
6. Lovering and Wark, *Suppl.* **2,** 157.

*X-Ray Data*

Preliminary X-ray study of tranquillityite also indicates it is a new mineral unrelated to any terrestrial one. Tranquillityite is hexagonal with

$$a = 11.69 \pm 0.05 \text{ Å}$$
$$c = 22.25 \pm 0.12 \text{ Å}$$

and a calculated density of $4.7 \pm 0.1$.[1] X-Ray powder data are given in Table 11-2.

On the basis of an analysis of tranquillityite from ophitic basalt 12051, it was suggested that the phase was essentially an (Fe) $(Ti,Si,Zr)O_3$ analog of perovskite ($CaTiO_3$).[2] This suggestion must be rejected, since perovskite is not a silicate, and since X-ray analysis of terrestrial perovskite shows it to be very different from tranquillityite. Perovskite is pseudocubic, with $a = 15.26 \pm 0.01$ Å, and the X-ray powder data of perovskite are quite distinct from those of tranquillityite.[3]

*References*

1. Lovering et al, *Suppl.* **2,** 42–44.
2. Brown et al, *Suppl.* **2,** 595.
3. Murdock, *Am. Mineral.* **36,** 57.

*Unidentified Fe-Ti-Zr Silicate*

In the mesostasis of fine-grained basalt 10017, clusters have been observed of prismatic or bladed crystals, often twisted. The largest of these crystals is only $4 \times 24$ $\mu$.

**Table 11-2 X-Ray Powder Data* for Tranquillityite**

| *hkl* | $d_{obs}$ | $d_{calc}$ | Intensity |
|---|---|---|---|
| 105 | 4.08 | 4.074 | 2 |
| 114 | 4.04 | 4.030 | 5 |
| 212 | 3.65 | 3.618 | 2 |
| 301 | 3.34 | 3.335 | 4 |
| 302 | 3.23 | 3.229 | 10 |
| 214 | 3.18 | 3.152 | 4 |
| 116 | 3.13 | 3.132 | 4 |
| 107 | 3.04 | 3.033 | 3 |
| 220 | 2.922 | 2.922 | 4 |
| 118 | 2.512 | 2.512 | 4 |
| 109 | 2.408 | 2.404 | 2 |
| 412<br>308 | 2.155 | 2.167<br>2.146 | 6 |
| 413 | 2.116 | 2.117 | 2 |
| 228 | 2.015 | 2.015 | 3 |
| 318 | 1.977 | 1.976 | 2 |
| 420 | 1.913 | 1.913 | 2 |
| 423 | 1.851 | 1.852 | 2 |
| 334<br>505 | 1.843 | 1.843<br>1.839 | 4 |
| 328 | 1.781 | 1.783 | 7 |
| 604 | 1.615 | 1.614 | 2 |
| 0.0.16 | 1.390 | 1.391 | 2 |
| 624 | 1.361 | 1.361 | 3 |
| 446 | 1.359 | 1.359 | 2 |
| 529 | 1.356 | 1.356 | 3 |
| 706 | 1.348 | 1.347 | 4 |
| 4.0.15 | 1.281 | 1.280 | 2 |
| 5.2.12<br>4.4.10 | 1.221 | 1.221<br>1.220 | 2 |

* Data normalized from Weissenburg singles crystal photographs about two directions oblique to each other.

*Reference.* Lovering et al, *Suppl.* **2,** 44.

The phase is red-brown and semi-opaque, appearing in its optical properties to be similar to tranquillityite.

The unknown phase is compositionally similar to tranquillityite also, except that its $SiO_2$ content is double that of tranquillityite, its Ti:Zr ratio is 2:1 instead of 1:1, and Y is undetected (Table 11-1; [10]).[1] The analysis does not fit the tranquillityite formula, and until X-ray study of the unknown phase is made, its exact nature cannot be determined.

Where the unknown phase coexists with whitlockite in the residual liquids, it has been noted that the U preferentially enters the unknown phase and the REE and Y occur in the whitlockite.[1]

*Reference*

1. Rice and Bowie, *Suppl.* **2,** 162–164.

## ZIRCON $ZrSiO_4$

*Occurrence and Form*

Zircon has been found in the nonmagnetic high-density fractions (those which sank in 3.8 Clerici solution) of both Apollo 11 and Apollo 12 soil samples.[1] Zircon has been detected in almost all examined chips of breccia 12013. Most of the zircon grains are only a few microns across, although some grains have a diameter of 50 μ.[2] In the darker lithology of 12013, some zircon grains are as large as 80 μ, but zircon occurs in relatively small crystals (<10 μ in diameter) in the light portions of the breccia.[3]

**Table 11-3 Zircon Analyses**

| | [1] 12034,3 | [2] 12013,10 | [3] 14321/19 | [4] 12070 | [5] 12032 |
|---|---|---|---|---|---|
| $SiO_2$ | 35.5 | 32.41 | 32.41 | 32.20 | 32.11 |
| $Al_2O_3$ | 1.3 | — | 0.20 | 0.09 | — |
| $TiO_2$ | 0.2 | — | 0.19 | — | — |
| FeO | 0.8 | 0.35 | 0.02 | 0.32 | — |
| MgO | 0.1 | — | 0.19 | 0.18 | — |
| CaO | 0.05 | — | 0.26 | — | — |
| $ZrO_2$ | 61.5 | 63.48 | 66.93 | 64.70 | 67.23 |
| $HfO_2$ | 0.6 | 3.01 | — | — | 0.88 |
| Total | 100.05 | 99.25 | 100.31* | 97.77* | 100.22 |

*References*

[1] Anderson et al, *Suppl.* **2,** 433. From a "gray mottled" basalt.

[2] Albee et al, *EPSL* **9,** 142. From a breccia.

[3] Gay et al, *Suppl.* **3,** 354. Crystal from an anorthositic xenolith in a breccia. *Analysis includes $Cr_2O_3$ = 0.03%, $NiO_2$ = 0.06%, and CaO = 0.02%.

[4] Keil et al, *Suppl.* **2,** 333. A single grain from a breccia fragment in a fines sample. *Analysis includes $Na_2O$ = 0.09%, $Ce_2O_3$ = 0.02%, and $Y_2O_3$ = 0.17%.

[5] Brown et al, *Suppl.* **2,** 595. From a fines sample.

One grain, from fines sample 12070, measured about 140 × 140 × 160 $\mu$. In noritic fragments in the soil, zircon is an interstitial phase of apparently late crystallization.[1]

### *Optics*

The zircon is colorless, brown, or hyacinth pink. Colorless zircon from soil sample 12070,35 is uniaxial with $\epsilon$ = 1.9 and $\omega$ = 1.925.[1] From soil sample 10084 zircon brown, which is biaxial, with 2V ~ 5°, $\epsilon$ = 1.853, $\omega$ = 1.838, is partially metamict.[1]

### *Chemical Composition*

The metamictization of the brown zircon probably is due to its U content, which is considerably higher (U = 167 ±19 ppm) than that of the colorless zircon (U < 24 ppm).[1] Zircon from the light-colored lithologic portion of breccia 12013 has a U content somewhat less than 100 ppm. The U content of the zircon from mare basalt 14072 is approximatley 500 ppm, and zircon from various clasts in breccia 14305,77 contains U up to 1300 ppm.[4]

It was noted that zircon from breccia sample 12013,10 showed almost no variation in the Fe or Hf content (Table 11-3; [2]).[2] Some of the zircon analyses show minor Al, Cr, Me, Ce, and Na (Table 11-3; [1], [3], [4]).

### *X-Ray Data*

Cell parameters of the zircon appear to vary with its color. Data are given in Table 11-4.

### *References*

1. Wood et al, *Spec. Rep.* **333,** 163–164, 168.
2. Albee et al, *EPSL* **9,** 143.
3. Drake et al, *EPSL* **9,** 122.
4. Lovering et al, *Suppl.* **3,** 284.

**Table 11-4 Cell Parameters of Zircon from Lunar Soil Samples**

| Sample | Color | $a_0$ (Å) | $c_0$ (Å) |
|---|---|---|---|
| 10084 | brown | 6.663 ±0.005 | 6.080 ±0.006 |
| 12037,20 | pink | 6.635 ±0.001 | 5.988 ±0.005 |
| 12037,20 | pink | 6.626 ±0.001 | 6.006 ±0.002 |
| 12037,20 | pink | 6.616 ±0.001 | 6.003 ±0.005 |
| 12070,35 | colorless | 6.606 ±0.001 | 5.991 ±0.002 |

*Reference.* Wood et al, *Spec. Rep.* **333,** 168.

## TITANITE $CaTiSiO_5$

### *Synonym*

Sphene; Gay et al, *Suppl.* **1,** 482.

### *Occurrence and Form*

In porphyritic basalt 12063/41, an ilmenite grain had a reaction rim that looked like titanite, but the rim was too thin for microprobe determination.[1]

### *Chemical Composition*

From basaltic fragment in breccia 14321,200, an average of six titanite analyses yielded:[2]

| | | |
|---|---|---|
| $SiO_2$ | = | 30.19% |
| $Al_2O_3$ | = | 0.91% |
| $TiO_2$ | = | 38.93% |
| FeO | = | 1.51% |
| CaO | = | 27.03% |
| $ZrO_2$ | = | 0.62% |
| REE | = | 0.85% |
| Total | | 100.04% |

### *X-Ray Data*

In a devitrified glassy spherule from fines sample 10085, a titanite-like structure was identified tentatively, by X-ray powder method, as one of the crystalline components.[3]

### *References*

1. Ramdohr, *Fortschr. Mineral.* **48,** 50.
2. Grieve et al, *Lun. Sci. Conf. '72, Abstr.*, 302.
3. Gay et al, *Suppl.* **1,** 482.

## THORITE $ThSiO_4$

Thorite reportedly has been identified from soil sample 14259,97. The chief importance of this rare accessory mineral is that it contains 21.6% of the uranium in the soil sample.[1]

### *Reference*

1. Haines et al, *Lun. Sci. III, Rev. Abstr.*, 350–352.

## AMPHIBOLE GROUPS

The occurrence of lunar amphiboles, at first doubted, appears to be established definitely from the descriptions of three different phases from three separate Apollo missions. However, the amphiboles are very rare very minor constituents of the lunar rocks, which are essentially anhydrous.

### MAGNESIOARFVEDSONITE $(Na,K,Ca)_3$ $(Mg,Mn,Fe)_5$ $([OH]?,F)_2$ $(Si, Al, Ti)_8$ $O_{22}$

*Synonymy*

Clinoamphibole; Gay et al, *Suppl.* **1,** 483–484.
Eckermannite-arfvedsonite; Charles et al, *Suppl.* **2,** 647.
Ferrorichterite; Charles et al, *Suppl.* **2,** 647.
Richteritic amphibole; Dence et al, *Suppl.* **2,** 292.
Sodic richterite; Brown, *J. Geophys. Res.* **75,** 6481.

*Occurrence and Form*

A crystalline material was chipped from a vug in medium-grained basalt 10058. The crystal was 0.5 mm long and had well-developed prism faces.[1]

*Color and Optical Properties*

The amphibole, described both as blue and dark green, is pleochroic. It has strong dispersion and fails to extinguish in most directions in the prismatic zone. The indices of refraction are:

$$n_\alpha = 1.638 \pm 0.002 \quad \text{green}$$
$$n_\beta = 1.642 \pm 0.002 \quad \text{yellow-green}$$
$$n_\gamma = 1.645 \pm 0.002 \quad \text{pea-green}$$
$$n_\gamma - n_\alpha = 0.007, \quad Z > X > Y$$

The $2V_\alpha$, determined only approximately, is 60°, with the optical axial plane ⊥ (010); Z = [010] and X Λ [001] = 20°.[1]

*Chemical Composition*

Microprobe analysis shows the mineral to be a low-alumina Na-amphibole. The analysis (Table 11-5; [1]) calculates to the rare magnesioarfvedsonite with some affinity to richterite. Because of its low alumina and high iron contents, ferrorichterite would be an acceptable name also. The presence of water has been postulated because of the low total of the analysis,[2] but

**Table 11-5 Chemical Analyses of Amphibole**

| | [1] 10058 | [2] 12021,22 | [3] 14163,42A |
|---|---|---|---|
| $SiO_2$ | 54.51 | 42.1 | 39.9 |
| $TiO_2$ | 0.16 | 0.1 | 4.0 |
| $Al_2O_3$ | 0.75 | 16.7 | 12.3 |
| FeO | 12.20 | 14.2 | 19.6 |
| MnO | 0.16 | 0.2 | 0.4 |
| MgO | 16.35 | 9.9 | 8.2 |
| CaO | 2.13 | 12.0 | 11.6 |
| $Na_2O$ | 8.69 | 1.3 | 2.7 |
| $K_2O$ | 1.49 | 0.6 | 1.8 |
| F | $1.2 \pm 0.3$* | 0.4 | — |
| Cl | — | 0.2 | — |
| Subtotal | 97.64 | 97.7 | — |
| less 0 ≡ F,Cl | 0.51 | 0.2 | — |
| Total | 97.13 | 97.5 | 100.5 |

* Uncertainty due to low counting rates and difficulties with correction factors.

*References*

[1] Gay et al, *Suppl.* **1,** 483–484. From a coarse-grained basalt. Magnesio-arfvedsonite.

[2] Dence et al, *Suppl.* **2,** 292. From a porphyritic basalt. Aluminotschermakite.

[3] Mason et al, *Lun. Sci. Conf. '72, Abstr.*, 460. From a thin- section of a breccia particle in fines sample. Kaersutite.

in an investigation of some meteoritic richterites, no compelling evidence for the presence of structural water in the amphiboles was found. It should not be assumed that insufficient fluorine indicates the presence of hydroxyl in the amphibole.[3]

*X-Ray Data*

Measurements of the single crystal gave:

$$a = 9.84 \pm 0.01 \text{ Å}$$
$$b = 18.03 \pm 0.02 \text{ Å}$$
$$c = 5.03 \pm 0.01 \text{ Å}$$
$$\beta = 103.25 \pm 0.25°$$

Space group $C2/m$

These data are consistent with a magnesioarfvedsonite-richterite.[2]

In grains of tridymite from Apollo 11 fines, hairlike inclusions of a greenish birefringent mineral with inclined extinction were noted.[4] The inclusions possibly may be the amphibole just described.

## Hornblende

General formula:[5]

$$(Ca,Na,K)_{2\text{-}3}\,(Mg,Fe,Al)_5\,([OH],F)_2\,(Si,Al)_2\,Si_6O_{22}$$

Two of the lunar amphiboles—aluminotschermakite and kaersutite—could be classified as hornblendes.

## Aluminotschermakite

Approximately: $(Ca,Na,K)_{2.5}\,(Mg,Fe,Mn)_4Al_{1.5}\,([OH]?,F,Cl)_2Al_2Si_6O_{22}$.

### *Synonym*

Ferrotschermakitic amphibole; Charles et al, *Suppl.* **2,** 657.

### *Occurrence*

An amphibole was found as loose grains in a package containing a fragment of porphyritic basalt 12021.[6]

### *Color*

This amphibole was described as gray-blue.[7] No optical measurements were reported.

### *Chemical Composition*

The microprobe analysis, from 12021,22 (Table 11-5; [2]),[7] yielded low F and Cl contents as well as a low total, and it was suggested that the phase possibly might be hydrous. The phase has been called both aluminotschermakite[7] and a ferrotschermakitic amphibole.[6]

### *X-Ray Data*

Calculations from the X-ray powder pattern yielded cell parameters:[7]

$$a = 9.81\ \text{Å}$$
$$b = 18.51\ \text{Å}$$
$$c = 5.32\ \text{Å}$$

KAERSUTITE $(Ca_{1.88}Na_{0.79}K_{0.35})(Mg_{1.84}Mn_{0.05}Fe_{2.47}Al_{0.18}Ti_{0.45})Al_2Si_6O_{23}$

*Synonym*

Titanian pargasite; Mason et al, *Lun. Sci. Conf.* '72, *Abstr.*, 460.

*Occurrence*

A few grains of hornblende were found in thin section of breccia 14319,13 and also in the fines sample 14163,42A.[8]

*Color and Optical Properties*

The hornblende is dark brown, nonpleochroic, and isotropic, with a relief and birefringence similar to the abundant pyroxenes. Its color is similar, also, to much of the glass in the breccia, making detection of the hornblende difficult. Refractive indices, determined by immersion method are:[8]

$$\alpha = 1.700$$
$$\gamma = 1.725$$

*Chemical Composition*

The chemical formula, calculated from the microprobe analysis (Table 11-5; [3]), is in good agreement with a general hornblende formula. From the chemical composition, as well as the optical properties, the phase could be described as a titanian pargasite or a kaersutite. It is noted that few if any terrestrial kaersutites have an Fe/Mg ratio as high as in this lunar specimen.[8]

*References*

1. Agrell et al, *Suppl.* **1,** 108.
2. Gay et al, *Suppl.* **1,** 483–484.
3. Olsen and Huebner, *Am. Mineral.* **58,** 872.
4. Frondel et al, *Suppl.* **1,** 469.
5. Strunz, *Tabellen* (1966), 370.
6. Charles et al, *Suppl.* **2,** 657.
7. Dence et al, *Suppl.* **2,** 292.
8. Mason et al, *Lun. Sci. Conf.* '72, *Abstr.*, 460–461.

MICA GROUP (tentative)

BIOTITE $K_2(OH)_4(Mg, Fe, Al)_6(Si, Al)_8O_{20}$ and (OH) may be replaced by F

*Occurrence and Form*

Two small crystals of micaceous appearance were found in fines sample 10084. Although the grains were believed to be indigenous to the sample, the possibility of their being a contamination could not be eliminated.[1]

*Chemical Composition*

The microprobe analysis of the grains is essentially that of a biotite:[1]

| | | | | | |
|---|---|---|---|---|---|
| $SiO_2$ | = 32.54% | | $Na_2O$ | = | 0.23% |
| $TiO_2$ | = 2.86% | | Cl | = | 0.06% |
| $Al_2O_3$ | = 17.51% | | F | = | 0.20%* |
| FeO | = 26.58% | | Subtotal | = | 95.03% |
| MnO | = 1.25% | | | | |
| MgO | = 4.90% | | less O, F, Cl | = | 0.11% |
| CaO | = 0.05% | | total | = | 94.92% |
| $K_2O$ | = 9.05% | | | | |

*X-Ray Data*

Preliminary single-crystal photographs of a part of one of the grains have the appearance of a mica pattern.[1]

*Reference*

1. Gay et al, *Suppl.* **1,** 483.

## Unnamed Mica

*Synonymy*

Material A; Drever et al, *Suppl.* **1,** 341.
Dioctahedral phyllosilicate; Drever et al, *Suppl.* **1,** 343.

*Occurrence and Form*

The fine material washed from grains in fines sample 10084-30 yielded more than 20 euhedral hexagonal flakes.[1]

*Chemical Composition*

Because the flakes were so small, only a quantitative probe analysis was possible: Si and Al are the major constituents. No Fe, Ti, Mg, Ca, Na,

* The uncertainty in F determination is due to low counting rates and difficulties with correction factors.

and K have been detected, but the extreme thinness of the flakes makes the microprobe technique insensitive, and it is possible that several weight percent of these undetected elements are present.[1]

### *X-Ray Data*

X-Ray diffraction patterns yielded a unit cell, measured on orthogonal axes, of

$$a = 5.159 \pm 0.555 \text{ Å}$$
$$b = 8.982 \pm 0.047 \text{ Å}.$$

The *b* parameter is consistent with that of various dioctahedral phyllosilicates such as muscovite or kaolinite. Under the electron beam the diffraction pattern weakened and disappeared.[2]

### *Reference*

1. Drever et al, *Suppl.* **1,** 341–344.

## TRIOCTAHEDRAL PHYLLOSILICATE

### *Synomyn*

Material B; Drever et al, *Suppl.* **1,** 342.

### *Occurrence and Form*

Less than 1 $\mu$ wide, rare flakes with irregular outlines were found on the surface of fine-grained basalt 10017. It is believed that these flakes are not contaminants from the gloves or the laboratory, nor could other contaminants have formed in the brief time of study. Similar flakes have been recovered from the approximately 2-$\mu$ fraction of fines sample 10084-30.[1]

### *X-Ray Data*

Under the electron beam the flakes decomposed in a manner similar to the above-described dioctahedral phyllosilicate. One of the diffraction patterns, however, was of sufficient intensity to permit the following cell parameter measurements:

$$a = 5.30 \pm 0.1 \text{ Å}$$
$$b = 9.25 \pm 0.1 \text{ Å}.$$

The *b* parameter is consistent with that of biotite, chlorite, or serpentines.[1]

*Reference*

1. Drever et al, *Suppl.* **1,** 342–344.

## Colorless Micaceous Grain

### *Occurrence and Form*

A colorless micaceous grain was found in the mesostasis of medium-grained basalt 10058-23.[1] The manner of occurrence appears to support an indigenous lunar origin.

### *Chemical Composition*

After short exposure to the electron beam, the grain decomposed (as do micas) to a dark brownish-black material. Only a partial analysis was obtained.[1] It may represent a mica.

| | | | | | |
|---|---|---|---|---|---|
| $SiO_2$ | = | 45.0% | MnO | = | 0.3% |
| $Al_2O_3$ | = | 7.8% | CaO | = | 12.2% |
| $TiO_2$ | = | 1.6% | $Na_2O$ | = | 2.8% |
| FeO | = | 3.5% | $K_2O$ | = | 0.2% |
| MgO | = | 0.4% | | | |

*Reference*

1. Brown et al, *Suppl.* **1,** 203.

## Garnet Group

## Almandine (predominantly) $Alm_{70.7}Gro_{25.0}Sp_{2.7}Pyr_{1.6}$

### *Occurrence and Form*

Three grains of garnet were found in rock fragments of porphyritic basalt 12021,22. The grains, 100 to 300 μ long and less than 40 μ thick, are anhedral and somewhat flattened. Although they were not observed in contact with other minerals in the rock fragments or in thin section, the anhedral form of the garnet suggests that it occurred interstitially rather than in vugs or as phenocrysts. The garnet probably formed late in the crystallization sequence.[1]

### *Optics*

The garnet grains have a gemmy appearance; they are pale brown with a pink tinge, and inclusion-free. Their index of refraction is 1.81.

*Chemical Composition*

No chemical zoning was observed in the grains. Microprobe analysis yielded:[1]

| | | |
|---|---|---|
| $SiO_2$ | = | 36.1% |
| $TiO_2$ | = | 0.1% |
| $Al_2O_3$ | = | 21.4% |
| FeO | = | 31.5% |
| MnO | = | 1.2% |
| MgO | = | 0.4% |
| CaO | = | 8.7% |
| Total | = | 99.4% |

No Cr, Y, K, or Na were detected.

*X-Ray Data*

X-Ray study yielded a unit cell edge of 11.624 ±0.005 Å.[1]

*Reference*

1. Traill et al, *Sci.* **169,** 981–982.

SPESSARTITE $Mn_3Al_2(SiO_4)_3$ (tentative)

*Synonymy*

Manganese garnet; Connell et al, *Suppl.* **2,** 2083.

*Occurrence*

The existence of spessartite as a possible phase in porphyritic basalts 12002,72 and 12021,18 was presumed on the basis of Auger electron spectroscopy of the rocks. No other data are given.[1]

*Reference*

1. Connell et al, *Suppl.* **2,** 2083, 2091.

MELILITE GROUP (tentative)

General formula: $(Ca,Na)_2(Mg,Zn,Fe,Al,Ca,Mn)(Si,Al)_2O_7$.[1]

*Occurrence and Form*

It was suggested that the yield of $SiO_4^{4-}$ derivative in trimethylsilylation of fines sample 10085,48 might be due in part to the presence of undetected gehlenite-åkermanite group minerals.[2] The existence of these melilite group minerals has been suggested again by the discovery of two groups of minute (10–20 μ diameter) grains in fines sample 14003,27.[3]

### *Optics*

Positive identification was not possible, but the optics of the two groups of mineral grains seem to indicate melitites.

1. Six grains are uniaxial (−) with birefringence about 0.010 and *n* about 1.67.[3] This is consistent with the optics for (synthetic) gehlenite.[1] Three of these grains had cleavages at an angle of 90°.[3]
2. Five grains are uniaxial (+), with birefringence about 0.010 and *n* about 1.64.[3] This is consistent with the optics for (synthetic) åkermanite.[1]

### *References*

1. *Winchell*, 473–474.
2. Masson et al, *Suppl.* **2,** 968–969.
3. ———, *Suppl.* **3,** 1031.

## Unidentified Silicate Phases

## Colorless, Birefringent Mineral

### *Occurrence and Form*

In breccia 10021-28 (303), an unidentified silicate grain was found. The grain consists of lamellae less than 50 μ long.[1]

### *Optics*

The mineral is colorless and it is possible that it is only finely devitrified glass, but the larger lamellae are moderately birefringent (0.005–0.010) and have approximately parallel extinction.[1]

### *Chemical Composition*

Probe analysis of the grain gave:

| | | |
|---|---|---|
| $SiO_2$ | = | 48.1% |
| $Al_2O_3$ | = | 25.5% |
| $TiO_2$ | = | 0.55% |
| FeO | = | 5.8% |
| MgO | = | 7.5% |
| MnO | = | 0.11% |
| CaO | = | 14.6% |
| $K_2O$ | = | <0.1% |
| $Na_2O$ | = | 0.56% |
| Total | = | 102.72% |

It is suggested that this analysis might yield the formula:[1] $[Fe,Ca,Mg,(Na,Al)]_3Al_2[Si,Al]_5O_{\sim 16}$.

*Reference*

1. Fredricksson et al, *Suppl.* **1,** 420, 427–428.

## Unknown Fe-Silicate

### *Occurrence and Form*

Three lathlike crystals were found in the $< 37\text{-}\mu$ fraction of 10 Apollo 14 soil samples.[1]

### *Optics*

The crystals are brown and strongly birefringent.[1]

### *Chemical Composition*

A semiquantitative probe analysis gave:

| | | |
|---|---|---|
| $SiO_2$ | = | 65% |
| $TiO_2$ | = | 0.5% |
| FeO | = | 36% |
| MgO | = | 2% |
| MnO | = | 0.5% |
| CaO | = | 1% |
| $K_2O$ | = | 0.5% |
| Total | = | 105.5% |

This calculates to formula $FeSi_2O_5$.[1]

### *X-Ray Data*

Attempts to obtain X-ray diffraction patterns were unsuccessful.[1]

*Reference*

1. Finkelman, *Suppl.* **4,** 183–184.

# APPENDIX I BIBLIOGRAPHY


| Abbreviation | Reference |
|---|---|
| *Am. Mineral.* **36** | *American Mineralogist* **36,** 1951. |
| *Am. Mineral,* **57** | *American Mineralogist* **57,** 1972 |
| *Am. Mineral.* **58** | *American Mineralogist* **58,** 1973 |
| *Apollo 14 preprint* | Preprint for *Third Lunar Science Conference, Abstracts*, 1972. |
| *Apollo 15 Lunar Samples* | Lunar Science Institute; *Apollo 15 Lunar Samples*, 1972. |
| *Apollo 17 Prelim. Sci. Rep.* | *Apollo 17 Preliminary Science Report, NASA SP-330.* |
| *Bull. Soc. Mineral.* **96** | *Bulletin de la Société française de Minéralogie et de Cristallographie* **96,** 1973. |
| *Carnegie Inst. Yearb,* '*69* | *Carnegie Institute of Washington, Yearbook* **69,** (1969–1970). |
| *Carnegie Inst. Yearb.* '*70* | *Carnegie Institute of Washington, Yearbook* **70,** (1970–1971). |
| *Contrib. Mineral. Petrol.* **30** | *Contributions of Mineralogy and Petrology* **30,** 1970 |
| *Contrib. Mineral. Petrol.* **33** | *Contributions to Mineralogy and Petrology* **33,** 1971. |
| *Contrib. Mineral. Petrol.* **41** | *Contributions to Mineralogy and Petrology* **41,** 1972. |
| *Dana VII* **1** | Palache et al, *Dana's System of Mineralogy, Vol. 1,* Seventh edition, John Wiley & Sons, New York, 1944. |
| *Dana VII,* **3** | Frondel, *Dana's System of Mineralogy, Vol. 3,* Seventh edition, John Wiley & Son, New York, 1962. |
| *EPSL* **8,** | *Earth and Planetary Science Letters* **8,** 1970. |
| *EPSL* **9** | *Ibid* **9,** 1970. |
| *EPSL* **10** | *Ibid* **10,** 1971. |
| *EPSL* **11** | *Ibid* **11,** 1971. |
| *EPSL* **12** | *Ibid* **12,** 1971. |
| *EPSL* **13** | *Ibid* **13,** 1971–72. |
| *EPSL* **14** | *Ibid* **14,** 1972. |
| *EPSL* **15** | *Ibid* **15,** 1972. |

| Abbreviation | Reference |
|---|---|
| *EPSL* **16** | *Ibid* **16,** 1972. |
| *EPSL* **18** | *Ibid* **18,** 1973. |
| *EPSL* **19** | *Ibid* **19,** 1973. |
| *EPSL* **21** | *Ibid* **21,** 1973–1974. |
| *Econ. Geol.* **62** | *Economic Geology* **62,** 1967. |
| *Econ. Geol.* **66** | *Economic Geology* **66,** 1971. |
| *EOS* **51** | *Transactions of the American Geophysical Union* **51,** 1970. |
| *EOS* **54** | *Transactions of the American Geophysical Union* **54,** 1973. |
| *Fortschr. Mineral.* **48** | *Fortschritte für Mineralogie* **48,** 1971. |
| *Geochim. Cosmochim. Acta* **37** | *Geochimica Cosmochimica Acta* **37,** 1973. |
| *Geology* **2** | *Geology* **2,** 1974. |
| *Geophys. Res. Lett.* **1** | *Geophysical Research Letters* **1,** 1974. |
| *GSA '71, Abstr.* | Geological Society of America; *Abstracts with programs*, 1971. |
| *GSA '73, Abstr.* | Geological Society of America; *Abstracts with programs*, 1973. |
| *J. Geol.* **81** | *Journal of Geology* **81,** 1973. |
| *J. Geophys. Res.* **75** | *Journal of Geophysical Research* **75,** 1970. |
| *J. Geophys. Res.* **79** | *Journal of Geophysical Research* **79,** 1974. |
| *J. Petrol* **12** | *Journal of Petrology* **12,** 1971. |
| *L. and T.* | Levinson and Taylor; *Moon Rocks and Minerals*, Pergamon Press, New York, 1971. |
| *Lunar Sample Information Catalog, Apollo 12* | *Lunar Sample Information Catalogue, Apollo 12,* NASA–MSC–01512, 1970. |
| *Lunar Sample Information Catalog, Apollo 15* | *Lunar Sample Information Catalogue, Apollo 15,* NASA–MSC–03209, 1971. |
| *Lunar Sample Information Catalog, Apollo 16* | *Lunar Sampie Information Catalogue, Apollo 16,* NASA–MSC–03210, 1973. |
| *Lunar Sample Information Catalog, Apollo 17* | *Lunar Sample Information Catalogue, Apollo 17,* NASA–MSC–03211, 1973. |
| *Lun. Sci. III, Rev. Abstr.* | *Third Lunar Science Conference, Revised Abstracts,* 1972. |
| *Lun. Sci. IV, Abstr.* | *Fourth Lunar Science Conference, Abstracts,* March 1973. |
| *Lun. Sci. V, Abstr.* | *Fifth Lunar Science Conference, Abstracts,* March 1974. |
| *Lun. Sci. Conf. '71, Abstr.* | *Second Lunar Science Conference, Abstracts,* January 1971. |
| *Lun. Sci. Conf. '72, Abstr.* | *Third Lunar Science Conference, Abstracts,* January 1972. |
| Lunar Science Institute; *Post Apollo Lun. Sci.* | Lunar Science Institute; *Post Apollo Lunar Science,* July 1972. |
| *Meteorit.* **6** | *Meteoritics* **6,** 1971. |
| *Meteorit.* **7** | *Meteoritics* **7,** 1972. |
| *Meteorit.* **9** | *Meteoritics* **9,** 1974. |
| *Mineral. Rec.* **2** | *Mineralogical Record* **2,** 1971. |

| Abbreviation | Reference |
|---|---|
| *Moon* **3** | *Moon* **3,** 1971. |
| *Moon* **5** | *Moon* **5,** 1972. |
| *Nature* **225** | *Nature Physical Science* **225,** 1970. |
| *Nature* **234** | *Nature Physical Science* **234,** 1971. |
| *Nature* **235** | *Nature Physical Science* **235,** 1972. |
| *Nature* **237** | *Nature Physical Science* **237,** 1972. |
| *Nature* **240** | *Nature Physical Science* **240,** 1972. |
| *Nature* **242** | *Nature Physical Science* **242,** 1973. |
| *Naturwiss.* **57** | *Naturwissenschaften* **57,** 1970. |
| Preprint; *Lun. Sci. IV, Conf.* | In a preprint for *Fourth Lunar Science Conference*, March, 1974. |
| Preprint; *Metal Silicate Relationships in Apollo 17 Soils.* | In a preprint for *Metal Silicate Relationships in Apollo 17 Soils*, Journal not indicated. |
| Preprint; *Niobian Rutile in an Apollo* 14 *KREEP Fragment*, submitted to *Meteoritics.* | In a preprint for *Niobian Rutile in an Apollo 14 KREEP Fragment*, submitted to *Meteoritics.* |
| *Sci.* **167** | *Science* **167**, 1970. |
| *Sci.* **173** | *Science* **173,** 1971. |
| *Sci.* **175** | *Science* **175,** 1972. |
| *Sci.* **179** | *Science* **179,** 1973. |
| *Sci.* **181** | *Science* **181**, 1973. |
| *Sci.* **182** | *Science* **182,** 1973. |
| *Spec. Publ. #3* | University of New Mexico, Institute of Meteoritics; *Special Publication #3*, 1971. |
| *Spec. Rep.* **333** | Wood et al; Smithsonian Astrophysical Observatory, *Special Report* **333,** 1971. |
| *Suppl.* **1** | *Supplement* **1,** *Geochimica Cosmochimica Acta*, 1970. |
| *Suppl.* **2** | *Supplement* **2,** *Geochimica Cosmochimica Acta*, 1971. |
| *Suppl.* **3** | *Supplement* **3,** *Geochimica Cosmochimica Acta*, 1972. |
| *Suppl.* **4** | *Supplement* **4,** *Geochimica Cosmochimica Acta*, 1973. |
| *Tabellen* | Strunz; *Mineralogische Tabellen*, Akademische Verlagsgesellschaft, Geest und Portig, Leipzig, 1966. |
| *Trans. AIME* **233** | *American Institute of Mining, Metallurgical, and Petroleum Engineers, Transactions* **233,** 1965. |
| *Winchell* | A. N. Winchell and H. Winchell, *Elements of Optical Mineralogy*, Part 2, *Fourth Edition*, John Wiley & Sons, New York, 1951. |

# APPENDIX II | SAMPLE NUMBERS AND THEIR NAMES

## Apollo 11 Samples

| | |
|---|---|
| 10003 | Medium-grained basalt |
| 10003–37 | Fine-grained basalt |
| 10003–44 | Medium-grained basalt |
| 10017 | Fine-grained basalt |
| 10017–20 | Basalt |
| 10017–50 | Fine-grained basalt |
| 10019–22 | Microbreccia |
| | Breccia |
| 10020 | Fine-grained basalt |
| 10020–40 | Fine-grained basalt |
| | Fine-grained olivine basalt |
| 10021–28(303) | Breccia |
| 10021–30 | Microbreccia |
| 10022–4 | Fine-grained basalt |
| 10022–28 | Fine-grained basalt |
| 10022–41 | Basalt |
| | Fine-grained basalt |
| 10024 | Basalt |
| 10024,23 | Coarse-grained basalt |
| | Basalt |
| 10024–33 | Medium-grained basalt |
| 10025 | Fine-grained basalt |
| 10029–1 | Olivine diabase |
| | Coarse-grained basalt |
| 10040–20 | Coarse-grained basalt |
| 10044 | Coarse-grained basalt |
| | Microgabbro |
| 10044–30 | Coarse-grained basalt |
| 10044–43 | Coarse-grained basalt |
| 10044–50 | Basalt |

| | |
|---|---|
| 10045 | Fine-grained basalt |
| | Fine-grained olivine basalt |
| 10045–29 | Fine-grained basalt |
| | Medium-grained basalt |
| | Fine-grained olivine basalt |
| 10045–35 | Coarse-grained basalt |
| | Basalt |
| 10045–35–5 | Fine-grained basalt |
| 10046–20 | Mare-basalt fragment in breccia |
| 10046–48–2 | Breccia |
| 10047 | Ophitic basalt |
| | Coarse-grained basalt |
| 10047,13 | Ophitic ilmenite basalt |
| 10047,20 | Mare basalt |
| 10047–44 | Coarse-grained basalt |
| 10047,68 | Mare basalt |
| | Coarse-grained basalt |
| 10049–29 | Fine-grained basalt |
| 10050 | Coarse-grained basalt |
| 10050,37 | Fine-grained gabbro |
| 10056–46 | Breccia |
| 10057–67 | Fine-grained basalt |
| 10058 | Basalt |
| | Medium-grained basalt |
| | Coarse-grained basalt |
| | Ophitic basalt |
| | Medium-grained cristobalite basalt |
| 10058–22 | Basalt |
| 10058–23 | Coarse-grained diabasic rock |
| | Medium-grained basalt |
| 10059 | Breccia |
| 10060–20 | Microbreccia |
| 10060–30 | Breccia |
| 10061 | Breccia |
| 10061–3 | Microbreccia |
| 10061–28 | Microbreccia |
| 10061–37 | Microbreccia |
| 10062–35 | Coarse-grained basalt |
| | Ophitic olivine basalt |
| 10065 | Microbreccia |
| 10065,15–1a | Breccia |
| 10067–8 | Breccia |
| 10069–26 | Basalt |
| | Fine-grained vesicular basalt |
| 10069–30 | Fine-grained basalt |
| 10072 | Fine-grained basalt |
| 10072–23 | Fine-grained basalt |

| | |
|---|---|
| 10072–33 | Basalt |
| 10072–49 | Fine-grained olivine basalt |
| | Fine-grained basalt |
| | Basalt |
| 10083–12–53 | Coarse fines sample |
| 10084 | Fines sample |
| | Soil sample |
| 10084–30 | Fines sample |
| 10084,75 | Fines sample |
| 10084,96 | Fines sample |
| 10085 | Fines sample |
| 10085–1–11 | Fines sample |
| | Coarse-grained basalt fragment in fines sample |
| 10085–4 | Fines sample |
| 10085–4–10a | Microanorthosite fragment in soil sample |
| 10085–4–14 | Fines sample |
| 10085–12–34 | Fines-grained basalt fragment in soil sample |
| 10085–16 | Soil sample |
| 10085–17M | Fines sample |
| 10085–25 | Fines sample |
| 10085–42 | Coarse-grained basalt fragment in fines sample |
| 10085–48 | Fines sample |
| 10085–LR No. 1 | "Gray mottled" basalt fragment in fines sample |
| | "Luny Rock" fragment in soil sample |
| | KREEP-basalt fragment in soil sample |

## Apollo 12 Samples

| | |
|---|---|
| Apollo 12, Double core sample, Unit IV | Gabbroic fragment |
| 12002 | Medium-grained picritic basalt |
| 12002,7 | Picritic basalt |
| 12002,72 | Picritic basalt |
| | Porphyritic basalt |
| 12003,17 | Fines sample |
| 12004 | Porphyritic basalt |
| | Porphyritic olivine basalt |
| 12004,11 | Porphyritic basalt |
| 12008 | Porphyritic olivine basalt |
| 12009 | Porphyritic olivine basalt |
| 12013 | Breccia |
| | "Black and white" rock |
| 12013, fragment #06A1 | Breccia |
| 12013,10 | Breccia |
| | Mixed light and dark breccia |
| | Variegated breccia |
| 12013,11 | Variegated breccia, light portion |

| | |
|---|---|
| 12018 | Basaltic rock |
| | Porphyritic basalt |
| 12018,49 | Porphyritic basalt |
| 12019 | Basalt |
| 12020 | Porphyritic basalt |
| 12020,8 | Porphyritic basalt |
| 12021 | Porphyritic basalt |
| | Ophitic basalt |
| | Basalt |
| 12021,18 | Porphyritic basalt |
| 12021,22 | Porphyritic basalt |
| 12021,29 | Porphyritic basalt |
| 12021,51 | Porphyritic basalt |
| 12021,134 | Porphyritic basalt |
| | Ophitic basalt |
| 12021,135 | Porphyritic basalt |
| 12022 | Porphyritic basalt |
| | Porphyritic olivine basalt |
| 12022,22 | Porphyritic basalt |
| 12028 | Fines sample |
| 12032 | Fines sample |
| 12032,44 | Fines sample |
| | Fines sample |
| 12033,23 | KREEP-glass fragment |
| 12033,97–2B | Glassy KREEP breccia |
| 12034,3 | "Gray mottled" basalt |
| 12035 | Granular basalt |
| 12035,9 | Basalt |
| 12036 | Ophitic basalt |
| | Feldspathic peridotite |
| 12036,9 | Ophitic basalt |
| | Feldspathic periodotite |
| | Coarse-grained basalt |
| 12037,20 | Soil sample |
| 12038 | Ophitic basalt |
| 12038,22 | Ophitic basalt |
| 12038,63 | Coarse-grained basalt |
| 12038,67 | Coarse-grained basalt |
| | Ophitic basalt |
| 12039 | Ophitic basalt |
| | Microgabbro |
| 12039,3 | Coarse-grained basalt |
| | Ophitic basalt |
| 12040 | Ophitic basalt |
| | Picritic basalt |
| 12040–39 | Picritic basalt |

| | |
|---|---|
| 12051 | Fine-grained basalt |
| | Ophitic basalt |
| 12051–36 | Ophitic basalt |
| 12051,59 | Ophitic basalt |
| 12052 | Coarse-grained basalt |
| | Porphyritic basalt |
| 12052,10 | Porphyritic basalt |
| | Basalt |
| 12052,57 | Porphyritic basalt |
| 12053 | Porphyritic basalt |
| 12057,14 | Basalt fragment |
| 12057,22 | Breccia |
| 12063 | Porphyritic basalt |
| | Basaltic rock |
| 12063,9 | Porphyritic basalt |
| 12063,15 | Basalt |
| 12063/41 | Porphyritic basalt |
| 12064 | Ophitic basalt |
| | Microgabbro |
| 12064,6 | Microgabbro |
| 12065 | Porphyritic basalt |
| | Fine-grained olivine basalt |
| 12065,8 | Porphyritic basalt |
| 12070 | Fines sample |
| 12070,35 | Microbreccia |
| | Breccia |
| | Soil sample |
| 12070,98 | Fines sample |
| 12071,6 | Chip |
| 12075,4 | Basalt |

## Apollo 14 Samples

| | |
|---|---|
| 14001 | Fines sample |
| 14001,7,2 | Fines sample |
| 14001,7,3 | Fines sample |
| | Basalt fragment in coarse-fines sample |
| 14002,7E–1–8 | Fines sample |
| 14003 | Soil sample |
| | Fines sample |
| 14003,27 | Fines sample |
| 14003–47 | Fines sample |
| 14003,79 | Fines sample |
| 14053 | Basalt |
| | Non mare basalt |

| | |
|---|---|
| | Medium-grained basalt |
| | Coarse-grained basalt |
| | Coarse-grained dolerite |
| | Coarse-grained gabbroic igneous rock |
| 14053,5 | Coarse-grained basalt |
| 14053,6 | Coarse-grained gabbro |
| 14053,17 | Basalt |
| 14055,7 | Microbreccia |
| 14063 | Breccia |
| 14063-20 | Breccia |
| 14063-28 | Breccia |
| 14066 | Breccia |
| | Microbreccia |
| 14068 | Breccia |
| 14068,7 | Breccia |
| 14072 | Basalt |
| | Mare basalt |
| | Crystalline clast |
| | Subophitic basalt |
| 14073 | Basalt |
| 14161 | Fines sample |
| 14161/20 | Fines sample |
| 14162 | Soil sample |
| 14162,12 | Microbreccia fragment in soil sample |
| | Breccia fragment in soil sample |
| 14162,41 | Norite fragment in soil sample |
| 14163 | Fines sample |
| 14163,39 | Fines sample |
| 14166,6E2 | Fines sample |
| 14257 | Coarse-fines sample |
| 14257,2 | Microbreccia fragment in coarse-fines sample |
| 14257,3 | Fines sample |
| | Soil sample |
| 14258 | Soil sample |
| 14258,28(1419-7) | Anorthositic fragment in fines sample |
| | Anorthosite |
| 14259,17 | Soil sample |
| 14259,33 | Fines sample |
| 14259,97 | Soil sample |
| 14261 | Fines sample |
| 14276 | KREEP basalt |
| 14276,13 | Basalt |
| | Gabbroic anorthosite chip |
| | Nonmare basalt |
| 14301/9 | Microbreccia |
| 14301/19 | Microbreccia |

|  |  |
|---|---|
| 14303 | Breccia |
| 14303-53B | Breccia |
| 14305,4 | Microbreccia |
| 14305,77 | Breccia |
| 14306 | Breccia |
| 14306-53 | Polymict breccia |
| 14306,53-20A | "Granitic" fragment in breccia |
| 14306,55 | Rhyolite fragment in breccia |
| 14310 | KREEP basalt |
| 14310,4 | KREEP basalt |
| 14310,5 | KREEP basalt |
| 14310,6 | KREEP basalt |
| 14310/20 | KREEP basalt |
| 14310/20A | KREEP basalt |
| 14310/20B | KREEP basalt |
| 14310,22 | KREEP basalt |
| 14310,30 | KREEP basalt |
| 14310,90 | KREEP basalt |
| 14310,145 | KREEP basalt |
| 14310,1159 | KREEP basalt |
| 14313 | Breccia |
| 14315,9 | Breccia |
| 14315,55 | Breccia |
| 14318 | Breccia |
| 14319 | Breccia |
| 14319,11a | Breccia |
| 14319,11b | Breccia |
| 14319,13 | Breccia |
| 14321 | Microbreccia<br>Breccia |
| 14321,21 | Microbreccia |
| 14321,42 | Feldspathic basalt |
| 14321,184-55 | Basaltic clast in polymict breccia |
| 14321,200 | Basaltic fragment in breccia |

## Apollo 15 Samples

|  |  |
|---|---|
| Apollo 15, rake sample, 116 | Pyroxene basalt fragment |
| 15016 | Porphyritic vesicular basalt |
| 15024,4 | KREEP basalt |
| 15065 | Basalt<br>Gabbro |
| 15071 | Soil sample |
| 15074,1 | Mare basalt |
| 15076 | Gabbro |
| 15076,12 | Mare basalt<br>Basalt |

| | |
|---|---|
| 15076,55 | Gabbro |
| 15085 | Coarse-grained basalt |
| 15085/14 | Basalt |
| 15102 | Soil sample |
| 15102,12 | Soil sample |
| 15116 | Gabbro |
| | Pyroxene-phyric basalt |
| 15125 | Pyroxene-phyric basalt |
| 15256 | Basalt |
| 15261 | Soil sample |
| 15264,4 | KREEP basalt |
| 15271 | Fines sample |
| 15301,116 | Fines sample |
| 15402 | Breccia |
| 15404,3 | KREEP basalt |
| 15404,5 | KREEP basalt |
| 15415 | Anorthosite |
| | "Genesis rock" |
| 15415,18 | Anorthosite |
| 15415,22 | Anorthosite |
| 15418 | Metamorphosed breccia |
| | Breccia with vitreous material |
| 15427 | Green glass in pyroclastic rock |
| 15434,4 | Anorthositic gabbro |
| 15434,7 | Anorthositic gabbro |
| 15445,10 | Peridotite |
| | Peridotite clast |
| | Breccia |
| 15455 | Black and white breccia |
| 15459 | Breccia |
| 15459,38 | Anorthositic gabbro |
| 15471 | Fines sample |
| 15473,3,2 | Anorthositic gabbro |
| 15474,1 | Mare basalt |
| 15474,4 | Olivine-pyroxene vitrophyre |
| 15475 | Basalt |
| | Mare basalt |
| 15475/125 | Basalt |
| | Mare basalt |
| 15495 | Porphyritic basalt |
| 15514,3 | Mare basalt |
| 15532,11 | Soil sample |
| 15534,1 | Mare basalt |
| 15534,2 | Mare basalt |
| 15538,5 | Mare basalt |

| | |
|---|---|
| 15555 | Basalt |
| | Olivine-cristobalite basalt |
| 15555/39 | Basalt |
| 15555,105 | Vuggy basalt |
| | Basalt |
| 15564,7 | Mare basalt |
| 15596 | Pyroxene vitrophyre |
| | Basaltic vitrophyre |
| 15601,113 | Fines sample |
| 15602 | Soil sample |
| 15602,29 | Soil sample |
| 15604,3 | Mare basalt |
| 15604,4 | Mare basalt |
| 15604,5 | Mare basalt |
| 15666 | Pyroxene-phyric basalt |
| 15682 | Porphyritic basalt |

## Apollo 16 Samples

| | |
|---|---|
| 60015 | Anorthosite |
| 60015,126 | Anorthositic breccia |
| 60016,95 | Polymict breccia |
| 60025 | Anorthosite |
| 60215/13 | Anorthositic cataclasite |
| | Cataclastic anorthosite |
| 60315 | Poikiloblastic noritic breccia |
| | Recrystallized polymict breccia |
| 60315,63 | Poikilitic basalt |
| | Poikiloblastic diabasic igneous rock |
| 60335 | Feldspathic basalt |
| | Recrystalized breccia |
| | Plagioclase-rich troctolite |
| 60335/75E | Xenoclastic basalt |
| 60601 | Fines sample |
| 60615 | Melt rock |
| 61016 | Shocked breccia |
| 61016–215 | Basalt |
| 61076 | Breccia |
| 61156,5 | Plagioclase-rich clast |
| | Feldspathic basalt |
| | Cataclastic anorthosite |
| 61156,31 | Annealed breccia |
| 61222,3 | Basaltic clast in breccia |
| 61281,8 | Fines sample |

| | |
|---|---|
| 61568 | Poikilitic rock fragment in rake sample |
| 62235 | Poikilitic basalt |
| 62295 | Norite |
| | Microtroctolite |
| | Spinel troctolite |
| | Spinel-microtroctolite |
| | Crystalline rock (possibly effusive) |
| 63501 | Soil sample |
| 63549 | Diabase rock fragment in rake sample |
| 64455 | Basalt fragment |
| 64585 | Mesostasis-rich rock fragment in rake sample |
| 65015 | KREEP-rich rock |
| | Polymict KREEP-rich rock |
| | Metamorphosed KREEP-rich breccia |
| | Ba-KREEPUTH-rich rock |
| 65785 | Spinel troctolite |
| 66055 | Breccia |
| | Microbreccia |
| 66081 | Soil sample |
| 66081,5 | Soil sample |
| 66095 | Metamorphosed breccia |
| | Breccia |
| 66095,78 | Breccia |
| | Metamorphosed breccia |
| 66095,80 | Metamorphosed breccia |
| 66095,87 | Breccia |
| 66095,89 | Metamorphosed breccia |
| 67075,45 | Brecciated anorthosite |
| 67435,14 | Microbreccia |
| | Partially melted breccia |
| 67455 | Breccia |
| 67455,8 | Polymict breccia |
| 67701,26 | Fines sample |
| 67712,16 | Fines sample |
| 67955 | Breccia |
| 68415 | Anorthositic gabbro |
| | Gabbroic anorthosite |
| | Porphyritic basalt |
| | Feldspathic basalt |
| 68415,8 | Anorthositic gabbro |
| | Gabbroic anorthosite |
| 68415,37 | Porphyritic basalt |
| | Anorthositic gabbro |
| | High grade metamorphic rock |
| 68416,77 | Porphyritic basalt |
| | Gabbroic anorthosite |

| | |
|---|---|
| 68501 | Soil sample |
| 68503,16 | Coarse-fines sample |
| 68503,16,6 | Poikiloblastic rock |
| 68815 | Metamorphosed breccia |
| 68841 | Fines sample |
| 69941 | Soil sample |
| 69941,13 | Soil sample |

## Apollo 17 Samples

| | |
|---|---|
| 70017 | Basalt |
| | Mare basalt |
| | Coarse-grained basalt |
| 70035 | Basalt |
| | Coarse-grained basalt |
| 70215 | Fine-grained basalt |
| 70215,149 | Fine-grained basalt |
| 71055 | Mare basalt |
| | Medium-grained basalt |
| 72415 | Dunite fragment in breccia |
| 72415,11 | Dunite fragment in breccia |
| 72415,12 | Dunite fragment in breccia |
| 72416 | Dunite fragment in breccia |
| 72417 | Dunite fragment in breccia |
| 72418 | Dunite fragment in breccia |
| 74220 | Fines sample |
| | Orange soil sample |
| 74241 | Basalt |
| 74242 | Fines sample |
| 74243 | Fines sample |
| 74275 | Mare basalt |
| 75055 | Coarse-grained basalt |
| 75081 | Basalt |
| 76055,10 | Breccia |
| | Micronorite |
| | Fine-grained noritic hornfels |
| | Impact-generated melt rock |
| 76315,11 | Breccia |
| 76535 | Troctolitic granulite |
| 77017 | Brecciated anorthositic gabbro |
| | Coarse-grained anorthositic gabbro |
| 77115 | Vuggy breccia |
| 77215 | Microbreccia |
| 79155 | Coarse-grained basalt |

## Luna 16 Samples

| | |
|---|---|
| Luna 16 | Fines sample |
| Luna 16, FQM–151 | Fragment from soil sample |
| Luna 16, G 38/a | Fine-grained basalt fragment from soil sample |
| | Basaltic fragment from soil sample |
| Luna 16,638/2 | Fragment from soil sample |

## Luna 20 Samples

| | |
|---|---|
| Luna 20 | Fines sample |
| Luna 20, B-A2 | Basalt fragment from fines sample |
| Luna 20, K–1 | Soil sample |
| Luna 20, K–4 | Soil sample |
| Luna 20, 509–10 | Anorthositic troctolite fragment from soil sample |
| Luna 20, 515–23 | Breccia particle in soil sample |
| Luna 20, 516–1 | Breccia fragment from soil sample |
| Luna 20, 528–34 | Mare-basalt fragment from soil sample |
| Lune 20, 532–3 | Breccia fragment from soil sample |
| Luna 20, 532–6 | Crystalline particle from soil sample |
| Luna 20, 808–35 | Anorthositic fragment from soil sample |
| Luna 20, 863 | Rock fragment from soil sample |
| | Olivine-plagioclase rock fragment |
| Luna 20, 926–3 | Anorthositic fragment from soil sample |
| Luna 20, 928–1 | Anorthositic fragment from soil sample |
| Luna 20, 22001,16 | Fines sample |
| | Soil sample |
| | Fine-grained soil sample |
| Luna 20, 22001,17 | Fine-grained fragment from soil sample |
| Luna 20, 22001,17–A16 | Fine-grained fragment from soil sample |
| Luna 20, 22002,2,1 | Handpicked crystal from fines sample |
| Luna 20, 22002,2,4c | Fine-grained soil sample |
| Luna 20, 22002,2,5a | Fine-grained soil sample |
| Luna 20, 22002,2,5b | Soil sample |
| Luna 20, 22002,2,7d | Handpicked albite crystal from an anorthositic fragment from soil sample |
| Luna 20, 22002,3 | Soil sample |
| Luna 20, 22003,1 | Soil sample |
| Luna 20, 22003,1,3a | Anorthositic fragment from soil sample |
| Luna 20, 22003,1–505–2 | Soil sample |
| Luna 20, 22003,1–512–2 | Fragment from soil sample |
| Luna 20, 22003, 1–515–23 | Soil sample |
| | Glass-rich breccia particle from soil sample |
| Luna 20, 22003,1–517–7 | Glassy breccia particle from soil sample |
| Luna 20, 22003,2 | Coarse-grained soil sample |
| | Spinel-troctolite fragment from soil sample |

| | |
|---|---|
| Luna 20, 22006 | Recrystallized polymict breccia fragment from soil sample |
| | Recrystallized anorthositic fragment from soil sample |
| Luna 20, 22007 | Troctolitic recrystallized polymict breccia fragment from soil sample |
| | Recrystallized vesicular fragment from soil sample |

# APPENDIX III | NOTE ON SYSTEM OF SAMPLE NUMBERING

Each lunar sample was photographed at the Lunar Receiving Laboratory in the Johnson Space Center in Houston, and was then assigned a generic number of five digits. If the sample was split or subdivided in any manner (e.g., splits of fines samples and chips, sawn sections, or thin sections of rocks), it received an additional specific number. Some samples may have a third part to the number assigned to them by the individual investigative teams. Thus:

1. A sample directly after photographing—14306.
2. A sample section—14306-53.
3. A further division of the sample by Principal Investigators—14306, 53-20A.

Some investigators separate the generic and specific parts of the sample number by a comma (,), others by a stroke (/), and still others by a dash (–). The lack of uniformity is retained in this volume to facilitate recognition of the samples in the cited references.

The Apollo 11 samples were assigned five digits in the ten thousands (10001, etc.), the Apollo 12 samples in the twelve thousands (12001, etc.), Apollo 14 samples in the fourteen thousands, and Apollo 15 samples in the fifteen thousands. This system was changed for Apollo 16 and Apollo 17 samples to provide enough digits for the numerous samples collected on the last two missions. Hence Apollo 16 samples have been assigned numbers in the sixty thousands and Apollo 17 samples numbers in the seventy thousands.

The Academy of Sciences of the USSR kindly made available to NASA some material from their Luna 16 and Luna 20 missions. Numbers were assigned to splits of these samples by the NASA lunar sample curator.

# INDEX